FOOD INTAKE
AND
ENERGY
EXPENDITURE

Edited by

Margriet S. Westerterp-Plantenga, Ph.D.
Department of Natural Sciences
Open University of The Netherlands
Heerlen, The Netherlands

Elisabeth W.H.M. Fredrix, Ph.D.
Department of Natural Sciences
Open University of The Netherlands
Heerlen, The Netherlands

Anton B. Steffens, Ph.D.
Department of Animal Physiology
State University of Groningen
The Netherlands

Editorial Advisor

Harry R. Kissileff, Ph.D.
Associate Professor of Clinical Psychology in Psychiatry and Medicine
Columbia University
College of Physicians and Surgeons
New York, U.S.A.

CRC Press
Boca Raton   Ann Arbor   London   Tokyo

Cover photograph by Pascal Tournere, Paris. With permission.

Library of Congress Cataloging-in-Publication Data

Food intake and energy expenditure/edited by Margriet S. Westerterp-Plantenga,
   Elisabeth W.H.M. Fredrix, Anton B. Steffens; editorial advisor,
   Harry R. Kissileff.
      p. cm.
   Includes bibliographical references and index.
   ISBN 0-8493-9228-4
   1. Nutrition. 2. Food habits. 3. Energy metabolism.
  I. Westerterp-Plantenga, Margriet S. II. Fredrix, Elisabeth W. H.
M. III. Steffens, Anton B.
   (DNLM: 1. Eating. 2. Body Weight. 3. Energy Metabolism.
4. Feeding Behavior. WI 102 F686 1994)
QP141.F666 1994
612.3¢9—dc20
DNLM/DLC
for Library of Congress   93-46315
                                              CIP

No claim to original U.S. Government works
International Standard Book Number 0-8493-9228-4
Library of Congress Card Number 93-46315
Printed in the United States of America 1 2 3 4 5 6 7 8 9 0
Printed on acid-free paper

To the memory of Eliot Stellar 1920–1993

Contents

Preface

Chapter 4

Chapter 5

Part II   Physiology and Endocrinology of Food Intake Regulation

Chapter 6

Chapter 7

Chapter 8

Chapter 9

Chapter 10

Chapter 11

Chapter 12

Chapter 13

Chapter 14

Chapter 18

Chapter 19

Chapter 20

Chapter 21

Part IV   Food Intake, Energy Expenditure and Evolution

Chapter 22

Chapter 23

Preface

Food Intake and Energy Expenditure has been developed for a course taught by the Open University of the Netherlands. We have tried to incorporate feedback from students and teachers, to produce a textbook that not only covers current courses in human nutrition, but also conveys a flavor of current research and an integrated approach to understanding food intake and energy expenditure.

This textbook is concerned with the relationship between the consumption of food by humans, and some other animals, and its utilization by the body, in various ways.

This idea is realized in four different parts, each with its own character.

Food Intake and Energy Expenditure is an integrated collection of essays on current problems in the field of ingestive behavior. Its strength and uniqueness lie in the selection of topics brought together under one cover by a group of authors who understand the importance of an interdisciplinary approach to the problem of energy cycling at the level of the individual animal or human subject. A complete understanding of this problem involves not only behavioral and metabolic processes occurring in the experimental subject as a whole, but also in the processes occurring at two distant ends of the organismic axis, i.e., the cellular/molecular coupled with the evolutionary. Because behavior of individual subjects frequently occurs in a social and environmental milieu, the special problems of studying food intake and energy expenditure in a social context and customary environment are incorporated.

Although topics in the individual chapters enable the reader/student to pursue a particular area in depth the editors have shown the interconnections among them so vital for a full understanding of the problems. Thus complete study of this book can provide the basis for a course on the regulation of energy balance that would be appropriate to a number of disciplines, such as biology, physiology, psychology, and nutrition.

This book is divided into four sections: psychobiology of ingestion, body weight regulation, energy expenditure, and evolution. Human and animal studies often treated separately in other contexts are integrated in the present context so that the application of animal studies to the human condition can be readily seen.

The emphasis in part I is on human eating behavior, and is unique in its conceptualization of analysis of behavior as an input-output system. Rather than summarizing debates about the use of psychological constructs, the approach is focused on the precise manipulation of quantifiable influences, such as variations in composition of a food stuff or gastric wall tension, on measurable behavioral outcomes such as reports of subjective feelings or amounts consumed. The criterion of precision is the ability of the manipulated input to predict the measured output. In the field of ingestive behavioral analysis, this approach represents a paradigmatic shift away from verbal explanation by means of constructs and intervening variables such as hunger, satiety, and palatability to residual variability in functional relational relationships and just noticeable differences. Adoption of this new approach will revolutionize thinking in this discipline.

The approach is based on the fundamental idea that energy input is matched to energy output so that manipulations in energy density of a diet or administration of energy before a meal ought to be compensated by change in voluntary consumption.

The theme of energy balance is continued in part II, but the emphasis is focused more on the animal than the human with body weight considered as a regulated variable in a system where energy input and output are controlled to conserve body weight. Aberrations in body weight and extrapolation back to disturbances in human body weight regulation and ingestive disorders such as bulimia nervosa and anorexia nervosa link parts I and II.

In part III the focus is primarily on the expenditure side of the energy balance equation, but the role of food intake on energy expenditure draws connections back to parts I and II. Because energy output is continuous while input is periodic, the issue of the time frame for their integration receives critical attention, and temporal patterns of ingestion and expenditure are carefully considered. The utilization of nutrients for both growth and locomotion links this section back to the selection of diets and its effects on behavior discussed in parts I and II.

Finally in part IV, the longer term effects of natural selection which adapt an organism to its niche are considered from the standpoint of how such effects would be incorporated into a genome and modified by means of biochemical effects expressed through the endocrine system. The involvement of energy balance in the control of reproduction is a key issue in this context. The relationship between food intake and evolution is reflected in the size of natural populations in the long term, as an ultimate effect of feeding throughout life over several generations.

Food Intake and Energy Expenditure represents four different research lines, clearly separated into the four different parts of the book, in a way that research findings, conclusions, interpretations and discussions are translated into education, with different views on the same topics. We believe this approach to be more valuable and more direct for students, rather than the more traditional method of teaching concepts and facts.

Usually, the four different parts are taught separately, and often by separate departments. In this book not only the significance of the psychobiology of food intake for the physiology of food intake regulation is discussed, but also the implications of the physiological aspects for the psychobiological approach come into focus. Secondly, the regulation of body weight is not only highlighted from food intake regulation, but the important role of energy expenditure is taken into account. Thirdly, we not only discuss body weight regulation in the short term (days, weeks or months) but also the longer term evolutionary implications. An endocrinological approach, as well as an approach from the field of animal ecology allow us the possibility of translating short-term or proximate functions into long-term or ultimate functions.

This book is suitable not only for intermediate students in biological, behavioral, and nutritional sciences, but will provide advanced investigators in each of these fields with insights into linkages with the others. Because the emphasis is on a selection of current research problems rather than a global historical review of the energy balance as whole, it opens up challenging questions and will be fruitful in providing suggestions for interdisciplinary research.

Harry Kissileff
Margriet Westerterp

Acknowledgements

Chapter 2 is partly based on: D.A. Booth, Summary: Part VI, The Psychobiology of Human Eating Disorders: Preclinical and Clinical Perspectives, Annals of the New York Academy of Sciences, 575, 1989.

Chapter 3 is partly based on: D.A. Booth: Sensory Influences on Food Intake, Nutrition Reviews, 48, 2, 1990.

Chapter 4 is partly based on: D.A. Booth, M.T. Conner and E.L. Gibson, Measurement of Food Perception, Food Preference, and Nutrient Selection. Nutrition and the Chemical Senses in Aging: Recent Advances and Current Research Needs. Annals of the New York Academy of Sciences, 561, 1989.

Chapter 5 is partly based on: D.A. Booth, Mood- and Nutrient-Conditioned Appetites, Cultural and Physiological Bases for Eating Disorders. The Psychobiology of Human Eating Disorders: Preclinical and Clinical Perspectives. Annals of the New York Academy of Sciences, 575, 1989.

We wish to thank Hanneke Drijver for editorial assistance, Ruud Hoefakkers for developmental testing, Lex Hendrix for logistic management, Evelin Karsten and Jo Hendriks for the final preparation of the layout of the manuscript.

I

Psychobiology of Human Food Intake

1
Current issues in intake psychobiology

The psychobiology of human food intake involves a great variety of interesting and important phenomena. Some of these are effects of eating on physiology. Also, our physiology can affect eating. Often effects in both directions are involved. This chapter gives some examples of each sort from recent research.

1   EFFECTS OF INTAKE ON PHYSIOLOGY

The quantity and composition of our food affects our physiology in many important ways. Food is of course necessary to sustain life. In addition, however, many of the nutrients in food are known to have specific effects on the physiology of humans and other species. A major aspect of the psychobiology of food intake is the impact of eating behavior on human physiology.

1.1   CEPHALIC RESPONSES

The sight, flavor and texture of food can have profound physiological consequences. Food stimulates the secretion of saliva, gastric and pancreatic juices from the intestine and other organs (i.e. insulin from the pancreas) and a variety of hormones. Heart rate may increase, pupil size may change, blood sugar content changes, and the metabolism of body fuels shifts. These responses occur before the actual digestion and absorption of food, sometimes even before food ingestion. They have a very short response time, usually within five minutes after the start of exposure to the food. Most of the changes have disappeared within 10 minutes.

Because such responses result from actions of food on the senses located in the head, such as taste, smell and sight, they are known as *cephalic responses*. That is, they have their origin in a stimulation of sensory receptors whose neuronal output is processed by the central nervous system to produce physiological changes, rather than from stimulation of the intestines or other tissues after the food has been swallowed.

It has been supposed that the size of the cephalic responses is related to the pleasantness of a food, the so-called *hedonic value* (Bruce *et al.*, 1987).

The physiological functions of cephalic responses are not yet clear. Four effects have been proposed:

1 Cephalic responses would enable the organism to process the food more appropriately to its quantity and nutritional composition.

2 Cephalic responses would delay satiation and enable the organism to ingest more food.

3 Cephalic responses would protect the body against the stress imposed by digestion, absorption and storage of nutrients. Food intake periodically exposes the body to a large

quantity of digested nutrients. These nutrients need to be processed and stored within a relatively short period of time. Storage is needed since food intake occurs in bouts, and yet metabolism proceeds continuously. Rapid storage of nutrients is mediated by the action of insulin. A strong cephalic insulin response would stimulate storage very promptly. Powley and Berthoud (1985) observed on the one hand a correlation between the size of the cephalic insulin response and the release of insulin once absorption started, and good control of the level of blood sugar on the other hand: hence the cephalic response may lead to longer-term good functioning.

4 A cephalic response may be explained as a conditioned physiological reflex. The organism learns to associate food ingestion with a number of centrally monitored, unconditioned physiological responses of the gastro-intestinal tract. As a result, eating by itself evokes the conditioned response, such as neurally triggered insulin release.

The *third* function seems most likely and is not incompatible with conditioning as described under 4. Further research will be necessary to determine the variation in the various cephalic responses among different persons and the stability of a cephalic response within a person. However, most work has been done on rats. Indeed, the cephalic insulin response in people has not yet been completely established. Another topic that needs to be investigated more closely are the relationships among the size of the cephalic responses, individual food preferences, food choices and the hedonic value of food.

1.2   THERMIC EFFECT OF FOOD

During and after food consumption the rate of energy expenditure increases independently of any changes resulting from movements of the limbs and torso. This metabolic increase is caused by energetic costs of food ingestion, digestion, absorption, storage and metabolism of nutrients in the body. The postprandial (= after the meal) rise in energy expenditure is influenced by the *quantity* and the *type* of food, by the *physiological characteristics* of the subject and by several other factors such as perhaps the *hedonic value* and *familiarity* of the food.

1.2.1   Definition and background

The rate of energy expenditure can be measured as an increase in oxygen consumption by the body and an increase in carbon dioxide production. On average, after a meal the metabolic rate increases by about 10 per cent. The increase in metabolic rate above the resting level (distinct from any changes in muscular activity) is known as the thermic effect of food, also denoted diet-induced thermogenesis.

The thermic effect of food increases rapidly within 1 hour after a meal to a peak value that is about 30 per cent above the premeal value. After one hour the thermic effect slowly diminishes to about 0-5 per cent after 4 hours. The pattern of change in energy expenditure after a meal is influenced by meal size and composition. After large meals the rise is larger and present for up to 8 hours. However, the total thermic effect of a mixed meal expressed as a percentage of the energy that is ingested is usually about 10 per cent. Fat tends to be digested more slowly than protein and carbohydrate and this causes the peak value of the thermic effect of fat to be later than that of carbohydrate and protein.

Interest in the thermic effect of food rose in the 1980s because it was found in some studies that currently and previously obese people had a lower thermic effect of food than normal-weight individuals. Several later studies, however, failed to confirm these findings. Current evidence is that some groups of overweight and obese people, in particular those that are insulin resistant, may indeed have a lower thermic effect of food than normal-weight people.

1.2.2   Thermic effect of food and eating behavior

Sensory characteristics of food may influence the thermic effect of food. An interesting question is whether relatively palatable food causes a relatively high thermic effect and whether such an effect is due to increased cephalic responses.

LeBlanc *et al.* (1985) were the first to report a 67 per cent increase in the thermic effect of food after a highly palatable meal compared to an unpalatable meal. However, in this study the meals differed in more characteristics than palatability. There were also differences in water content, weight, texture and appearance, for example. Such differences may have contributed to the result. Weststrate *et al.* (1990) found no difference in thermic effect between liquid meals that were identical in overall composition but differed in palatability. The issue of the effect of palatability on the thermic effect of food is still unresolved and more rigorously controlled studies are needed.

In 1990, Westerterp-Plantenga *et al.* observed a relationship between the size of the thermic effect of food and subject characteristics. It was found that the thermic effect of food was higher in normal-weight unrestrained eaters who had a decelerated cumulative food intake than in dietarily restrained normal-weight and post-obese eaters with a constant rate of food intake. This observation raises the question, does the reduced thermic effect of food in restrained subjects affect the satiating effects of eating in the way predicted by a thermostatic theory of food intake control, i.e. does heat production inhibit intake?

2   EFFECTS OF INTAKE ON MOOD AND PERFORMANCE

Interesting research opportunities have developed recently in the study of the impact of food on mood states. Does consumption of such a food enhance an existing mood state or may it actually alter the mood? Are certain foods preferred in some mood states because they improve those states? This is an area that is extremely difficult to investigate because of methodological pitfalls.

In one preliminary experiment in schoolchildren, when ingestion of a flavored drink was associated with feeling relief of anxiety, liking for the flavor was reinforced. Alcohol is of course well-known for its ability to relieve anxiety and 'drown sorrows'. One of the attractions of caffeine drinks may be their effect on the ability to concentrate and hence an improvement in mood when facing intellectual work. However, physiological effects by themselves seem unlikely to lead to powerful 'mood foods'; the relatively mild and perhaps non-specific effects of food and drink constituents have to meet important subjective needs before consumption of those materials become at all compelling.

Wurtman *et al.* (1987) suggested that obese persons, women suffering from premenstrual syndrome and persons afflicted with seasonal affective disorder would all ingest more carbohydrate-rich foods during periods of depression. In rats, high-carbohydrate, low-protein foods can cause an increase in synthesis in the brain of the neurotransmitter serotonin. Serotonin plays a role in sleep and distress. Hence, more serotoninergic activity and one might feel soothed or sedated; less and one might feel tense or anxious. Nevertheless, the evidence so far is against an effect of high-carbohydrate foods on brain-serotonin concentrations in people (Leathwood, 1984). Indeed, the existence of a phenomenon of true carbohydrate craving is questionable: snackfood-eating habits and their physiological and social reinforcers could explain the observed emotional eating of items high in starch, sugar and fat (Booth, 1989). The taste and feel of sweet and fatty foods may release endogenous opiates, which can affect mood and increase feelings of wellbeing (Drewnowski, 1989).

Recently, observations have been made that various alterations of the diet in children can affect misbehavior or the development of intelligence via physiological effects on vitamin or mineral deficiency or intolerance of additives or natural allergens. However, it is an open question, whether the observations are caused by biological or social effects of the dietary manipulation. Changing the foods served in an institution is bound to have a considerable impact on the spirit of the place. Even vitamin capsules might be distinguished from placebo capsules and provoke different reactions in children. Applications of the psychological effects of food intake could have a promising future. A prerequisite is, however, that research into the mechanisms of such effects is scientifically sound and informative. This is an incentive to design experiments capable of falsifying theories on the processes by which food intake affects mental states and behavior, rather than merely demonstrating effects that require further investigation to establish their reliability or reality, let alone the evidence for interpretable and usable mechanisms.

3   EFFECTS OF PHYSIOLOGY ON INTAKE

3.1   FOOD CRAVINGS

Carbohydrate craving has already been mentioned as the second loop of a proposed cycle of effects of carbohydrate intake on mood via brain serotonin activity (Wurtman & Wurtman, 1987). Those physiological processes in the brain are said to induce a selective appetite, specifically for food rich in starch or sugar. This loop back to food intake has proved at least as controversial as the serotonin idea. Drewnowski (1989) pointed out that the foods selected in Wurtman's experiments are even richer in fat than they are in carbohydrate. Booth (1987, 1988) noted that the food selections that were high in carbohydrate (and fat) were of conventional packet products for eating without plate and cutlery (snackfoods). Therefore those people who had a habit of snacking would be tempted to eat such items as an extra dessert or between meals, as was observed in those experiments. Furthermore, their mood is likely to be affected by succumbing to such a temptation to eat what they might regard as 'junk food', especially if they are obese or depressed. Thus the felt compulsion to eat conventional snackfoods need not have anything to do with their carbohydrate content.

Scientists and laypeople alike have generally assumed that cravings for foods are based on some nutritional deficit or imbalance or perhaps an addictive substance in the food. Does pica (eating earth) provide some of the minerals that are needed in small amounts, such as magnesium? Do food cravings during pregnancy help to meet the requirements of the growing foetus? Is 'chocolatism' caused by neurotransmitter-like actions of the methylxanthines or phenylethylamine in chocolate?

Rozin *et al.* (1967, 1968, 1971) were quite sceptical about these physiological theories of common food cravings and their own and others' subsequent evidence has cast further doubt on them (Rodin, 1990). Possibly, hormonal states during pregnancy or physical or emotional stresses during dieting or anxiety attacks create a state of mind that seeks varied or strong sensations. Foods that have been enjoyed at some time but are not frequently eaten may become very attractive when one has an excuse to indulge oneself and this disposition is described as *craving*. Maybe the desire to eat that food is augmented by the belief or hope that it might do some good to the baby, the bodily discomfort or the emotional distress, as the case may be. Careful research is needed to disentangle these cultural and cognitive processes from possible physiological effects.

3.2   WISDOM OF THE BODY

Davis (1935) found that infants remained healthy and even recovered from rickets when allowed to select their diet from an array of natural foodstuffs. However, it has been pointed out that random variety-satisfying selections from these foods were likely to have been healthy.

Rats prefer strongly salty tastes when they are sodium-deficient with no prior experience to tell them of the benefits. It has been supposed that human beings also have an innate appetite for sodium. A tragic case is always cited of a young boy with undiagnosed adrenocortical insufficiency and consequential sodium deficit who ate rock salt at home but was prevented from doing so in hospital and died. However, the child could have found for himself that salt prevented him from feeling ill. Careful experiments in adults have yet to establish clear evidence for an innate sodium appetite, like that seen very clearly in the rat.

Given the great variety of flavors and textures of the foods available to our ancestors and the unreliability of relationships between sensory characteristics and nutrient contents, any specific hunger shown by mammalian omnivores would probably have to be learned. Rats clearly can learn to associate the texture or flavor of an artificial diet with physiological effects of its protein content and to select that dietary flavor only when they need protein. Gibson & Booth (1986) have some evidence of learning a protein-specific appetite in human adults. If these complicated and carefully controlled learning experiments can be repeated for other essential nutrients, might we find that people have a range of hidden capacities for wisdom of the body? If so, how strong is such learning compared to other physiological and social influences on food preferences and intake?

3.3   SOCIOBIOLOGY OF DIET

Some anthropologists have proposed that certain ethnic cuisines and food taboos are physiologically based. The theory is that the particular human group and its culture survived because their dietary practices provided nutrients that were limited or lacking in their crops and hunting and food gathering from the wild. For example, it is proposed that the origins of the Mexican tortilla lay in the effects of cooking maize with lime: the vitamin niacin and the essential amino acid lysine are then released. If the bean crop had failed, these nutrients would be lacking in a diet subsisting on corn.

Another example is lactose intolerance. Most human adults (the exception are those of European stock) have lost the enzyme lactase in the intestinal wall that enabled them to digest milk sugar in infancy. Hence the lactose in milk cannot be digested and absorbed and so it passes to the lower intestine where it is fermented by bacteria which creates discomfort and illness. These consequences of malabsorption of lactose, it is suggested, have excluded milk and its products from the diet of some groups. Or, it resulted in their consuming only fermented products in which the lactose has been hydrolysed to glucose and galactose or converted to other compounds which can be absorbed without previous digestion.

3.4 HUNGER AND SATIETY

Like other animals, human beings seek and consume a wide variety of materials that provide metabolic energy. This desire for food *(hunger)* is quite general to sources of energy (see also Chapter 11 and 13). It is therefore assumed that it is an effect of incipient if not actual lack of energy supply to some innervated tissue that has been signalled to (or within) the brain. However, it has been very difficult to track down hunger signals in people because experimental subjects are inevitably aware of any food deprivation that is imposed on them (Wooley *et al.*, 1972). Also, as the diet industry knows all too well, it is technologically possible to provide apparently normal foods that are in fact energy-free, and so to carry out naturalistic experiments on the duration of hunger-suppression effects of ingested energy.

Some exciting possibilities have recently been developed, however, by applying the technique of social and physical isolation from temporal cues. Relationships can be investigated between the initiation of intake or request for food and physiological measures, such as blood level of glucose and other substrates, and also hormones involved in energy metabolism such as insulin. The first study of this kind failed to find a relationship between blood glucose and hunger (Foltin *et al.*, 1990). However, use of shorter time intervals is beginning to give positive results. In the human species, as in the rat (Newman & Booth, 1981), the slowing of absorption from the digestive tract and/or timing mechanisms in the brain are likely to produce a shift in flows of substrates. This may temporarily leave tissue glucose uptake in arrears of glucose absorption, mobilization and production, resulting in a dip in blood-glucose level just before the meal.

Satiety mechanisms have been easier to investigate than hunger mechanisms, in other species as well as in humans. Food components or digestion products can be intubated to the stomach or infused into the intestine or blood vessels. The chemical composition of foods (particularly calorific fluids) can sometimes be altered in ways that are not detectable to the external and oral senses and so any effects on subsequent intake can only be attributed to physiological actions. If this disguising of post-ingestionally effective differences is done properly, this 'preloading' technique can test for and even measure hypothetical sources of satiety signals. The clearest results on human satiety mechanisms so far have employed intestinal infusion.

Human satiety is considered in more detail in the next section, in connection with the health issue of weight control.

4   SOME HEALTH-RELEVANT EFFECTS OF INTAKE ON EATING

Effective research on influences on food intake in humans is not very extensive, compared to that on the visceral physiology and neuro-endocrinology of feeding in rats and other experimental animals. This chapter ends with a selection of psychobiological issues that are of particular interest to nutritionists, not already mentioned.

4.1   NUTRIENTS, SATIETY AND HUNGER

A major area for psychobiological studies relevant to nutrition and health is the role of physiological factors in hunger and satiety on the total cumulative food intake.

Such research has been started in obese and lean persons (to gain insight into the etiology of obesity), in slimmers (to determine how permanent weight loss is achieved), in diabetics (to deal with some of the problems they may have with keeping to their diets) and in the elderly (to understand why they sometimes undereat).

The food industry would like to know the satiating power of a food so that they can ensure that healthy products are satisfying. So, much recent psychobiological research has been concentrating on which factors determine the satiety value of a food: its calorie content, its nutritional composition, its weight (volume) or its fiber content? Can the satiating power of foods at a given caloric content be improved for those who want to slim or control their weight?

This work is founded on the concept of behavioral compensation for energy intake (Booth, 1972). *Energetic compensation* can be defined as the tendency to adjust food intake to cancel out variations in the energy content of meals. Eating and drinking occasions vary in duration, amount of food and energy density. (The *energy density* of a food is the amount of metabolic energy derived from the food per unit of weight, e.g. MJ per kg or kcal per g). These factors, as well as what preceded the meal, may all affect the activation of satiating processes after the meal. Also, the timing and nature of subsequent opportunities to eat will affect the role of satiety mechanisms in suppressing subsequent intake. Studies on energetic compensation are important, because they can be designed to reveal the mechanisms of the physiological regulation of food intake, and the extent to which such processes operate.

One issue is whether these mechanisms are different for the various macronutrients, for instance with respect to effects on substrate oxidation and gastro-intestinal action. Is it possible to fool the compensation of food intake, e.g. by use of reduced-calorie foods? If so, to what extent is an effect helpful to slimming influenced by the type of calorie reduction? The usefullness of reduced-calorie foods in weight control and prevention of obesity has simply been assumed but is in fact a highly controversial scientific issue.

Increasing numbers of research workers have investigated energetic compensation in human subjects. The degree of energetic compensation of intake found in these studies varied between full compensation and even overcompensation to almost no compensation (e.g. 10 per cent, Lissner *et al.*, 1987). The reasons for these different outcomes are most likely to be related to differences in the interactions between the composition of the ingested foods, satiety mechanisms and subsequent eating occasions. Hence, the observed degree of compensation will depend on the experimental design, the nature of the experimental manipulation, the characteristics and number of subjects and their eating habits, and the duration of the experiments. A food or a nutrient has no fixed satiety value.

One possibility being pursued is that the degree of energy-intake compensation is higher when sugar is replaced by non-caloric or low-caloric sweeteners than when high-fat products are replaced by low-fat products. This would imply that on current eating habits, fat reduction is more useful then sugar reduction in weight control. In Lissner's study, 24 normal-weight women consumed in random order a low-fat (15-20 per cent), a medium-fat (30-35 per cent) and a high-fat (45-50 per cent) diet during three successive periods of two weeks. The records of spontaneous food intake indicated remarkable differences in energy intake between these diets. On the low-fat diet the women were estimated to ingest on average 8,738 kJ per day, on the medium-fat diet 9,848 kJ and on the high-fat diet 11,364 kJ. Body weight after two weeks on the high-fat diet was 0.7 kg higher than it was after two weeks on the low-fat diet, which is consistent qualitatively (allthough not quantitatively) with what would be expected from the trend in the dietary records (see also Chapter 20). In contrast, covert replacement of sugar by aspartame in a diet, including a good deal of sweet drinks and desserts, resulted in 60 per cent compensation within two weeks (Porikos *et al.*, 1982).

Another research approach is to investigate whether ingestion of equicaloric amounts of proteins, fats or carbohydrates have different short-term effects on rated appetite or food intake. Geliebter (1979), Driver (1988) and Hill & Blundell (1986) did not observe differences between the macronutrients and their effect on subsequent intake. However, in other studies, evidence has been found that proteins have more prolonged satiating efficiency than carbohydrates (Hill *et al.*, 1986). The long-term implications will, however, be critically dependent on meal patterns and on learning processes.

4.2   COGNITIVE EFFECTS ON ENERGY INTAKE

The effect of expectations on food intake have seldom been adequately investigated. If people believe a food to be rich, it could make them expect to feel fuller for longer. Such 'cognitive satiety' effects were observed by Wooley (1972), Wooley *et al.* (1972), Booth *et al.* (1982) and Rolls *et al.* (1982), but not by Wardle (1987) and Ogden & Wardle (1990). However, investigators have seldom measured the expected time-course of the satiating effect of a meal. Moreover, very few experiments on the suppression of intake by the nutrient contents of a prior ingestion have shown that variations in the contents of the preloads were effectively disguised and so did not produce differential expectations.

The effect of cognitive restraint of eating, i.e. dieting behavior, is being studied intensively. Restrained eating is characterized by a conscious effort to limit food intake, both in quantity and frequency. In general, no differences have been observed in spontaneous eating behavior between the overweight and the normal weight, or between non-dieters and restrained eaters. However, Booth & Toase (1983) found better learning of satiation in dieters who have been more successful in losing weight. Early experimental findings by Herman, Polivy and students, have influenced much behavioral research on dietary restraint. Paradoxically, a high-energy preload or even just believing it was high in energy was followed by an immediate increase in food intake in dieters, relative to non-dieters. In other words, if the believed energy content of a snack is outside the range of normal consumption, it may lead to disinhibition of the restrained eating behavior of a dieter. It has been suggested that some give up on their diet (the 'what the hell!' effect) and others may become so anxious or guilty about the breach of their diet that they eat for comfort (emotional overeating).

4.3   FOOD-SPECIFIC SATIETY: HABITUATED AND CONDITIONED

Rolls and colleagues (Rolls *et al.*, 1981, 1982, 1988a, 1988b) have extensively demonstrated short-term satiety specific to a food that has just been consumed by human subjects, as observed earlier by several groups of researchers in rats (Le Magnen, 1954, 1956a,b; Booth, 1969, 1972, 1980, 1983; Deutsch, 1983; Treit *et al*, 1983). They found that the consumed foods had a lower palatability and were eaten in smaller amounts for many minutes than identical types of non-consumed foods. Also, after consumption of other tastes, there was an increase in palatability and intake, the *variety effect*, or a restoration of interest similar to dis-habituation of the orienting response. This type of food-specific satiety is called a natural desire for variety, or a form of habituation or boredom which is known to be highly specific in stimulus and response to recently repeated behavior.

Booth *et al.* (1976, 1982, 1983) also observed in humans a type of food-flavor-specific satiety that they had shown in rats (Booth 1972, 1980; Booth & Davis, 1973; Gibson & Booth, 1989). Unlike the satiety and variety effects found by the Rolls group, this food-specific satiation could appear long after previous consumption, because it was based on associative learning from a super-satiating effect of the meal which ended with the distinctively flavored food. Also, unlike habituation satiety and the variety effect, this conditioned hesitancy about a dessert depends on a certain degree of fullness of the digestive tract in combination with the food flavor which has been learned to be predictive of the supersatiation shortly following such a meal (Booth & Toase, 1983; Gibson & Booth, 1989).

4.4   FIBER CONTENT OF FOOD

Various books and the popular press have been advocating fiber-rich foods or fiber supplements as an aid in slimming. One mechanism by which dietary fiber might have a role in weight reduction is the control of appetite. Are feelings of hunger reduced on a fiber-rich diet compared to a fiber-poor diet? Burley *et al.* (1987) and Levin *et al.* (1989) observed increased feelings of satiety after a fiber-rich meal compared to a fiber-poor meal, but no effect on actual food intake during a subsequent test meal. The texture and other sensory characteristics of high-fiber foods could create expectations of satiety. Booth (1987) mentions evidence that the rich aroma of malt fiber reduced intake in unrestrained eaters and increased it in dieters.

Large amounts of bran are unpalatable but may be needed to produce a long-term increase in feelings of fullness sufficient to help in weight reduction. In addition, some rapidly satiating macronutrient such as readily digested starch is needed in order to maintain the learned satiety effect of moderate fullness of the stomach (Booth & Toase, 1983) and the satiety signals in the upper intestine.

4.5   SWEETNESS AND ENERGY CONTENT ON LATER APPETITE

A considerable research effort has been directed at effects of sweetness and energy content of a preload on subsequent food intake and feelings of hunger and satiety. Sweetness is a potent innate stimulus of intake of sweet food, but the issue is whether the taste has inhibiting or stimulating after-effects on intake, separate from post-ingestion actions of sugars. Such studies (Brala & Hagen, 1983; Blundell *et al.*, 1988; Rogers *et al.*,

1988; Rodin, 1990; Mattes, 1990) therefore compare the effects on later eating behavior and food ratings of sweet energetic preloads based on caloric bulk sweeteners such as sucrose, and of sweet preloads without energy, containing low- or non-caloric sweeteners such as aspartame or saccharin. The results of such experiments are contradictory. Sometimes more hunger and a higher food intake have been found after sweet rather than non-sweet low-energy preloads, sometimes not.

This type of research has received a good deal of attention from the media, particularly after reports by Blundell *et al.* (1988) and Rogers *et al.* (1988) indicating that artificial sweeteners could actually stimulate appetite. Saccharin has such a stimulatory effect on intake in rats through a variety of physiological mechanisms (Tordoff & Friedman 1988, 1986). This has suggested to some that artificial sweeteners might not be effective in controlling appetite or reducing food intake. However, experiments in people have to contend also with expectations of satiety created by sweetness after experience with sugar consumption. There are two other reasons for doubting the efficacy of sugar substitutes in weight control. As mentioned earlier, taking a snack may break a dieter's self-restraint and so sweet drinks and snacks may have after-effects on eating and are not limited in their effects to their own intake. Also, the fattening effect of sugar consumption may be crucially dependent on timing relative meals (Blair *et al.*, 1989): sugar and other sources of energy in and with drinks are less likely to inhibit intake at meals and substantial snacks.

4.6 AFTER-EFFECTS OF PALATABILITY OF FOOD

Do differences in palatability between earlier meals also affect food intake in later meals? If so, is such an effect caused by physiological effects on feelings of hunger or by some cognitive process independent of physiological differences in after-effect? Hill (1984) found a more rapid rise in rating of desire to eat after consumption of a highly palatable lunch than after a less palatable lunch. The lunches were identical in energy and macronutrient contents but their effect on attitudes to later eating were not assessed. The authors conclude from their findings that it may be prudent for those who wish to control their energy intake to realize that palatable food may give a delayed increase in feelings of hunger, in which case food intake would be stimulated and control of energy intake jeopardized.

5 SUMMARY

Current issues in intake psychobiology are:
1 effects of intake on physiology;
2 effects of physiology on intake;
3 health-relevant effects of intake on eating.

Effects of intake on physiology encompass:
1 cephalic responses (that protect the body against the stress imposed by digestion, absorption and storage of nutrients);
2 the thermic effect of food (which, of a mixed meal expressed as a percentage of the energy that is ingested is usually about 10 per cent; the pattern of change in energy expenditure being influenced by meal size and composition);
3 effects on mood and performance that have been suggested as enhancing or altering an existing mood state.

Effects of physiology on intake encompass:

1 food cravings (that may be based on some nutritional deficit or imbalance);

2 so-called wisdom of the body (such as an innate appetite for sodium);

3 sociobiology of diet (surviving because of dietary practices that provide nutrients limiting or lacking in crops, hunting and food gathered from the wild);

4 hunger (desire for food as a source of energy as an effect of incipient or actual lack of energy) and;

5 satiety (food component-dependent post-ingestional effects).

Health-relevant effects of intake are:

1 nutrient, hunger and satiety effects (with respect to expected energy-density of the food);

2 cognitive effects on energy intake such as shown in restrained eating behavior;

3 habituated and conditioned food-specific satiety effects as a variety effect;

4 effects of fiber content of food (creating expectations of satiety);

5 effects of sweetness and energy content on later appetite;

6 after-effects of palatability of food.

2
Concepts and methods in the psychobiology of ingestion

The previous chapter illustrated a range of current issues facing research into the psychobiology of food and water intake. The following four chapters are systematic reviews of the state of the evidence provided by research on theory and practice with regard to human eating and drinking behavior and its biological bases. The reviews include relevant evidence from other species, which is particularly abundant from the laboratory rat. They emphasize the requirements for analysis of the causal structure of the behavior resulting in intake. Biological determinants and consequences of ingestion are included, as in Chapter 1. However, areas of biology that cause these factors are not treated systematically in these reviews. Such matters are dealt with in later chapters.

Psychobiology is essentially psychological as well as biological. This implies that the theoretical concepts and research methods of the psychobiology of ingestion must address the causal processes that are operating in the influence of the external and internal environments on the overt activities of eating or drinking. Indeed, in the human case, the verbalizations and other symbolic behavior can be shown to illuminate how the environment influences the acts and reactions of ingestion.

The pattern of stimulation from the diet, the body and the physical and social surrounding acts on movements and words through the brain of course. However, the approach of psychological science is to hypothesize about the causal structure and dynamics in such influences in terms of the externally observable stimuli and responses. That is, psychology takes a 'black box' approach to the transmission of information from the environment through the organism and into the environment again.

As in any other science, there is a tension within psychology between a minimalist approach to theory and investigation, known historically as *behaviorism*, and an abstract and speculative approach to theory construction and experiment, now increasingly known as part of *cognitive science*. The ordinary person's way of talking about behavior, in 'mentalistic' language, is consistent with either extreme. At present a careful use of ordinary words like 'appetite', 'satiety', 'preference' and so forth is sufficient for analyzing the causal processing within ingestive behavior, and for designing and interpreting experiments on the psychobiology of food and water intake in humans and other mammalians. Care is necessary, however, to make sure that words like 'hunger', 'thirst', 'desire to eat' and so on are used to interpret tests between different processes that organize ingestion.

This chapter will therefore outline the behavioral meaning of the necessary concepts and the basic requirements of the design of investigations that tell us something about information processing in elements of eating or drinking behavior or by talk relevant to it by the eater or drinker.

1   EATING BEHAVIOR: THEORETICAL CONCEPTS

Eating behavior is a psychological process resulting in the ingestion of food. The results of this behavior are seen in the total amount of food or nutrients ingested during a day, the timing of periods of intake throughout the day, qualitative and quantitative selection between foodstuffs, the duration of eating, the eating rate, and the possible variations in these aspects of food intake. Eating behavior consists of such ingestive responses to sensory input as are mediated through thoughts and feelings. Food intake can only be studied fruitfully if this causal structure of eating behavior is clearly analyzed and described.

Drinking behavior and fluid intake are not excluded by the above statements. Unless specified otherwise, food consumption or ingestion includes the intake of water as well as energy and all other nutrients. Meals typically include fluids as well as solid foods. Sometimes we consume fluid by itself but even then a drink is often more than just water and may be calorific.

Food and water intake accumulate as a result of series of acts involved in eating and drinking, i.e. ingestive behavior. This behavior of ingesting solids or fluids is a set of oral movements that are under the control of the sensed physicochemical characteristics of the foods and drinks, e.g. soft, warm and meaty flavored or sweet and crisp. However, the eater's or drinker's reaction to the sensory characteristics of available items of diet also depends on bodily state, e.g. how much is in the stomach or what nutrients are stimulating the wall of the upper intestine or supplying substrates to be oxidized in the liver. The ingestive reaction also depends on cultural and interpersonal features of the situation, e.g. what time of day or course in the meal it is and who is present and what their social relationships are to the eater or drinker. Hence, to explain the choices and amounts of food and fluid that are consumed, we will have to take into account each successive decision to accept or reject a mouthful in terms of the sensory, somatic and social influences bearing on the person or animal at each moment in time.

Ingestive behavior is thus a causal relationship between stimuli from the food and the external and internal environment and the motor patterns of sucking or biting, chewing and swallowing.

First then, let us look at the patterns of intake that have to be explained as the cumulative result of changing causal structures of ingestive activity. Next, we consider the terms used to refer to different aspects of this causal structure of eating and drinking. Thirdly, we look briefly at the history of types of mechanisms assumed to be involved, before turning at the end of this chapter to the fundamental method of science as applied to analysis of ingestive behavior.

1.1   TERMINOLOGY

Terms such as *hunger, appetite, preference, satiation, satiety, sensory-/nutrient-specific satiety* and *satiating efficiency/power* are frequently used to describe aspects of eating behavior. Some of these terms, such as hunger, are well-known and appear clear to everybody. In scientific literature, however, there is frequent disagreement on the exact meaning of such terms and hence on the measurement of eating behavior.

All constructs or concepts to be used in science need to be translated into measurable variables, a process called 'operationalization'. This translation can be difficult and cause controversy among researchers, especially when they do not attend to the problems of distinguishing the causal processes underlying the phenomena they are investigating. Especially for behavioral terms a scientific understanding is sometimes difficult. This is partly because we use such language to talk about ourselves and other people in everyday life. It is also because there are long traditions of more or less philosophical discussions about humanity in that kind of language that do not have to be tied down to the testing of causal explanations of systematic empirical observations. For over 150 years, experimental psychologists have found that, as soon as they sought to understand the basis of a simply named sensation, emotion or motive, unexpected complexities rendered definitions of ordinary vocabulary quite pointless. Thus, all we should expect of words like 'appetite' is a pointer to a set of behavioral phenomena we may then seek to understand mechanistically.

An *appetite* is a disposition or orientation in behavior towards some object or objective, such as earning money, becoming famous or eating some food. The disposition may be weak, in which case anyone can observe that behavior is readily reoriented to something else. The appetite may be strong and dominate activity. Subjectively, the appetite may be a faint wish or an overwhelming compulsion respectively.

When we want to separate appetite for food from appetite for water, we can use the word hunger for the motive to eat, and the word thirst for motivation to drink. *Hunger* is the motivation to acquire and ingest food, i.e. the observable tendency or the conscious desire to eat. However, the word hunger is secondarily used to name bodily sensations associated with a relatively empty stomach.

Similarly, the word *thirst* is sometimes used to refer to sensations such as dryness of the mouth. Nevertheless, hunger and thirst are fundamentally the publicly observable behavioral tendencies to eat food and drink water, not the subjective experience of bodily sensations, the existence of a physiological deficit of energy or water, or a signal of it such as an empty stomach or low blood volume.

The effect of sensory qualities of a food, such as its appearance, color, odor, mouthfeel and taste, on the disposition to eat is often called preference, or palatability, or liking. *Preference* is a simple concept for beginning to describe the organization within observed behavior of selecting specific foods from alternatives. However, the study of sensory preferences cannot be confined simply to observing the effects of the food stimuli. *Palatability* has its origins. There are some inborn sensory preferences but mostly they are learned by assocation of sensed food qualities with effects of the energy content and macronutrient composition of the food, and by social learning of menu structure and the suitability of appearances to the eating occasion.

Satiation is the process of the inhibition of food intake as a consequence of eating and the tendency to refuse food as a result of that process. *Satiety* for food is this state or degree of inhibition of food intake as a result of ingestion. Satiety specific to a particular food occurs as well as inhibition of intake of foods more generally. Recent sensing of food during its consumption and what is left on the plate contribute to satiation. Processes in the gastro-intestinal tract (stomach distension, stimulation of chemoreceptors, release of intestinal hormones, e.g. cholecystokinin) also contribute to the process of satiation, i.e. to the early states of satiety towards the end of the meal and for a while afterwards. A variety of processes triggered during the period of absorption of the digested food constituents usually contribute more to the later stages of satiety. This sequence of activation of signals is depicted in Figure 2.1. Memory of the timing and content of the meal also prolong or shorten the inhibition of hunger.

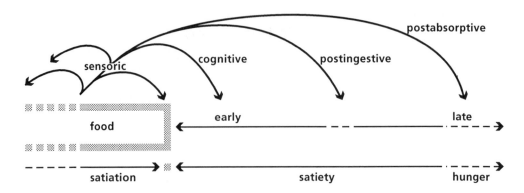

FIGURE 2.1
Outline of the sequence of activation of satiety signals during and after a meal
(Blundell, 1990)

The arrows indicate only the initial activation of a source of satiation during and after meals. Cognitive effects, for example, could last as long as memory and gastric effects while the stomach remains considerably distended.

All these effects of food ingestion, digestion and absorption on eating behavior are mediated by internal signals such as peripheral metabolism and neuro-endocrine responses that are integrated with external cues by complex processes in the brain, into characteristic aspects of behavior. That is, hunger, food preferences and their satiation are multifactor processes in eating behavior. Cognitive mechanisms integrating information from foods, the body and the surroundings play a role before, during and after eating. The interaction of two sources of influences, the body and the food, has been analyzed in a line of experiments in rats, monkeys and human beings by Booth and colleagues (1972, 1976, 1985), paralleled by work in rats by Deutsch (1983) and in young children by Birch (1980, 1982, 1984, 1986, 1987, 1989).

2 PHENOMENA OF EATING BEHAVIOR

The purpose of the scientific study of food and water intake is to understand and explain eating and drinking behavior. It is therefore important to agree on the phenomena that need to be explained and understood. There are at least six phenomena that research on ingestive behavior has addressed.

1 The broad pattern and outcome of eating behavior is relatively stable. Most people eat, if possible, several times a day, usually at the same times of the day. Between successive days, food intake may vary considerably but in the long term food intake seems to be rather constant. To what extent is this relative constancy in pattern and total food intake based on biological controls and how far can it be explained by social factors, e.g. contraints on the habit of timing and size of meals?

2 Eating behavior is susceptible to changes in the energy density of food. Some investigators have found such effects in children and adults, and they are clearly seen in animal experiments. The human response to known manipulation of the energy density of food is an important topic to study for reasons of health. If the energy density of food is lowered, will people compensate for the caloric deficit by ingesting more food? If so, to

what extent? Is the degree of compensation dependent on the nature of the manipulation of the energy density and on the timing of eating occasions?

3 Energy intake usually changes when the level of physical activity is permanently altered. What changes in eating behavior mediate these effects and are they physiological or cultural in origin?

4 Eating behavior is not continuous, but occurs in bouts that are often placed periodically over the day. To what extent do physiological processes terminate bouts of eating (meals and snacks) and influence the different intervals between, say, breakfast and lunch and lunch and supper? Do the social factors in human meal sizes and timing have some biological bases?

5 The rate of eating typically decreases during food ingestion if it changes systematically at all. Is this the beginning of physiological, sensory or social processes that end that bout of eating? Are there other signs of inhibitory processes developing during a meal, whether or not eating does slow down?

6 Most people find it extremely difficult to force themselves over a prolonged period of time to eat substantially more or less than they would normally do. Can we find long-term physiological signals of excess or deficit behind the social influences, for example from how much other people eat and/or expect one to eat, and from the availability of foods and occasions to eat them?

3 HOMEOSTATIC FEEDBACK THEORIES OF EATING BEHAVIOR

The conceptual models developed over the last century to explain physiological influences on eating behavior and the resulting food intake have been based on the theory of homeostatic feedback. In homeostatic theories, eating behavior serves to regulate a certain variable within the body. This regulation is effected by a set of autonomic and behavioral controls by which increases and decreases in that variable are corrected, returning the variable towards its original value. Such a corrective control is said to be providing *negative feedback*. (Positive feedback intensifies the change instead.) Based on the theory of negative feedback, engineers developed mathematical techniques, called control theory, for analyzing the performance of self-stabilizing and goal-directed machinery, such as air conditioning and guided missiles.

A very simple model of eating control based on physiological feedback is depicted in Figure 2.2. This model assumes that an organism starts eating as a consequence of hunger sensations. Food intake results in the release of satiety signals causing the termination of feeding. After some time these signals fade away and the organism is stimulated to eat once again by emerging hunger sensations. This conventional feedback model of eating behavior may be extended to allow for intake-inhibitory signals from different parts of the body, each affecting different kinds of food (Mook, 1988).

Two types of homeostatic theory have been proposed. In some people, a particular substance or type of substance is focal, such as glucose, fat or protein. In others, the metabolism of all these substances for energy is regulated by eating behavior. A proxy for this metabolism signalled to the brain would be body temperature, rate of energy expenditure or energy production in the liver, depending on the theorist.

Homeostatic theories on food intake control are named according to their regulated variable, thus:

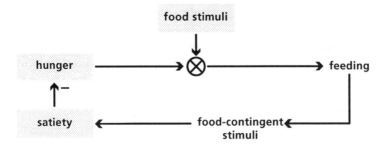

FIGURE 2.2
Simple model of eating control based on physiological feedback (after Mook, 1988)

1 *Lipostatic theory*. This proposes that the amount of fat in the body, particularly the triglyceride stored in adipose tissue, is regulated by food (energy) intake and by energy use by the body.

2 *Glucostatic theory*. As its core, this theory assumes regulation of the use of glucose by certain privileged cells. If the use of glucose by those cells is low, food intake would be stimulated, whereas food intake would be suppressed if the glucose metabolism by these cells is enhanced.

3 The *aminostatic theory* states that eating behavior regulates a deficit or surplus of amino acids in the blood plasma.

4 The *thermostatic theory* considers metabolic heat production as the variable regulated by the control of food intake. The signal to the brain may be body temperature. In what is termed the ischymetric theory, it reflects body heat production as well as the cost of movement.

5 The *energostatic theory* postulates that the energy produced by the liver from oxidative metabolism of all the absorbed and mobilized nutrients is monitored to control food intake. This theory is a refinement or combination of the other four theories on food intake (Booth, 1972; Friedman & Stricker, 1976).

All homeostatic theories assume that the level of a certain variable is monitored by a 'detector' in the brain via signals from the body and that this monitoring leads to an appropriate activation of effectors that control eating behavior. Identification of the physiological or biochemical factor serving as a signal has proved to be an extremely difficult task. Some progress has been made since this type of research started in the 1950s, but this issue is not likely to be resolved satisfactorily until the homeostatic feedback idea is extended with the consideration of responses adapted by learning and integration of physiological signals with nonphysiological influences (see also Chapter 9).

4   HOMEOSTATIC THEORY OF DRINKING

The body continuously loses water through the lungs and at least a minimum is used in excretion, as well as in sweating, panting and other body-cooling strategies. These losses can sometimes be replenished by water produced by metabolism as well as by water obtained from moist foods. However, the amount of water in the body is regulated to balance intake and losses. The antidiuretic hormone vasopressin reduces the amount of water allowed to pass through the kidneys into urine. However, if and when despite this effect a deficit still develops, thirst is aroused and water is sought and consumed.

The brain monitors the amount of fluid inside and outside cells. The intracellular fluid level reflects the osmotic pressure of ions and molecules in the blood that are excluded from cells. Sodium is usually the most important solute that varies in concentration. The extracellular fluid level is signalled via the pressure of the blood in large vessels and by hormones such as rennin, which causes release of angiotensin.

A deficit in either intracellular or extracellular fluid triggers thirst in adults of the mammalian species tested, including humans. However, newborn mammals have yet to be shown to distinguish watery from non-watery materials in consumption aroused by a bodily water deficit, osmotic loads or administration of angiotensin. So, although it is generally assumed that water-specific appetite is an innate part of a homeostatic system, this still remains to be established in the human or other species.

Drinking by humans is probably not usually triggered by a water deficit, though. Sometimes it seems that we drink because it is a conventional time or place to consume fluid. Another reason may be that we want a refreshing feeling in the mouth, the effects of the caffeine, alcohol or sugar in the drink or just to join in the fun of which drinking is a part. Nevertheless, much drinking in humans as well as in other species is associated with eating or physical exertion, both of which make demands on the water balance. Rats drink much of their water before meals on a dry diet. Most of the rest of their drinking immediately follows meals, before much of the water is required for digestive secretion and urinary excretion of waste products. Eating triggers drinking in the rat by several physiological mechanisms. It remains to be seen how much meal-associated drinking in people is obligatory from demands on body water, how much is anticipation pre-established in the species and how much is set up by experiences of the individual.

Hence, drinking not triggered by water deficit is not necessarily non-homeostatic. Even drinking for refreshment or fun might turn out to be learned or innate behavior that has functioned in some circumstances to prevent water deficit. It is therefore arbitrary to divide particular bouts of drinking and fractions of water intake in a bout into homeostatic and non-homeostatic categories. Of real importance are the actual influences on drinking from moment to moment and how those influences came into being. 'Homeostasis' and 'regulation' are only words that were useful to short-circuit posing questions about mechanisms. They must not be allowed to substitute for causal analysis and block any progress in our understanding of ingestive control processes.

Because drinking as well as eating is a normal part of bouts of ingestion, this part of the book will focus on eating but most of what is said is readily extended to fluid intake under non-deprived conditions.

5 NON-HOMEOSTATIC THEORIES ON EATING AND DRINKING BEHAVIOR

Non-homeostatic theories on eating behavior emphasize factors other than physiological requirements. The most popular of these is the *hedonic* theory of eating behavior. It postulates that eating behavior does not necessarily compensate a deficiency in a homeostatic manner, but may originate from the motivation of the organism to find pleasure, reward and satisfaction. Eating behavior, regardless of how it started, is continued because certain sensory characteristics of the food stimulate pleasure systems in the brain.

This approach is confused. Eating and drinking triggered by deficit is instrumentally rewarding and emotionally satisfying and the food and drink can give great pleasure. Indeed, it has been suggested that pleasure is a homeostatic phenomenon (Cabanac, 1971).

The hedonic view may even be circular. It is the motivation to eat food and drink fluid that is to be explained and palatability is just another way of looking at this motivation. To say that food is pleasant may be no more than to want to eat it. The scientific issue is, what sources of influence other than energy deficit or water deficit make food and water palatable, rewarding, satisfying and a source of sensual pleasure? Obviously, social factors are important in the timing of meals and drinks, and the amounts consumed. The challenge for the behavioral sciences is to find out what the relationships are between the effects of social influences and biological factors insofar as these are operative.

Most scientists now prefer a synthesis of homeostatic and non-homeostatic conceptions of eating behavior (Kissileff & Van Itallie, 1982) and drinking behavior. The modern theories on food-intake regulation imply a heterogeneous feedback system. Various signals derived from the gastro-intestinal tract, liver, circulating energy nutrients, depot fat, or metabolizing cells are thought to transmit information to a central system, i.e. the brain. This integrates information from the various peripheral components and the environment and responds to the integral pattern. Regulatory influences of physiological signals have to be adapted by evolution or learning to variations in foodstuffs and demands from the environment (Booth, 1980). Hence, while regulatory mechanisms may on occasion be overwhelmed by non-homeostatic influences, many potentially disregulating influences on eating and drinking may abide by biological imperatives while at the same time serving social or personal objectives.

6 RESEARCH METHODOLOGY

6.1 MEASURING EATING BEHAVIOR

To assess the processes of eating behavior, various methods are available which can be subdivided in experiments and observational studies. The latter are invaluable when the data can be analyzed to test hypotheses against each other.

In an experiment the investigator changes the level of one or more potential influences (the independent variables) in a controlled way. A measured change in an effect of interest (the dependent variable) compared to a control or starting condition, or to the effects of other levels of the influence(s), allows conclusions to be made on causes. Dependent variables include weight of food consumed, relative amounts or probabilities of choice among foods, eating rate, and, in the human case, statements on foods, the body or the situation as perceived by the eaters.

Usually investigators study eating behavior under the influence of various factors. Even when the intention is to vary only one factor, it is often difficult to avoid varying other influences at the same time. For example, the energy content is generally related to a food's weight and volume. Macronutrient contents vary with texture, smell, taste and appearance; in real foods, they vary with assumed nutritional functions and habitual culinary roles. Thus, many experiments and observational studies require co-varying factors, to be distinguished by statistical techniques such as multiple regression or further analysis.

The development of the psychobiology of food and fluid intake depends on acknowledgment of the unmistakable truism that science is the identification of causal processes. The only way to begin to understand eating behavior, therefore, is to characterize the immediate causally effective input to the momentary output from the organism that we are observing, whether it be in action or in words. The observations and measurements might be carried out around or during an ordinary meal, for example, with reference to the rules determined by a dieter's eating, or the mental state of a bulimic patient as she starts a binge.

The methodology for the study of eating behavior comprises analysis of influences on food intake. The experimental designs required to study eating behavior assess the strength of causal effects of sensory, bodily, and external stimuli on responses that result in observed selection and consumption of food.

6.2   OUTPUT PATTERNS: WHAT EFFECTS ARE THERE TO BE INFLUENCED?

We can analyze the rat's microbehavior (e.g. licking rate, lick bout duration), or people's food ratings or their selection of food intakes, and attempt to pick out patterns in these dependent variables. Identification of pattern of output is the first step, or rather, part of the very first step, towards characterizing the mechanisms of any phenomenon. One output measure, such as an intake or a rating, by itself does not say anything about the causal processes that produced it. One should dissociate the multiple measures as a first step towards finding any effect that is under some influence(s).

For example, some microbehaviors in the effects of a drug on licking in the rat are dissociated from others. Faced with such a pattern of data, we must allow that they contain evidence of the drug affecting one eating mechanism and not another. It remains to be determined, however, which mechanisms are differentially affected.

Verbal information on food preference and its changes during feeding can be used to indicate appetite, as well as ratings of bodily sensations and of desire to eat food generally. This is also possible with the help of food preference lists, which may contain photographs of foods. A procedure where a subject is forced to choose between pairwise presented foods is another possibility. Assessing the palatability or the hedonic value of foods is possible by using questionnaires with a line to be marked for each food, anchored with the terms 'would always choose' to 'would never choose'.

The output could be one set of various kinds of ratings of appetite, which range from abdominal sensations to the general question of how hungry one is, the pleasantness of several staple foods, or how much energy is in the food that people say they would order from a menu. It was shown that all these ratings tend to show a high correlation, and indeed that before/after meal differences also intercorrelate even better (Booth et al., 1982). This is the same phenomenon that Halmi (1989) and others noted, namely that many of these ratings lean heavily on one single factor, both in normal eaters and in sufferers from eating disorder. All of these ratings also correlate with measures of rated preferences, of actual food choice, and of the cumulative results of eating behavior, such as food caloric intake. That is why we should say that all these ratings are of appetite (including its inhibition by eating, i.e. satiety).

Food intake as an output measure has the limitation of being the cumulative result of much behavior. Hence, intake is at best an indirect index of causal output. More direct output measures are momentary eating rates, food choosings and ratings of the disposition to eat. Despite their verbal distinctions (no doubt expressing genuine experience), these ratings are in fact all highly correlated with the tendency to eat. Hence, each rating is no more than another measure of overall appetite. It is therefore logically fallacious and contrary to the evidence to assume that different somatic influences on eating behavior are reliably identified by ratings of sensations of hunger and satiety, or dietary factors by the pleasantness of foodstuffs or of their sensory characteristics. This also makes a nonsense of the supposition that differently worded ratings measure different experiences or distinguish between the different kinds of objective influences that they refer to, for example, different gastro-intestinal processes or metabolic versus sensory influences on eating. Insofar as these output measures all share common variance, and insofar as they are all intercorrelated, they cannot be effective discriminators of different influences. Such different causes of appetite may well exist but the ratings provide no evidence of it under the conditions of testing that have confused, for example, cognitive and physiological influences. One way to put it is that they are all the same rating, even though the words are different.

By the same token, however, when we do dissociate outputs, most interesting and productive questions can be asked. For example, in Halmi's results, ratings expressing preoccupation with food were an output measure that was not associated with most of the other direct measures of eating (Halmi, 1989) . The question now arises: what is this distinct behavior by these individuals? The evidence is that what the persons are saying by their dissociated rating of preoccupation with food is different from 'I have an appetite' or 'I don't have an appetite'. Is it guilt because of the food they have just eaten? Is it the respondent appealing to the experimenter to consider what is being done to the long-suffering subject of investigation? That is a genuine scientific and clinical question that obviously could be important to follow up.

Is it money-saving? Is it perhaps stereotyping the binge as consumption of 'junk' food? In other words, is the binger picking out the low-protein foods because he or she feels that what he or she is doing is just eating 'rubbishy' foods in a 'rubbishy' way? There is a real question of what is the cause underlying this particular output measure, which arises only because that output is separate from the other responses. Because it works differently from the other measures, it is telling us a great deal more than masses of data that are all redundant with each other.

So, those are examples of real scientific findings and further questions that can arise from multiple output measures. If we show that we do have dissociations, we can ask what is peculiar about the cause of each coherent and independent set of indices. If the relevance of such an issue to the aspect of eating that interests us needs to be established, the output measure in question has to be associated with another output measure that is characteristic of that aspect of eating. We need a valid measure of bingeing, dieting, palatability, satiety or whatever, before we can see whether our new measure is related to it.

6.3   DIAGNOSIS OF MECHANISMS OF INFLUENCE ON EFFECTS:
      INPUT/OUTPUT ANALYSIS

Yet, this approach still gives us no information on the causes or controls of the differential effect on which for instance drugsact. Output analysis never identifies the origin of the effect. To do this, we must have input data that relate to the output dissociation.

The nature of the influence on a distinct output can only be identified by input-output analysis. This is the nearest we can reach in the way of an 'algorithm' or a paradigm for psychobiological research into food and water intake. We do not need scientifically indeterminate arguments about the definitions of words. Nor do we need conceptual diagrams littered with boxes and arrows that have no definite relation to data gathering or data analysis. This paradigm, algorithm, or model is the specification of a mental mechanism in appetite in a way that implies how to measure it and how to interpret any relevant observations. This is indeed the approach that has yielded generalized laws of physical, biological and economic causation.

We must therefore measure distinct *causal inputs* to eating. Furthermore, the identification of *mediating causative processes* requires the double dissociation of responses to these inputs, be they gastric wall tension, gustatory stimulation, tactile or visual sensing of skinfold or muscle thickness, etc. Then the *quantitation of identified causal influences* on orderly or disorderly eating would, in addition, require unconfounded and unbiased psychophysical (dose-response) analysis: only two levels of a factor cannot establish linearity.

No diet or rating by itself can be a 'test' of palatability or preoccupation with food. The test for a mechanism is a discrete set of input-output relationships within behavior. There is no way that we can interpret an intercorrelated set of outputs that has been dissociated from other outputs, unless we have some input measure that tightly correlates with the output factor.

The mental mechanism generally found in this context is the appetite triangle (Figure 2.3).

When differences from maximum motivation to ingest are unfolded around the theoretical peak level of output, an appetite mechanism is measured by the linear psychophysical relationship between any kind of input and any output measure that relates to eating, be it appetite rating or ingestive behavior.

Note also that it is not the intercept or the slope of the input-output function, nor even its correlation coefficient, that most effectively measures the strength of a causal factor or the sensitivity of the eating to the influence (and its modulation by an intervention). The more unexplained scatter there is in the relationship between the input measure and the output measure, the less we know about what is going on in the situation. Thus, the strength of the observed mechanism is the scatter of the input-output function relative to the slope; the smaller that ratio, the more difference the influence makes to the output. Psychologists have used this measure of causal strength for over 150 years, under the name of the *just noticeable difference* or *difference threshold*. It measures the sensitivity of the output to that particular input over the tested range.

In the *preload/testmeal* design of satiety experiments a subject receives a fixed amount of food or drink - the preload - with a known nutritional composition and energy content. The subject is requested to refrain from additional eating for a certain period of time after the ingestion of the preload. During this time ratings of, for example, hunger and satiety may be obtained.

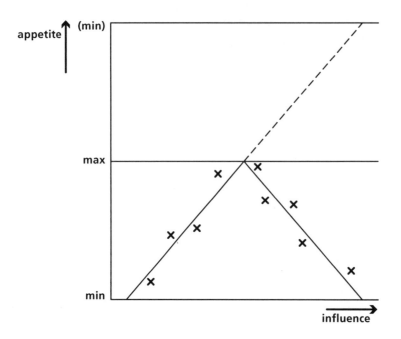

FIGURE 2.3

Quantitative relationship between learned appetite for food and an influence on it in an individual organism

Differences from the most motivating level of the influence are unfolded and a linear regression calculated through the raw data.

The subject subsequently receives a test meal and may eat as much as he/she wants. This procedure is then repeated under identical circumstances, with the same amount of the preload having a different nutrient composition but identical sensory characteristics. If, at the time of the test meal or of the ratings the two nutrient compositions are known to be acting on different receptors, then the effects can be used to characterize the roles of those receptors in satiety. When enough is known of the passage of food through the gastro-intestinal tract and circulation, of what the satiety signals are and what and when people habitually eat, generalization to everyday life may be possible.

However, the usual difficulty of the input/output analysis, is dissociating relationships between different sets of inputs and outputs.

An example of this comes from flavors of food (Table 2.1). Sweetness ratings for sweet, white coffee appeared to correlate very tightly (showed very little residual variance) with the sugar content of the coffee, whereas they correlated very poorly with the creamer (whitener) content. On the other hand the whiteness ratings correlated rather poorly with the sugar content but did correlate very tightly with the creamer content. This is a *quantitative double dissociation*, using psychophysical techniques. It shows that when people talk about sweetness they are actually talking about the sugar or the sweetener in what they are tasting. Anyone would hope that this could be shown, but without data like these we are merely assuming it to be so.

TABLE 2.1

Median (range) percentage residual variance of mean linear regressions between log concentrations of constituents and ideal-relative intensity ratings on freely described off-ideal characteristics of hot coffee drinks, in nine habitual drinkers of sweet, white coffee*

	constituent	
ratings	*sugar*	*whitener*
sweetness ratings	3 (18)	48 (93)
whiteness ratings	28 (81)	2 (3)

* $p < 0.01$, one tailed Wilcoxon test, in all comparisons between constituents and ratings.

6.4   WHICH MECHANISM IS A MANIPULATION ACTING ON?

As soon as one faces the basic reality of the mechanistic and quantitative scientific approach, it becomes obvious that to identify and measure any influence on eating behavior we must measure the inputs, and distinguish the inputs that affect the relevant outputs from those that do not. Furthermore, without this we are not justified in claiming what mechanism it is that a nutrient, a drug, a part of the digestive tract, social influence or a therapy affects (Figure 2.4).

Figure 2.4 applies this basic logic for scientific analysis to the effects of manipulated conditions on dissociated mechanisms. Each panel in the figure gives a pair of input-output functions under a particular condition. There are three levels of each input, which is the absolute minimum to see whether there is a straight-line relationship between input and output. The output can then be whatever measure of appetite is suitable, whether behavioral or verbal.

The rows of panels in Figure 2.4 represent two kinds of conditions. They may be two drug treatments, for example, D1 and D2 dopamine antagonists, and each of three doses. They may also be different kinds of people, a restricting anorexic and a bingeing anorexic, an anorexic and a bulimic, or a bulimic and a control. Alteratively, the recovered bulimic, the moderately disordered bulimic and the acute ostentatious bulimic would be a kind of dose response, as with the drugs.

Whatever the overall design, it might be two different foods we are talking about, for example, tomato soup and a biscuit.

Alternatively, the two inputs could be constituents of one food. In that case, in each panel of the figure we have the effects on intake or preference of these constituents (sugar and salt, say) in the test food.

If we find that, for example, in the acute bulimic the responses to the two inputs behave differently, whereas in the normal they behave in the same way, this would point in the direction of a mechanism that operates differently in bulimics from normals. The bulimic's fear of sweetness might be greater than that of saltiness, out of all proportion to that seen in normal users of food.

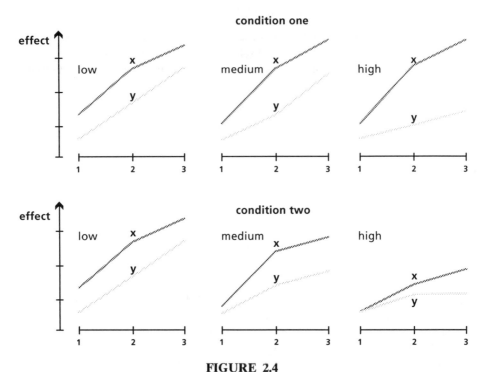

FIGURE 2.4
Identification of the differential effect of a condition on an identified and measured causal process

Examples of effective conditions could be two forms of eating disorder or two drugs with distinct receptor specificities. The causal effect (the ordinate in each panel) could be a measure of eating behavior, contrasted with other behavior. The causal influences (X or Y, each tested at three levels) could be dietary sugar and salt, gastric distension and hepatic oxidation, or nearness to mealtime and breaktime at work. The strength of a mechanism is measured by the slope and precision of its linear effect. The specificity of the effect of the condition on one mechanism rather than on the other is shown by differential effects on slope or intercept with increasing intensity of condition (i.e. severity of disorder or dose of drug). Condition One specifically weakens the effect of Y, whereas Condition Two generally disrupts the category of X and Y.

Take a psychopharmacological example. If a drug makes the same preference measure less sensitive to both saccharin levels and salt levels, then the drug may be affecting a palatability mechanism. If the drug affects only the slope for sweetness, then it would appear to be affecting a taste perceptual mechanism.

In other words, to obtain any sound evidence of the effect of an abnormality or an intervention on a mechanism, we have to hold together three kinds of variables and measure them all at once. The results of a single test, with no measure of differences in influences on the response, cannot advance our scientific understanding of the mechanisms involved. This is true whatever number of subjects are used, however sophisticated the statistics and however the significance of the p value for a difference. We not only need to know the differences between the conditions, for example, the kinds of people, but also the differences between the levels of the inputs and outputs, and, furthermore, we need at least two examples of influences that show dissociable input-output effects to work with (Table 2.1).

7   SUMMARY

A behavioral explanation of intake consists of theoretical concepts in eating behavior, including *terms* such as hunger, appetite, preference, satiation, satiety, sensory-specific satiety, satiating efficiency/power, and *phenomena* such as: pattern of eating behavior, susceptibility to changes in energy density of food, relationship to the level of physical activity, periodicity of eating and timing, eating rate and eating rate change, forcing oneself eating substantially more or less from normal.

Homeostatic (lipostatic, glucostatic, aminostatic, thermostatic, energostatic) and non-homeostatic theories on eating and drinking have been developed.

Modern theories on food-intake regulation imply a heterogeneous feedback system: various signals transmit information to a central system, i.e. the brain. This integrates information from the various peripheral components and the environment and responds to the integral pattern. The research methodology comprises analysis of influences on food intake.

3

Palatability and the intake of foods and drinks

This chapter focuses on the impact of foodstuffs on the senses during eating and drinking. It emphasizes that all control of food and water intake operates through the effects on ingestion of the sensed characteristics of the solid and liquid materials that the foraging animal finds to be available. Examples are given of sensory influences on choice and intake of foods and drinks in rats, monkeys and humans. The psychological measurement of sensory preference - often called palatability - is explained. A major example of health effects of palatability is then discussed in detail, both the theory and the practical implications: the role of snackfood consumption in obesity.

Eating and drinking are usually under the control of the currently sensed characteristics of foods and drinks. That is, the impact on the senses of the composition of the dietary alternatives that are available from moment to moment is the direct cause of accumulation of intake energy, water and other nutrients throughout.

Even though we might not be thinking about the flavor or texture of a food we are eating, it determines our ingestive behavior.

This is not a denial of physiological influences on our hunger, thirst, satiety and preferences, nor is it to discount the influence of culture on eating habits. The point is that the exact effects of the momentary somatic and social influences on dietary choices and intakes usually depend on what foods and drinks are present. That is, the influences of physiological state and external environment are cognitively integrated with sensory influences. Hunger and satiety signals, and culture and interpersonal factors do not just interact with sensory factors in an additive manner. The interactions are non-linear.

Furthermore, each of us has distinctive patterns of integrating sensory and contextual effects on ingestion. That is so because these patterns have been established by our personal past experiences of the significant sensory, social and somatic associations of eating. For example, sweetening a new flavor of tea might make it more pleasant next time, even without sugar. Eating or drinking a novel item given with a smile as a reward makes it better preferred subsequently. The smell, taste or texture of a food becomes more attractive when it is followed by nutritional benefits. These are all examples of *learned sensory preferences* or, in other words, the *experiential induction of palatability*. As we shall see, these acquired likings are also learned in relation to the moment and context experienced.

Thus, the actual controls of ingestion are *momentary*, *individualistic* and *situationalized*.

Let us therefore now consider, in illustrative detail, the processes of palatability and its contextualization.

1   SENSORY CHOICES DURING INTAKE

All food and fluid intake results from choices; choices of what to eat and drink, and how much or how little. Certain foods and drinks are chosen for particular occasions or people, such as breakfast tea, traditional Sunday lunch in Britain, Thanksgiving dinner in the United States, pea soup and sausage on a cold winter's day in the Netherlands, croissants at breakfast in France, snackfoods, party foods, dieters' products, children's foods, red meat for a macho male, and champagne for a celebratory toast.

This sense of appropriateness to context is very obvious when there are two or more courses on the menu. Choices change during such a meal. However, even in a laboratory experiment using a single foodstuff, cessation of intake is specific to the food that is being eaten. If a different food is presented, the chances are that eating will start again. In other words, eating induces a change in relative preference between eaten and uneaten foods. This is evidence for the stimulating effect of sensory variety as well as for a boredom or habituation component in satiation.

Intake is also affected by learned changes in relative food preference from early to late in the meal. Many of these shifts are picked up from conventions in a culture. Examples include the distinction between entree and dessert menus, the eating of salty, savoury or meaty foods before sweet foods in the West, or the follow-up of fish by sorbet in elaborate French meals. However, there is also a major influence of physiological context on choices among desserts and on choice of how much dessert to eat. An entree that is rich in readily digested carbohydrate produces a post-ingestional effect after a short delay. This effect may therefore arise after the eating of a dessert has started or even just been completed. It can then change the liking for that dessert, by a mechanism of *associative conditioning*. The sensory qualities of the dessert and signals of satisfaction of the digestive tract are together conditioned to inhibit ingestion by association with this briefly delayed effect. It appears that concentrated starch eaten in the early part of a meal can produce a mild duodenal distension at about the time the meal ends (Booth & Davis, 1973; Booth *et al.*, 1982). This entree-induced oversatiation conditions a lowered attraction to the dessert. Hence the dessert is less attractive during the next meal when it is offered and the eater is already half-full. Less may be eaten of this dessert or it may even be refused altogether. In either case, there is a *learned* reduction in the amount eaten.

Either facilitation or inhibition of eating can be conditioned by ordinary nutrients consumed. Readily assimilable energy (or a balanced mixture of essential amino acids) conditions increased liking for the foods associated with its absorption. Without any countervailing influence, this effect would encourage a person to eat more and more of those foods. Indeed, such learned de-satiation may be the foundation of food binges (see Chapter 5).

The conditioned decrease in liking for the dessert is specific for the state of a partly full stomach during which the dessert was eaten. Indeed, a strong enough 'bloat' will condition a fullness-dependent aversion to the dessert. Hence, this process of learning establishes a satiating effect of the combination of moderate gastric distension with the sensory characteristics of that dessert. A mild sense of fullness may well bring the meal to a conclusion but this is a learned physiological signal. It acts on ingestive behavior only through recognition of the sensory characteristics of the dessert that had also been followed by excessive satiation on one or more previous occasions.

The cue of mild gastric distension and the consequence of duodenal distension are both probably chemically non-specific internal signals. Hence, this learning mechanism induces appetites and satieties that are not orientated towards particular nutrients.

Nonetheless, a nutritionally specific state of need can become the learned internal cue. This, therefore, results in a nutrient-specific hunger. In both rats and people, a learned appetite for protein is a sign of the state created by omission of protein from a single meal. The organism learns, only when in that incipient state of need, to choose and ingest foods having sensory characteristics that were associated with repair of the need.

In such cases, both the *sensory* and the *physiological* control are complete: if either of them is missing, the learned control of ingestion does not appear. Thus, food preferences and aversions cannot be assumed to be independent of the momentary conditions of testing. This has crucial methodological implications that have been ignored by most research on dietary selection and food and water intake. If we are to understand the influences on intake, we must measure or control the sensed characteristics of each food and also test and retest choices between foods from the start to the finish of a meal, and afterwards.

It also follows that the usual concept of palatability as a fixed property of a food is not empirically viable, even for one person. Palatability seldom is, if ever, simply added on to physiological signals of hunger or satiety to determine meal size, except perhaps in unfamiliar, artificial and extreme situations. The proper empirical concept of the palatability of a food or drink is its momentary sensory facilitation of an individual's disposition to ingest in a specified context. This disposition is measurable as relative amount or probability of intake. In people, it can be expressed as a verbal degree of acceptance that is predictive of ingestion.

Therefore, the idea that one could make adults eat more by adding palatable constituents to familiar food misapprehends the nature of sensory influences on eating and drinking behavior.

It is a fallacy because everybody has learned to like the particular level of aroma, taste, texture, etc. that is familiar in a food. Adding more of what is liked is therefore likely to result in too much of a good thing. High levels of sweetness may be an exception because they stimulate innate ingestive reflexes. Even at the age of 4-6 months, human infants learn to consume more of whatever level of saltiness they have experienced a few times before. By 12 months of age, the acceptable level of saltiness is particular to each food among those foods that the infant is used to eating.

This induction of preferences by habit continues throughout life.

The palatability of a familiar food is thus the particular combination of levels of sweetness, crispness, coloring, etc. that the individual has learned to regard as ideal. Therefore, in principle, whether the average daily intake, for example, is increased or decreased by the net effect of all the choices that have been subject to a sensory influence (e.g. saltiness) is unpredictable. This is so, unless we allow for the whole variety of contexts of the different kinds of food choice involving that aspect of food composition. In other words, questions such as whether the palatability of sugar or fat causes overeating are incoherent. Also, it is not sensible to expect to predict intake of sugar, salt, fat or fiber from preferences in one or two tests, especially if the tests are not on habitual foods and menus.

Similar problems arise in assessing whether the intake-stimulating effect of sensory variety can cause greater long-term intake outside the laboratory. Any extra intake stimulated by a succession of differing foods may be compensated for by subsequent intake reduction, either directly or by learning. Furthermore, people may become used to a succession of alternatives and the stimulating effect of variety would then disappear. Providing a choice of foods to a group of individuals is always liable to provide each individual with an option that leads to a higher average level of palatability than if everybody had been offered only one food.

2   MEASUREMENT OF THE STRENGTH OF SENSORY INFLUENCES

The facilitating effect of a sensed factor on preference or intake will be strongest at the learned level of that factor in the learned-eating situation. This facilitation will decrease by degrees as any sensed feature - or indeed a contextual factor - departs more and more from its usual strength in that familiar or ideal configuration. The degree of this change in the effect that is produced by a given difference between foods is the measure of the strength of that sensory influence. This is a principle that applies to measurements of the strength of causation anywhere in quantitative science. If a small change in input produces a large change in output, then that input has a strong influence over the output.

The strength of the influence of a sensory (or other) factor on intake, preference, intensity or any other response is limited by the fineness of our perception of differences in that factor. It may not be easy to decide whether the level of, for example, sweetness or astringency of a cup of tea is exactly as learned or slightly stronger or weaker. Discrimination of such differences is also liable to be made difficult by the usual multiplicity of influences on appetite. Furthermore, attention is liable to focus more on some sensory factors than on others, which may reduce the sensitivity to differences in the less salient factors. Thus, there will be regions of doubt in which the response may be equivocal.

Moreover, even when the eating situation is clearly distinguishable from the most appetizing configuration, it may still evoke appreciable ingestion. As long as the food is sufficiently similar to the ideal, it may be liked to some degree. In other words, a person's tolerance of differences from the learned most motivating food and context may be less discriminating than the finest possible discrimination of differences. That is, the threshold for detection of a difference from the most preferred level may be lower than the threshold for tolerating a difference, i.e. for the sensed factor being too much or too little to be perfectly liked.

The simplest objective measure of the strength of an influence relates the variability in the individual's response to each stimulus level in the task facing the assessor. This relationship generates an estimate of sensitivity in that task, be it rating preferences, describing attributes, drawing distinctions without using words, or simply eating. So long as the testing has been carried out appropriately, a simple formula (Conner *et al.*, 1988) yields the traditional measure of a *just-noticeable difference (JND)*, which is a form of threshold for detection of a difference. This measure relates to suprathreshold sensitivity in psychophysics, the slope of the plot of intensity rating against amount of stimulus. However, the slope of a psychophysical function depends on the anchor points for the rating and what they refer to for each rater. The JND depends only on the direction set by the anchor's description.

The same calculation can be used to estimate a *just-tolerable difference*, i.e. the threshold for a difference large enough to affect preference. As predicted above, these tolerance thresholds tend towards a lower limit, probably the detection threshold. This limit tends to be lower in water than in a drink or food, e.g. for sugar in chocolate or salt in bread (Conner *et al.*, 1988), presumably because of masking effects.

3   SENSORY INFLUENCES ON CUMULATIVE INTAKE

The influences of a sensed characteristic on choice and intake will depend on context, in ways that vary from person to person. Thus, to predict the effects of sensed components on food intake we will have to measure individualized food- and situation-specific influences on momentary ingestion.

To approach a realistic calculation of the net effect of a sensory influence on average daily intake, the contribution of all main sensory factors to the individual's diet would have to be sampled. Also, the responsiveness of the sizes of all common portions, courses and meals would have to be measured, as well as any sensory effects on the timing of meals. Then, this immediate causation of eating would have to be weighted for the observed frequencies of the food choices that involve influential sensory factors for that person. Once a representative sample of people has been measured in this fashion, however, the general picture is available by simple addition, as has been demonstrated for several particular food-choice situations. It is the interaction for each individual of satiety with the diverse opportunities to eat at different delays that is so complicated. Because of this, no food or constituent has a determinate satiety value or satiating efficiency: the regulatory effect of intake-suppressant responses to eating depends on the timing and variety of the foods presented and the individual's habitual reactions to their sensory characteristics in those circumstances.

Sensory influences on normal food intake in humans have been identified qualitatively in experiments such as those on post-ingestional conditioning of meal size. Quantitation of the strength of sensory influences on any real-life intake occasion has hardly begun, however.

The strength of the influence of some particular sensory factors on preferences in eating or drinking situations has been measured, but mainly for foodstuffs eaten singly, and this work has been mostly confined to taste-makers.

Salt is the only nutrient with a characteristic taste. Therefore, it was one of the first candidates for preference measurement and for investigation of relationships between preference strength and daily nutrient intake. Individuals' preferences for salt in four diverse foods appeared to be substantially intercorrelated. However, these foods were all constituents of the early parts of meals. Preferences for saltiness in them applied primarily to the situation itself and only weakly to the addition of salt to foods in general. Preferred salt levels also remain related to choices between high- and low-salt foods, in the same way as has been shown for sweeteners (Conner & Booth, 1988). Moreover, much salt intake is caused by foods and drinks that do not taste salty, such as milk and bread. Such items are chosen under all kinds of quite different sensory and non-sensory influences. So, it should not be at all surprising that these very limited assessments of salt preference failed to predict the total salt intake.

For decades, a favourite culprit for obesity has been the supposed high palatability of sugar and an alleged difficulty in satiating its influence on intake. People's likings for sweetness in different foods correlate to some degree and are also partly correlated to choices of sweet over non-sweet alternatives. However, at least in British consumers, a liking for sweetness in snackfoods and desserts does not consistently go together with a liking for sweetness in vegetables and fruit. Often, sweetness in snackfoods coincides with a high fat content of those foods. It may, however, be the preference for fat that is expressed accidently as a preference for sweetness.

4   SNACKFOODS AND DRINK PREFERENCES

There are frequent occasions for ingestion when usually the sole point is to have a drink for refreshment or to be sociable, while there is no need or desire to have a meal, or even a snack. Nevertheless, the traditional beverages consumed during breaks from work often contain energy, e.g. in the form of sugar, milk constituents or alcohol. Many drink services also provide one or more types of small items of solid food to nibble with the drink. For example, any decently organized workplace in Britain, from the factory floor to the board room, serves biscuits with the mid-afternoon and mid-morning tea or coffee. Learning makes us expect and like to have these drinks and snackfoods together in these contexts. In many Western countries also, beer is often served together with chips potato or peanuts. Drinkers of cola and other sodas are encouraged by advertizing slogans to eat a filled and coated wafer bar or a piece of chocolate confectionery, even as part of the refreshment or social enjoyment, e.g. 'helps you work, rest and play'.

Habitual eaters of these combinations of starch, fat and sugar have learned to find their crispness and other sensory qualities such as saltiness or sweetness appealing. Those sensory factors thereby gain control of intake on those drinking occasions. Such a modest energy intake can have no adverse post-ingestional consequences to condition intake downwards. Indeed, the innately reinforcing sweetness and preference-inducing associative effects of the ready energy in them could well foster a strong temptation to have the between-meal snack with the drink. No doubt there is also the sociability factor.

Furthermore, food items having similar sensory properties, usually in larger and more elaborate formulations, are often served as desserts. Smaller items that are high in fat, sugar and/or starch are often served with the coffee or other drink after a meal. It is presumably as a result of using such items that some people develop a liking for the sugar in them at levels at the higher end of the available range. Some people who do not eat desserts and packetted snackfoods like the sweetness of plant foods, which is presumably because of a habit of eating fruit or the sweeter vegetables.

5   DRINK-BREAK ENERGY AND OBESITY

Thus, many people have acquired the habit of consuming a beverage and perhaps a solid accompaniment containing ≤ 500 kJ of energy twice or more times a day. The frequent ingestion of sweet, creamy or crisp food, especially 'when not really hungry', may sound self-indulgent. Yet it cannot just be assumed to cause obesity without a plausible mechanism, and preferably some decent evidence that the mechanism has the predicted effect in real life. In fact, the effect of between-meal kilojoules on the long-term energy balance is more predictable than that of high proportions of sugar and starch or even of alcohol and fats in the diet. Readily assimilated energy has an appetite-suppressant effect while it is being absorbed, especially at a high rate. However, more than 500 kJ can be absorbed within an hour by an adult. As a result, there will be no short-term moderating influence on appetite and intake at the next eating occasion. For some chocolate bars, it is even promised that it 'won't spoil your appetite'. In contrast, a substantial breakfast may moderate appetite at lunch and a decent lunch may prevent ravenous eating at suppertime.

In other words, the between-meal (and end-of-meal) energy intake will bypass the mechanisms for regulation of intake that we have within meals and between adjacent meals close enough together in time. These short-term effects are the only mechanisms of weight regulation that have been demonstrated in human beings. In such a case of habitual short-term non-regulation, there is no question of what the longer-term effect on energy balance will be. The only issue is how big the fattening effect is likely to be. This depends on how frequently and persistently the modest amount of energy slips in past the intake-moderating system and 'goes to waist'. The advertized nibble that 'won't spoil your appetite' promises to spoil your waistline as a direct result.

This prediction has been tested in self-report studies of 500 respondents from the general public in the English Midlands. The most effective strategies for getting weight off and keeping it off for 1-2 years are indeed cutting calories between meals, especially energy in high-fat snackfoods. But also cutting back on alcohol and on sugary foods which are also high in fat, helps. In Britain, a high proportion of many people's alcohol intake is between and after meals, especially in the evening, when it is metabolized many hours before breakfast and the satiety effect of alcohol has long gone.

There are other strategies associated with success in keeping weight off. These include cutting back on fat in meals and taking extra exercise. Although rich in energy, fats may be rather ineffective at satiating. They are a concentrated form of energy and a major contributor to palatable sensations such as crispness and creaminess. A habit of taking exercise makes only a small contribution to the energy balance. As with the small intakes of energy between meals, however, it is the persistence of the habit that has the cumulative impact on body weight. In a culture where mechanization has greatly reduced personal energy expenditure, and energy is usually ingested in and with drinks between and after meals, it is not surprising that there is a high prevalence of obesity in both men and women.

6   ENGINEERING THE SENSORY CONTEXTS

The implication is that one of the most effective ways to reduce and prevent obesity would be to engineer the sensory and social influences on drink-breaks and snacks so that eating habits are changed to minimize caloric ingestion between meals or at the ends of meals (Booth, 1988). Universally popular alternatives to alcohol, sugar and starch ingestion at these times seem to be essential.

Following this view, educational and clinical approaches to reducing obesity should focus on the self-managed timing of energy intake, which should occur as much as possible in the first half of meals. They should also recommend that the richer energy sources are to be avoided, i.e. fats, and should not put too much emphasis on the total amounts of sugars and starches. Indeed, high concentrations of readily assimilated carbohydrate early during the meal could help to moderate meal sizes by automatically reinforcing a hesitancy about desserts.

More broadly, health education, food technology and marketing should introduce a sharp distinction between two kinds of 'good' snacks. One is the drink break that contains no energy, not even in its solid accompaniment. The other is a quick meal that is based on either at least three types of food or a nutritionally complete drink or food item. At present, there are many sensory similarities between calorific drinks and their traditional accompaniments and both the zero-calorie drink breaks and the convenience meals. This is confusing and helps to maintain the temptation to consume unnecessary energy outside meals and therefore is damaging to health. Fruit and salad vegetables not containing sugar or starch can of course substitute for a drink when fluid is needed, or water.

7   SUMMARY

Palatability and the intake of foods and drinks is concerned with sensory choices during intake.

Choices during intake depend on appropriateness, relative preference and physiological control.

The usual concept of palatability as a fixed property of a food is not empirically viable, but its momentary sensory facilitation of an individual's disposition to ingest in a specified context.

Measurement of the strength of sensory influences relates the variability in the individual's response to each stimulus level. This yields the just noticeable difference: a threshold for detection of a difference, and the just tolerable difference: a threshold for a difference large enough to affect preference.

Habitual consumers of snackfoods and drinks have learned to find their sensory qualities appealing.

Using items high in fat and sugar can result in a liking for these at levels at the higher end of the available range. Persistence of a habit like this causes a cumulative impact on body weight, with obesity as a final consequence.

One of the effective ways to reduce and prevent obesity could be the engineering of sensory and social influences on drink-breaks and snacks so that eating habits are changed to minimize caloric ingestion between meals or at the ends of meals.

4
Recognizing and liking foods and the consequences for nutrition

This chapter illustrates the processes by which foods are recognized and selected, by looking at the nutritional implications of taste and smell deficits in older people, which may shed some light on these processes. Can these learned controls of food intake enable the selection of sources of a nutrient when that nutrient is needed? Would aging of sensory capacities be liable to interfere with such nutrient-specific appetites?

Food perception, preference and selection are all objectively measurable performances that discriminate between different levels of entities such as the sugar, thickener or protein in the food. A rating or an intake may not be controlled by the word the experimenter or the subject uses to describe it: protein-orientated food selection, thickness preference, or sweetness perception. These perceptions are proved only when the objective influence, isolated from any other influence, accounts for the observed variation in choice response.

Much research depends on finding a difference between an experimental condition and a control. A stronger design is to measure the effect of a difference between high and low levels of an influence. Such a design can be extended from merely two levels of an influence to a continuum over the natural range of levels. When properly designed, such tests evoke linear dose-response functions or psychophysical performance. From this straight line, the strength of the influence on the effect can be estimated.

Usually, at least a few influences are operative on a choice at the same time. Because the interactions may vary among people, the only valid causal analysis is at the individual level.

Examples of this approach to the psychobiology of ingestion are described in the following. They involve food-taste tolerance in the elderly, integration of sensory and attitudinal influences, and dietary choices that were supposed to reflect a craving for carbohydrate.

1   FOOD SELECTION

Selection of a food requires the food to be recognized and discriminated from other foods. Thus we should first consider the exact psychological processes involved in this biological function.

1.1   FOOD PERCEPTION

Any input from a food that affects an output from an individual has to be sensed or perceived to that extent. That is, the perception of a food attribute can be defined as the effect on any response that arises from differences specific for that attribute. This definition encompasses sensations generated by the effect on the senses of the inherent characteristics of the composition of the food itself. However, it also includes affective reactions , i.e. pleasure or emotions. Furthermore, perception is not merely a subjective experience; it is primarily an objective achievement. It can be made evident in the behavior of choosing, accepting or rejecting a food sample, as well as by description of the food. The description and choice may also be affected by attributes of the food, such as price, nutritional information or advertizing image.

Food perception, then, can be measured as the sensitivity of a detection, description or choice response to inherent or attributed aspects of the foodstuff. The food item has been correctly recognized if it is sufficiently similar to the standard version to receive the same response.

1.2   FOOD PREFERENCE

Preference is a word that has many uses. This section concentrates on the non-mediated sensory influences on choice.

A widely accepted theory is that the inherent or sensory characteristics of a food (such as flavor, texture and appearance) are what is liked, while conceptual or extrinsic attributes (such as nutritiousness or low price) are not really liked, even though they may influence the choice. This view presupposes that the effects on choice of the *sensory* attributes can be separated from the effects on choice of *non-sensory* attributes, such as a brand image and even of sensorily cued attributes, such as the calories expected from a sweet or creamy food.

The identification and measurement of such uninterpreted, inherent influences on food choice, however, are much more difficult than is generally assumed. To demonstrate that the liking really is for some sensed inherent characteristic, e.g. flavor, we would have to construct a model containing both non-mediated and sensation-mediated mental pathways from food composition to choice response. We would then have to test this model against a model containing those pathways plus the attitudes based on food composition. That kind of analysis uses a methodology that will be outlined below.

A *hedonic rating* is also often assumed to measure some affective process such as a thrill of pleasure. Yet, when people say they like something or that it is pleasant, all they may be doing is expressing a positive behavioral disposition towards it, i.e. saying that they would choose it if other things were equal. The only way to establish whether there is an emotional experience with a causal role in choice is to use an adequately designed experiment to test the contrary theory that there is no affective mediator.

Another scientific principle is that observing choices among a set of foods provides data only on those particular foods in the situation tested. What is needed to understand the determinants of food intake are preference responses in tests that identify and quantify which aspects of the foods influence the observed choices in the situation of interest. Then the preference data can be interpreted and, even better, generalized beyond the particular samples of foods tested.

1.3   NUTRIENT SELECTION

Similarly, the concept of selection of particular nutrients can only make scientific sense when we consider the psychological mechanisms required to achieve nutrient-specific choices among foods.

One mechanism by which the selection of a food for its actual nutrient content could occur is for someone to use correct information on the nutrient composition of the alternatives. Such performance is readily measurable by assessing the accuracy of the consumer's beliefs and showing that they explain the declaration of intention to choose.

Such a mechanism, however, is not what is meant by attributing 'carbohydrate craving', 'protein sparing', or other nutrient-orientated selection to food choices, even in people. The hypothesized performance is the choosing between foods under the control of post-ingestional effects that are specific to the carbohydrates and/or the proteins in these foods.

Indeed, beliefs and intentions concerning supposed nutrient contents or other aspects of using the foods might interfere with control of choice by actual nutritional after-effects. In any species, moreover, 'sensory' preferences for the foods could determine the observed choices rather than resulting from post-ingestional effects or presented and/or remembered information. Differences in nutrient intake could then result entirely incidentally from the control by nutritional effects, and the nutrients consumed would have had no causal role in the subject's behavior.

2   CAUSAL INFLUENCES ON BEHAVIOR OF THE INDIVIDUAL

Dietary habits and patterns as well as nutrient intakes need to be determined in order to monitor and evaluate the effects of food intake on health. Furthermore, this needs to be done in individual people, because patterns, intakes and the preferences behind them can differ greatly from one person to another, even within the same culture. It is widely assumed that food preferences are a strong influence on food choices and hence on nutrition. It can be argued from accumulating evidence that dietary preferences are virtually entirely determined by dietary habits. Hence, the systematic measurement of the main preferred selections and portion sizes of food of a single person could turn out to be as precise as the usual recording methods applied to estimate actual dietary habits and intakes in everyday life.

In the following we shall concentrate on methods for measuring food preferences, and on some recent research results obtained using methods applicable to individuals.

If there is a large perceived difference between two foods, a number of questions arise. What sensed or attributed factor or factors make the foods so different? Or, what is it about the food samples in an experimental set that determined the subject's preferences among them? Is the selection between diets varying in nutrient composition actually controlled by the nutritional effects intended by the experimenter, or instead, by sensory differences to which a subject has different attitudes?

Methods have recently been developed that can identify and measure the causal efficacy of cognitive processes mediating food recognition and liking in an individual eater. This research approach originated from a method of obtaining undistorted peaks in preference for a sensory characteristic and using them to discriminate differences in food composition. In this particular range of applications, the new approach has been called the *appetite triangle* (see Figure 2.3) or *tolerance discrimination measurement*. In a multidimensional form, this method is equally applicable to the combination of features that constitue a food or eating situation.

The methodology is illustrated here by a study of the effect of aging on the role of the taste of caffeine in the acceptability of instant coffee drinks. The use of the method to measure also multiple influences, including non-sensory factors, is illustrated below by an effect of nutritional attitude on the response to a taste. Finally, the approach is used to tackle the vexing question of the role of nutritionally controlled food preferences in protein intake.

3   SCIENCE OF PSYCHOLOGICAL PERFORMANCE

Properly construed, food perception, preferences and nutrient-oriented selection are observable cause-and-effect phenomena.

The empirical method of theoretical science is to design the conditions of observation and to analyze the results obtained in a manner that distinguishes between theories on the causal processes that are operative. In the case of food recognition and liking, the chain of causation is in the subject's mind. This causal chain is formed by the processes that transform the situational inputs into the eventual choice.

It is essential to record patterns of food intake or choice, or verbal data predictive of food-oriented behavior, as well as all causal influences of the situation in which a choice is made. Then we can estimate the quantitative relationships between the observed choices made in a certain type of situation, and all the influences on choice that have been presented to the subject under study. These quantities, relating situational input to personal output, are measures of psychological performance. From them we can obtain an objective and valid specification of what the individual has perceived, preferred or selected.

4   QUANTITATIVE ANALYSIS OF A SINGLE INFLUENCE ON FOOD CHOICE

Thus, the core of the scientific method - that is, the empirical study of causal mechanisms - is the recording of independent input variables, as well as of one or more output variables.

To build a secure understanding of the operative mechanisms, at least some of the observed inputs must prove to have a substantial relationship with the output. That is, for the chemical senses and nutrition, the data should provide evidence that they encompass the main influences on food choice.

This approach has been applied to preferences for salt levels in soups and breads and for sugar levels in a soup, chocolate, a lime drink and instant coffees. It has also been used to measure preferred aroma levels in flavored drinks, the preferred color of low-fat spreads, and the combined visual, tactile, and gustatory impact of whitener in coffee. A quite different kind of physical aspect that can also be examined quite well using this method is the ease with which margarines spread. As we shall see, the influence of the level of calories indicated by a label on choice can also easily be quantitated. The factors in price could be assessed as well, to give a consumer's elasticity function. Furthermore, it is feasible to scale influential aspects of nominal variables such as brand names. Nutrient-specific dietary selection has to be subjected to a quantitative multi-dimensional analysis.

5   PREFERENCE FOR FLAVOR STRENGTH IN THE ELDERLY

Sensory deficits might be expected in the elderly. Some absolute threshold data are available, i.e. individuals' *just noticeable difference (JND)*, *difference threshold* or *Weber's discrimination ratio*, all effectively the same concept but under a different name. The method below uses fewer data than the traditional methods that are usually applied to groups of people.

5.1   CAFFEINE IDEAL POINT AND TOLERANCE

The individualized, cognitive measurement methodology can be illustrated by a simple application to the possibility that deficits in perception can affect food preferences in the elderly. Discrimination between tastant (taste-maker) levels were analyzed by ratings of preference and of intensity from small numbers of healthy elderly, middle-aged and young people of both sexes.

The test assessed individually tolerated levels of caffeine added to the decaffeinated version of the subjects' usual brand of instant coffee, made up in the way and at the time the individual subjects would normally drink a cup of coffee. There were two sessions. Only ratings of likelihood of choice were taken in the first session, from 'Always choose' to 'Never choose'. In the second session, both these choice ratings and ratings of bitterness from 'Not at all' to 'Same as my usual coffee' and above were obtained .

Intriguingly, the taste of caffeine appeared to be integrated with the roasted coffee-bean flavor. The subjects did not realize that only the caffeine level was varied; they thought different types of coffee were tested that varied in quality.

In most of the subjects' ratings of bitterness in the second session were considerably less sensitive to caffeine (the bitterness discrimination ratios in Table 4.1) than were the choice ratings in the first session (tolerance discrimination ratios in Table 4.1). In this context the rating of bitterness may be a quantitative expression of the same phenomenon of unconsciousness of the taste of caffeine.

Only a few subjects showed similar sensitivities in the two ratings or even a slightly greater sensitivity using the word bitter.

Weber's discrimination ratio is a threshold for the detection of a difference between levels of an influence. In this case, it is that ratio of caffeine concentrations which are discriminated 50 per cent of the time - halfway between random or totally confused performance and perfectly discriminated or with certainty. The smaller the discrimination ratio, the smaller the ratio of caffeine levels that have been distinguished by the responses. This means, the finer the discrimination by that response, the more sensitive the response and the stronger the causal influence of caffeine on that response.

At the top of the range of discrimination ratios for caffeine in coffee the sensitivity was as high as that for caffeine in water; the lowest observed taste discrimination ratio was 0.13, slightly lower than the lowest individual value (0.17) reported for caffeine in water. Yet, individual differences in the ideal point, namely the most preferred level of caffeine estimated from the choice ratings, and in discrimination ratio were unrelated to taking sugar or milk in coffee. So, these constituents appeared not to have major masking effects (Schutz & Pilgrius, 1957).

Table 4.1 shows some interesting results. With regards to possible effects of aging, the most elderly male testees tended to prefer higher concentrations of caffeine, and they were more tolerant of deviations from the ideal point than some younger men. Also, their bitterness sensitivities were appreciably lower than, for example, those for the oldest women. That is, these old men liked their coffee strong, possibly because it tasted less strong to them.

TABLE 4.1

Ideal points and choice or bitterness intensity discrimination ratios difference threshold for caffeine in instant coffee as habitually drunk

age range (years)	sex	n	ideal (log g/ 100 ml)	logarithm to base 10 of discrimination ratios	
				tolerance	bitterness
< 35	F	11	−1.4 ± 0.3	−0.1 ± 0.4	0.8 ± 1.3
35-55	F	5	−1.2 ± 0.2	0.3 ± 0.7	0.7 ± 1.4
56-69	F	11	−1.3 ± 0.4	0.0 ± 0.2	1.8 ± 2.5
> 70	F	6	−1.3 ± 0.7	0.8 ± 0.9	0.7 ± 0.6[a]
< 35	M	8	−1.4 ± 0.3	0.2 ± 0.5	1.2 ± 1.9
35-55	M	3	−1.3 ± 0.2	0.0 ± 0.2	0.6 ± 0.6[b]
56-69	M	4	−1.1 ± 0.5	0.3 ± 0.4	1.4 ± 1.7
> 70	M	4	−0.8 ± 0.3	0.2 ± 0.4	1.8 ± 3.9

Note: each of the values in the three columns on the right is a mean ± *SD*.

[a] Based on four values: the two others were infinite.

[b] Based on two values: the third was infinite.

In contrast, the most elderly women were no different from women of other ages in their ideal caffeine level but may well have been more tolerant of deviations from the ideal than any other group. The elderly women may, however, have been somewhat sensitive to differences in bitterness.

In short, evidence that the preferred tastant level is dissociated from tastant sensitivity, and even from the precision with which the ideal level is preferred, may well have been provided by the first tests of that possibility in the elderly. This is consistent with the views discussed earlier that people begin to like what they habitually use, rather independent of how it tastes.

There were signs of preference for less caffeine in men and women under middle age. Also, the youngest women's choice ratings were considerably more sensitive to caffeine differences than were those of other groups. There is little reason to suspect that a decline in taste sensitivity sets in at middle age, and so these preferences might be secondary to culturally influenced habits such as avoiding strong coffee and measuring it out rather precisely.

5.2 UNDISTORTED MEASUREMENT OF FLAVOR PREFERENCES

The results discussed above illustrate an important principle. That is, possibly within quite broad limits of flavor balance, for example, between bitterness and milkiness or sweetness, the particular level a person prefers is most likely adapted by habit. Thus, a preference for the habitual formulation is likely to persist even though decline of the senses may make it taste less strong.

Indeed, the readapted ideal level with an increased tolerance of deviation from it, could generate the artifactual impression that stronger flavorings make foods better liked in old age. Only an experimental design including an objective choice of psychophysical parameters, carried out on individual subjects with a minimum distortion of the responses, can distinguish changes in peak preference from changes in tolerance of deviations from it. By doing so, one can ascertain whether aging does indeed create a demand for strong flavoring.

6   MULTIPLE INFLUENCES ON FOOD CHOICE

The above example considers only one sensory factor influencing a set of choices. Usually several discrete factors affect what a person prefers among the choices available.

6.1   INTEGRATION OF INFORMATION IN INDIVIDUAL EATERS

The process by which someone integrates two or more influences into perception and choice can be determined by comparing a calculated model of the integration with observed choices. One discriminable difference from the ideal level in any one of the factors could be modeled as decisive. Alternatively, discriminable or tolerable differences may interact in a Euclidean space: the combined distance is the hypotenuse of a right-angled triangle whose sides are the distances in the two separate dimensions; or in a city-block metric in which the distance down one street is added to that along the street around the corner to the opposite corner of the block. So far, it appears that Euclidean combinations of discriminable differences in distinct types of influence are the best predictive of food preferences.

The tastes of different species of fruit provide one of the simplest examples of these two forms of interaction between sensory qualities in the mind of the eater. Most ripe fruits contain both sugars and acids. However, the sugars and acids and their concentration vary between species and varieties, even at the same stages of ripening. For example, malic acid is predominant in apples while citric acid is the main contributor to the sour taste of citrus fruit such as oranges and lemons. Oranges contain higher levels of sugars than lemons, obviously. Even though the proportions of sugar and acids may vary among oranges, the range of variation is sufficiently different from that in lemons to be recognized after some experience of eating oranges. That is, we can learn to perceive the proportion of sweetness and sourness that is characteristic of an orange or its juice. If we are used to orange-flavored drinks, we may be able to judge the most realistic levels of sugars and acids.

This then provides a neat example for mapping the cognitive processes of integration of information from two 'channels' or patterns of taste - those entering the brain from receptors that are particularly sensitive to sugars, recognizable as sweet, and those from acid receptors, named sour.

If we provide mixtures of sugar and acids having the color and odor of an orange, we can test the idea that the perceptual effects of two sugars should add in a city-block metric. The same should be valid for the sourness of two acids. Also, the combined effect of the sugars should be one dimension and the combined effect of the acids should be another dimension perpendicular to it, with the distance of the orangey taste above or below the proper level forming the hypotenuse of a right-angled triangle. This is illustrated in Figure 4.1.

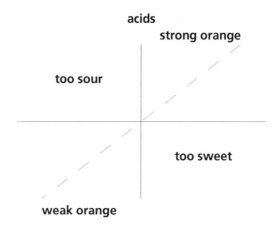

FIGURE 4.1
Mixtures scaled in just-noticeable differences (JND)
The origin is a mixture of sugar and acids that mimics the taste of normal orange juice.

6.2   INTEGRATION OF SENSORY AND ATTITUDINAL INFLUENCES

A multi-dimensional psychophysics of choice is needed to measure the combinations of the various influences on food choice and intake. The choice and intake of a particular food may be influenced, for example, by how the composition of the food is perceived, by the context in which the food is perceived, and by what the food is perceived to be yielding in the way of benefits, as compared to the other foods available. A quantitative account is needed of how these discrete influences are integrated and transformed to produce describable sensations and emotional experiences as well as verbally expressed and actual behavioral tendencies.

Studies on coffee drinkers have illustrated the multi-dimensional version of this approach. Various influences on choices between coffee drinks were considered: perception of coffee drinks, preference for particular constituents in coffee drinks, and reaction to labelling of coffee drinks. The implications of these factors varied with the attitudes of the subjects tested.

Are coffee-sweetness ratings really based entirely or mainly on sugar perception, as anyone would assume? It was shown that indeed they were, by obtaining evidence from linear regression analysis that sweetness and milkiness ratings of coffee usually are differentially sensitive to sugar and whitener levels (Table 4.1). Given this finding, we can use partial regressions of these two psychophysical functions to measure the subject's performance in discriminating differences in sugar and whitener concentrations by the ratings. From the regression equations the ideal points and different thresholds, Weber's discrimination ratio, can be calculated. Then the numbers of discrimination ratios from the ideal levels for sugar and whitener can be calculated for each drink and its two-dimensional Euclidean distance from the subject's ideal coffee estimated by Pythagoras's Theorem. The length of the hypotenuse for the right-angled triangle of sugar and whitener distances from the ideal does indeed correlate well with the rated overall preferences for the different coffee samples.

TABLE 4.2

Estimated percentage probability of choice of samples of coffee containing different levels of sugar or artificial sweetener and with different sweetener labels attached, for an individual drinker of sweet, white coffee

sugar equivalents (g/200 ml)	sweetener label		
	all sugar	midpoint	all artificial
0	36	36	35
2.8	60	55	49
3.5	78	76	74
4.0	73	83	92
4.5	53	72	90
5.5	23	45	67

More recently, each subject's attitudes towards sensed constituents of a food have been preliminarily identified and the sensory and image aspects of the food's design to that person's preference optimized. Illustrated in Table 4.2 is a drinker of sweetened coffee, who in prior personal scaling expressed a fear of the fattening effects of sugar in coffee. Not surprisingly, this person showed a preference for coffee with low-calorie sweetener. More interestingly, there was also some sign of a preference for less sweetness when the sweetener was known to be sugar.

Thus the multi-dimensional analysis of the causal strength of the most important influences on preference, intrinsic or extrinsic to a food, provides an entry into the eater's mind that can be both theoretically illuminating and practically useful.

7   CARBOHYDRATE CRAVING AND PROTEIN SPARING

Diets that differ in sensory as well as nutritional composition are liable to be eaten in different amounts. However, because of the sensed differences, it is unscientific to attribute the differential intake to particular nutrients. Observed intakes or preference ratings could be controlled entirely by sensory properties of the foods when these are not used as guides to nutritional effect.

7.1   NUTRIENT-INTAKE DIFFERENCES WITHOUT NUTRIENT SELECTION

Different food attributes, such as taste, texture and ease of consumption, may be liked differently on their own account. Alternatively, the food might be identified by its sensory qualities, name or compositional information as being of a type toward which the individual has particular attitudes and habits. One such type could be junk food. Hence obese people may have strong emotional reactions to eating what they think are poorly nutritious snackfoods. An appetite-suppressant, such as the serotoninergic drug fenfluramine, is likely to change people's choices of convenience foods by a cognitive process that is not related to the physiological effects of eating such foods.

Studies on macronutrient selection and effects of fenfluramine on carbohydrate craving in people have observed effects on intake after and between meals. People become adapted - by familiarization and associative conditioning - to liking the foods they usually have in the situations they normally have them in. Thus, use of a food in a particular situation tends to induce a craving in that situation for its distinctive combination of attributes. This craving will be particularly strong if this person is trying to refrain from eating such foods and feels deprived of them.

Indeed, in British consumers, preferences for sweetness in commonly consumed snackfoods and drinks tend to group together into a single factor. This constitutes a 'sweet tooth' that is separate from likings for sweet vegetables or fruit, which other people have presumably learned from eating fruit or carrots and the like as snackfoods, or juices as drinks. So, even though most snackfoods are high in fat, sugar and/or starch and hence somewhat low in protein, people who stop eating the usual sort of snack food between meals, are not likely to suppress a craving for carbohydrate or fat. More likely, they are cutting back because they believe that conventional snackfoods are not nutritious and are therefore unwise to eat if one's appetite is limited because of drug action or to avoid getting fat.

Conversely, a genuine protein appetite would be a preference and intake controlled by food characteristics that the individual has learned to be predictive of repair of protein need. It is not a specific hunger for protein if it results from other sensory or attitudinal attributions to what happen to be high-protein foods, such as the desire for a hearty meal of beef steak or the strong flavor of a cheese.

Therefore, an observed choice of foods rich in a particular nutrient over foods poor in that nutrient need have nothing to do with the nutritional effects of the foods. Actually, nutrient-oriented selection occurs only if the choice is based on a liking for sensory characteristics that are distinctive of the nutrient. This liking could be innate, based on the validity of the stimulus as a predictor of those nutritional effects in the previous history of the species. The liking could be learned, where the individual has come to like the sensory stimulus because it has been associated with nutritional improvements. In either case, however, to be a nutrient-specific hunger the liking has to be increased by some deficiency or incipient lack of the nutrient of which the sensory stimulus is predictive.

7.2   EVIDENCE FOR SPECIFIC HUNGER

Sodium appetite is the one proven example of genuine nutrient selection, at least in some mammals, and the mechanism is innate. A liking for strongly salty tastes appears when and only when the animal has a serious sodium deficit.

Notwithstanding continuing widespread assumptions to the contrary, there is no evidence for a sugar or glucose appetite, in which a liking for very sweet tastes serves to select foods because of the glucose they provide for the tissues that need it. There is simply an innate sweet preference. It may be made more vigorous by the arousal effects of food deprivation and weakened by the sedating effect of a large meal.

Nevertheless, the existence of a carbohydrate-selective appetite has been proposed. Rats and people choose from foods either rich in sugar or starch, or both under some conditions. These dietary choices are modulated by neurotransmitter activity of serotonin (5-hydroxytryptamine, 5-HT) or supply of its precursor. Such phenomena have been interpreted as indicating a liking for carbohydrates regulated by 5-HT in the brain, as distinct from proteins and fats.

A weakness in this evidence is that the differences in nutrient composition of the foodstuffs used in these tests are highly confused with sensory differences. The following alternative reasoning could be taken into consideration.

The nutrient preparations in the rats' diets differ greatly in texture. The involvement of 5-HT in brainstem oral motor control alters innate textural preferences. Hence the effects of drugs could arise from textural differences between diets, and not only their nutritional after-effects as is commonly assumed.

The high-carbohydrate foods in the human tests are immediately recognizable as conventional snackfoods. So their sensory characteristics are likely to trigger reactions such as conformity with conventional antagonism towards junk food, guilt about self-indulgence, or an educated awareness of foods having low densities of essential nutrients. Even if the sensory characteristics of these foods themselves are craved for, this liking will be the habituated and conditioned preference for familiar food between meals. A specific selection of carbohydrates mediated by 5-HT has to be distinguished from this non-specific snacking appetite.

A few people are unconventional and acquire the habit of eating main-meal foods in modest quantities on the occasions that they eat between meals. Someone who prefers to snack on a ham sandwich is liable to gain satisfactions that are quite different from those gained by someone who snacks on candy bars. So it is hardly surprising that mood differences between conventional and unconventional snackers are reported. Such effects are no evidence, though, that carbohydrate and fat craving is acquired by mood changes induced by post-ingestional effects of the foods, let alone on 5-HT systems.

7.3   MEASUREMENT OF NUTRIENT SELECTION

It is very likely that a carbohydrate appetite or a protein appetite would have to be learned. This is because there is no reliable natural dietary cue known for either nutrient, not even a cue of so limited reliability as the salty tast of sodium chloride. The heating of carbohydrate can produce a caramelized smell and the extraction of proteins may introduce a smell into the preparation used in the animal laboratory, but these cannot have evolved as innate cues. So, a test for macronutrient selection and its modulation requires the observation of the influence on food choice of an arbitrary sensory characteristic that has been learned to be followed by some relief of current lack of that nutrient. The influence of this learned nutrient-predictive cue is more easily assessed when it is isolated from other sensory characteristics, such as those of the nutrient itself. If it is not, the confounding effects of any other sensory influences on likings and attitudes must be monitored and analyzed out, because these effects could be the cause of the observed dietary choices.

Rats can learn protein-specific selection, as evidenced by the choosing of a protein-associated food aroma or texture by rats mildly deprived of protein. This protein-sparing selection is not modulated by the action of a drug on brain 5-HT, however. There is also preliminary evidence for one-trial learning in human subjects of greater intake of a food of which the flavor is paired with protein, when the protein intake has been experimentally reduced to a low level for 15-18 hours. Such a test of preference for nutrient-paired and -unpaired flavors between otherwise identical foods is essential to measure any genuinely protein- or carbohydrate-specific selection and its modulation by a brain transmitter.

Thus, in conclusion, it will be important to isolate genuine nutrient cues from other influences on food preference in further exploration of the possibility that poorly nourished elderly people have sensory preferences that might express a protein appetite. Signs of protein deficiency are associated with a greater preference for the taste and smell of hydrolysed casein. However, this meat or vegetable consommé flavor might be attractive to someone who is not feeling well and so is undereating. This possibility must be distinguished from a real preference for protein-associated cues when lacking in protein.

How might taste or smell relate to nutrition in the elderly?

In conclusion, it is far from clear how deterioration of the chemical senses might affect nutrition in older people. Adequate research techniques have yet to be applied to this issue.

We can suppose that biological aging involves some decrease in sensitivity to some tastants or odorants or to differences between them or their concentrations. Also, the sensory deficits induced by disease or injury become more likely as the environmental exposure lengthens and physical frailty increases. In addition, there have been suggestions that an inadequate diet itself can cause sensory deficit; that is, a micro-nutrient deficiency in the elderly might lower the absolute sensitivity to a tastant.

Absolute thresholds, however, need not be related to difference thresholds or to supra-threshold judgements of intensity. Still more important, although reduced absolute sensitivity might reduce the aversion to spoiled food, reduced sensitivity to differences in the level of a tastant or aroma in a food does not necessarily make any difference to the preferred level. This is because preferences for the characteristics of familiar foods, including their bitterness and saltiness and even their sweetness, are entirely acquired. So, preference can be continuously reacquired during a slow decline in sensation.

This malleability of food preferences to the exposure provided by eating habits makes even longitudinal incidence research impossible to interpret. Changes in chemical sensitivity will occur in parallel with changes in life-style. On a group basis this totally confuses the relationship between the chemical senses and eating habits and, hence, the relationship between food preferences and nutritional status. Only an individualized multivariate theoretical model could isolate the influences on food choice from non-experimental data.

Some researchers have well appreciated this possibility of the elderly adjusting their eating habits, and even their food preferences, to compensate for sensory losses in smell or taste. Smell appears to be particularly open to social suggestion. Demonstration by research of an olfactory deficit in a person could have the unintended disadvantage of changing for the worse that person's hitherto normal attitude towards food. On the other hand, if a smell deficit was a problem (particularly a general anosmia), counselling to attend to food names, colors, textures and tastes might be sufficient to sustain the pleasures of eating. After all, appearance is normally crucial to food liking, and yet blind people lack no enjoyment of eating. Augmented food flavors might also help some people use this suggestion effect. On the above considerations, however, flavor enhancement could hardly be relied on generally to encourage eating in the elderly or to discourage it in the obese, for a sudden increase above the familiar level can be as disagreable as a sudden decrease.

The conclusion therefore is that malnutrition in healthy older people is unlikely to have arisen from sensory deficits. Lack of mobility, financial resources and social interests are more likely to be contributory factors.

8   SUMMARY

Recognizing and liking foods and the consequences for nutrition is concerned with food selection including food perception (measured as the sensitivity of a detection), description or choice response to inherent or attributed aspects of the food, food preference with respect to sensory and non-sensory attributes, nutrient selection (with beliefs and intentions concerning supposed nutrients interfering with control of choice by actual nutritional after-effects).

Because these phenomena are observable cause-and-effect phenomena, quantitative analysis of a single influence on food choice is possible, as well as of a multi-dimensional influence on food choice.

Integration of sensory and attitudinal influences finally produces describable sensations and emotional experiences as well as verbally expressed and actual behavioral tendencies.

5
Physiological and cultural bases for eating disorders

This chapter presents an argument that might explain the psychotic eating disorders and also provide pointers toward more successful therapy from what we know about the normal psychobiology of eating and emotions, and about the effects of social culture on individual development. This argument serves the additional purpose of illustrating how science and its practice cannot make definite progress without systematic causal analysis. Such analysis involves the measurement of the effects of the experimental or clinical interventions on defined mechanisms in a programme of investigation directed by coordinated theory.

The main line of theory proposed here is that eating is controlled by learned responses to the physiological consequences and social contexts of food intake.

1   PSYCHOBIOLOGICAL AND SOCIAL APPROACH

1.1   PSYCHOBIOLOGICALLY NORMAL BASES

The syndromes of disordered eating diagnosed as *anorexia nervosa* and *bulimia* (nervosa) are undoubtedly psychopathological (see also Chapter 13). Yet, the mechanisms that establish normal appetite for food can explain the eating disorder when coupled with socially induced distortions of feelings about bodily shape. The syndromes can be the outcome of a process of cultural inveiglement of the sufferer into extreme operation of normal learning of appetite and normal physiological adaptations to starvation or vomiting.

That is, the causes of these syndromes do not necessarily involve anything organically or functionally wrong in the hypothalamus, the limbic system, the gastro-intestinal tract or any other part of the brain or body involved in the physiology of appetite, emotions or endocrine control. On this account, any biologically normal person may be susceptible to developing disordered eating behavior and abnormal attitudes to foods and to the body (Booth, 1988a).

1.2   TREATMENT SUPPORT AND SELF-MANAGEMENT

It follows from this approach that success in managing and indeed in preventing disordered eating will depend on better use of people's normal physiological, psychological and social capacities. The patient should be adapted back to more orderly psychobiosocial performance in accord with self-organized strategies. Drug treatment, in particular, should not be aimed at correcting some presumed underlying abnormality of the brain mechanisms of appetite, mood control or neurovisceral regulation. Rather, if drugs are to be used at all, they should be directed at assisting psychophysiological reintegration of the somatic, social and sensory influences on appetite (Booth, 1988a).

2   A PHYSIOLOGICAL AND COGNITIVE THEORY OF NORMAL APPETITE

A holistic theory of the mechanisms controlling normal eating behavior was developed nearly 20 years ago in several laboratories. It provided the basis for a computer model which is an integrated quantitative basis for a central and peripheral physiology of appetite. It also provides a cognitive psychological framework for the cultural anthropology of eating habits.

The concepts included in this theory were required by evidence from the laboratory and the field. All the postulated mechanisms had been observed independently of the operation of the system as a whole.

Thus, for example, the limited regulation of body weight that is actually observable, was achieved without set points in the computer-modelled theory. The behavior processes that integrate sensory and visceral signals and form a basis for cognitive aspects of appetite and satiety, were explicitly analyzed by laboratory experiments in adult and infant rats, lean and obese monkeys and human adults.

Appetite and its associated phenomena of hungers, satieties, palatabilities, cravings, likings and distastes are specified and measurable aspects of the overall performance of the behavioral system.

Also, the cognitive postulate that eating is a set of learned reactions to configurations of sensory, somatic and social cues has provided the basis for fundamental methodological progress in the scientific analysis of the influences on appetite in the individual: this takes the form of linear causal analysis of influences from, for example, sensory characteristics of the diet and their integration with the perceived bodily effects or socio-economic attributes of eating particular menus and foods.

3   CULTURAL BASES OF EATING DISORDER

There seems to be no convincing reason to doubt that we and our culture are entirely based on physical reality. This faith, however, in no way commits us to treating as biologically determinated or reducible the mental organization of behavior, let alone the organization of our culture, social institutions and economy. Cultural processes operate over and above the actions and reactions of persons toward one another. Also, a person's (and a rat's or a computer program's) conscious and nonconscious world of information processing works according to its own laws. Biological processes in the body and in the history of the species (and the laws of physics) merely constrain and influence the psychological and social processes. Neuronal network activity provides the physical 'engineering' on which the mental 'software' runs: the brain's structure does not decide the program, any more than a computer's structure does.

Thus, the slimming culture, individuals' weight control and the physiology of fat deposition are all real (and interlinked) causal networks. Moreover, they all have to operate within the first law of thermodynamics - that is, energy balance across the skin.

Thus, any realistic scientific approach to eating disorders will regard people as integral psychobiosocial systems.

3.1   THE CULT OF LEANNESS

The mannequin has become the representative of an institution or ideology. She is like the priestess of a religion whose gods are slimness and youth. As in the prophet Ezekiel's vision, her 'dry' bones come to life all too impressively when dressed in fine clothing.

From a sociological point of view it is readily understandable why a devotion to slimness should become prevalent during widespread affluence. Plumpness can be functional and gross obesity can be prestigious in times and places where most people barely have enough to eat. Such shapes are liable to become repellant, though, when food is generally abundant. This is because, under these conditions, fatness can be taken as a sign of personal weakness and not strength. Also, it may serve as a biological signal of infertility more than of capacity for reproductive nurturance.

In such an affluent and fatness-hating culture, the normal distribution of subcutaneous adipose tissue in females can become like possession with a demon. The slightest sign of fleshiness needs to be exorcised by suitable rituals. These practices are widely known as 'going on a diet'. They include exercising, fasts, taboos on 'danger' foods, self-denial of palatable and satisfying foods and the consumption of 'diet' or 'slimming' foods, beverages, potions and pills. Dietary restraint is thus not a personal trait; rather, it is allegiance to an informal institution containing a wide variety of behavioral, attitudinal and emotional options that are unlikely to form a homogeneous scale.

Adolescence is a time of search for one's identity, when the normal human capacity for ideological or religious commitment gains an extra fervor. It is hardly surprising that many young people become converted to this specialized form of asceticism. The sex who are sculpted to maturity by adipose tissue will be especially liable to conflicts about shape and eating. Normal youthful enthusiasm, coupled with ecological and emotional gains such as an alternative to social competiton and an escape from sexual predation, could carry dieting to the extreme of a complete and endless fast.

Genuinely theological or political motivation can produce a not totally dissimilar disruption of eating. Usually, religious fasts and hunger strikes are carried out by presumably psychobiologically normal adults.

Moreover, a ready justification for self-starvation is provided by the demonology that is implicated in the religion of leanness. An ultimate extreme of devotion to the rituals of dieting expresses a loathing for normal rounding of the body or even for the fleshiness of relaxed musculature. Such alliance of fasting with a morbid fear of body fat is diagnosed as *anorexia nervosa.*

Self-starvation of this or any other sort might be made easier by the normal physiological adaptations to lack of food. These include reduction in capacity of the stomach or in its rate of emptying.

3.2   FROM CYCLIC DIETING TO BULIMIA

In contrast, some recovering anorexics or ordinary dieters may exercise a more intermittent affiliation to the cult of leanness, with repeated short or long phases of observing the rituals. Such cyclical dieting carries the risk of developing into compulsive food abuse. Again, this is not necessarily because of any biological abnormality. It could be just a perversion of the normal mechanisms by which appetite develops and adapts.

We now turn to the basic mechanisms of normal food-intake control in order to see how the eating disorders could develop from them.

4   APPETITE FOR FOOD IS LEARNED

The key to this theory of eating disorder is the evidence that an ordinary appetite and orderly eating are a natural, healthy and of course life-sustaining addiction to the usual circumstances of ingestion. This acquired addiction is not just to the foods themselves. It is also to states of the body and to times of day, emotional states and interpersonal situations.

Even our smooth-brained commensal, the rat, provides evidence for this proposition. Indeed, the first evidence for the conditioning of appetites or satieties cued jointly by diet and viscera was from rats. More recent work on humans has confirmed that eating motivation is under learned control by food stimuli which have formed in configurations with internal contexts, such as physiological cues to the need to eat or to stop eating.

These appetite configurations (or *Gestalten*) do not depend on only one particular type of learning mechanism: combinations of cues that most motivate ingestive activity are partly formed by a process of mere habituation. An appetite or a satiety could also be a respondent (reflex pattern) evoked by a conditioned configural stimulus (classical conditioning), an operant (intentional act) emitted as a set of deliberate actions occasioned by a reinforcement-discriminative stimulus (instrumental learning), or a set of deliberate actions ruled by reasoned objectives. These theoretical aspects of the exact nature of the learned eating will not be further addressed here. What matters is the evidence that palatability, satiety and eating habits, including binges on foods, are all learned responses of some sort. However, the nature of the appetite-conditioning reinforcers will be considered briefly below, because this brings out how ubiquitous are the learning processes that gain control of eating.

As soon as controlled naturalistic designs were used instead of extremes of toxicity or nutritional deprivation and of aversiveness and chemical purity of the test diets, it became clear from animal and human experiments that dietary energy and also essential amino acids (Chapter 4) were strong reinforcers of dietary preferences. It was also clear that preferences were readily learned in a way that was tied to contexts such as states of repletion or depletion. Such nutritional conditioning of contextualized eating could be sufficient to explain the emergence of bulimia from cyclical dieting, as we shall see below. Nevertheless, sometimes socio-affective conditioning and sheer familiarization may also be operative. These latter mechanisms are considered first, before turning to reinforcing effects of nutrient, about which more is known.

5   AFFECTIVELY REINFORCED EATING

Receiving a food as a reward and the sight of one's peers eating a food increases that food's attractiveness to the preschool child. Either of these situations could be providing positively reinforcing social or emotional associations that condition preference to the food. It is not known yet whether this socio-affective conditioning depends on the slight nutritional reinforcements gained by eating the reward or the shared food choice.

5.1   EFFECT OF MOOD

It is widely assumed that improvements in mood and a sense of competence can be induced by food and drink constituents such as caffeine and glucose, and that these psychological benefits contribute to the attractiveness of those items. Reinforcing effects of caffeine on human drink preferences have indeed been observed.

It has also been proposed that the association of emotional gains with eating could contribute to food cravings. The emotional states of some stages of pregnancy or indeed hormonal imbalance or mild undernutrition might be relieved to some extent by indulgence in unusual or long-avoided foods. Similar attractions could be found in snackfoods by obese or depressed people.

Such processes have been proposed to explain the tendency to binge or indeed to refuse food in the eating disorders. Attempts to assess moods and their changes around the time of a binge have not clearly supported this hypothesis. Often, unless vomiting occurs (or even when it does), distress increases after a binge. However, this could be the feelings of anxiety and guilt about the consequences of eating. The sensory experiences of eating that occur beforehand might be sufficient emotional reinforcement to make a binge more likely in the future in similar circumstances.

The timing and sensitivity of the mood assessments to date may not have been adequate to pick up slight and transient elevations of mood during the binge itself. In theory, a merely transient but very prompt change can be a strong reinforcer. Thus, it could become increasingly tempting to eat and to keep eating certain foods in those circumstances in which they had been paired with sensual pleasure, had served as a distraction from distress or had simply sedated or anesthetized the emotional turmoil.

For instance relief of a state of mild anxiety induced in normal schoolchildren, after a single pairing with a distinctive drink flavor, increases the liking rating for that flavor in a majority of the girls and boys tested. Thus, no special susceptibility would appear to be necessary for repeated affective gains to create food cravings.

This experiment provided no evidence that the relief-conditioned preference depends on the presence of anxiety, however. Also, it has so far not proved possible to obtain the effect in students by similar procedures. However, considerable variations in the design of the experiment would be necessary before one could conclude that emotional conditioning of food preference and indeed its contextualization to a distressed state does not occur in adults. Therefore, this mechanism for the development of emotional overeating should still be considered a major possibility.

5.2   MERE EXPOSURE

There is a mechanism that might establish likings for foods or contextualized eating habits without caloric or affective associates. Repeated exposure to the unreinforced stimulus at least habituates neophobia, i.e familiarization destroys the fear of the novelty. Merely repeating the exposure to eating a food may induce some positive hedonic judgement. Indeed, many types of stimuli are rated as more pleasant after they have been experienced without obvious consequences.

Food preferences induced by familiarization have been extensively studied in animals. This is an effect that must be controlled for in caloric conditioning experiments. Familarity also increases food preference in adults, preschool children and 6-month-old infants.

Nevertheless, those experiments confused mere exposure with slight caloric and affective reinforcement. The test foods or drinks were caloric and, even though the amounts were small, a few calories can condition a preference. Also, there is an implicitly hospitable gesture in any offer of food or drink, which may be another contribution to positive affective associations.

The exposure-induced human food preferences steadily increased up to 20 exposures, and so the upper limits on this effect remain to be defined. Furthermore, it is conceivable that habituation is context-dependent, in which case mere exposure could induce an appetite *Gestalt*.

6   LEARNING FROM AMINO ACIDS

Amino acids that are essential to life and health were in fact first identified by the reduction of food intake that occurs when one of the essential amino acids is omitted from the diet. However, this is not just an aversion-conditioning phenomenon. As soon as the idea was tested, it became evident that the balanced mixture of essential amino acids provided by good-quality proteins helps to establish the normal, strong likings for distinctive sensory characteristics of protein-containing foods. Furthermore, this protein-induced palatability does not arise from the correction of protein deficiency. The omission of no more than one or two normal mealtime intakes of protein was sufficient to make the amino-acid balance preference-reinforcing.

6.1   POST-INGESTIONAL CONDITIONING OF ODOUR PREFERENCES

Protein eaten in a flavored diet, or a balanced amino-acid mixture administered via gastric intubation in association with an odorized diet, conditioned a strong preference for the purely olfactory stimulus (not just to taste).

The learning was strong enough to occur over delays of up to a maximum of about 30 minutes between the sensory cue and the post-ingestional consequence . This interval would be sufficient to generate signals of dietary provision for metabolic needs. Longer delays would be adaptive for slow-acting poisons, and indeed toxin-conditioned aversions occur at CS-US (conditioned stimulus-unconditioned stimulus) delays of 12 hours or more.

Furthermore, the amino-acid induced food preferences were fully acquired in one or at most two experiences with bland diets under mild food deprivation. The rapid acquisition of ingestive motivation for protein-associated dietary qualities was shown by healthy adult rats, fast-growing adolescent rats and rat pups who were just beginning to sample solid foods independently of the dam.

There is now also evidence in human subjects that, after a low-protein breakfast, protein disguised in the following lunch makes that menu more palatable. Protein-conditioned preferences are difficult to see, however, because they are easily buried in general hunger and in food likings from other sources. Nonetheless, this mechanism seems likely to contribute to the attractions of any cuisine, for virtually all traditional food mixtures are adequate in their amino-acid balance.

6.2   SOMATIC CONTEXTUALIZATION OF FOOD PREFERENCES

As pointed out above, acquired eating motivation is not always attached to sensory characteristics of the diet alone. The dietary preference reinforced by effects of protein in adult rats is not expressed unless there has been no protein intake for 4-6 hours. Our experiments on humans have also provided some evidence for learned protein preference only during the early stages of protein need that had been present along with the distinctive diet during conditioning, i.e. the repleting protein-conditioned ingestion to a combination of relevant dietary and physiological cues. The result is a genuine protein-specific appetite and a mechanism for learning which diets to select to obtain the protein needed.

In principle, external environmental cues might also be recruited to the protein appetite *Gestalt* if, for example, incipient protein needs were greater at certain times of day (Baker *et al.*, 1987; Booth *et al.*, 1988a).

7   LEARNING FROM CALORIES

It is usually assumed that ordinary hunger stimulates energy intake and that other needs, except perhaps water, are met automatically from the nutrients in the energy-containing foods. Yet, few psychobiologists have considered how it is that aversive or, much more commonly, neutral sensory characteristics of foods become so highly palatable in later life, especially some hours after a meal. Theorists have been distracted by the ingestive power of sweetness (and of fluidity) from birth in both rat and man.

7.1   CALORIC CONDITIONING OF SWEETNESS PREFERENCE

Caloric conditioning of flavor preferences was first demonstrated in rats in the early 1970s. Up to virtually 100 per cent preference for a tastant was acquired by association with glucose polymers or dilute glucose. Yet, the myth has persisted that only conditioned taste aversions can be so strong. Indeed, the effect was so powerful that it reversed the innate preference for greater sweetness.

It has also been demonstrated psychophysically that the liking for sweetness in familiar drinks and foods can be entirely acquired in human adults. When the sweet sensation is part of the hedonic *Gestalt*, a food's sweetness becomes no different from its creaminess, crispness, color and every other salient sensory characteristic. Causal analysis of the sensory preferences for common types of food or beverage shows that the congenital greater preference for strong sweetness is overwhelmed by an aversion to too much sweetness. This dislike for excessive sweetness is no less strong than the aversion to too little sweetness in a familiar sweet food or drink.

7.2   HEPATIC OXIDATION REINFORCER

The original conditioning for sweetness preference in rats was obtained with hypotonic sugars and with dilute or concentrated solutions of glucose polymers (maltodextrin, or Polycose®). Data obtained from gastric intubation, and from subsequent gastric and intravenous infusion, demonstrated that this was not just post-ingestional conditioning, as has since been shown in many experiments in Sclafani's laboratory using gastric maltodextrin infusion. It proved to be also specifically parenteral conditioning, as has recently been confirmed with intravenous glucose infusions.

Similar infusions or injections suppress the appetite for food in freely feeding rats by supplying energy to hepatic oxidation. This satiety mechanism of energy supply was in fact the second postulate of the original cognitive and physiological theory of appetite (see Chapter 2). In this theory a privileged position for glucose sparing by the liver was assumed to be likely. This was, however, subsequently discounted by others.

Such satiety from the energy flux into intermediary metabolism is most accurately named 'cytodynamometric', i.e. measuring the energy production of cells such as those of the liver, because this mechanism does not maintain a constant amount of energy. This liver-dependent mechanism also has to be distinguished from some manner of sensing the energy exchange across the skin (less physical activity), called *ischymetry*. In fact, these recorded variations in metabolism corrected by movement will reflect the absorption rate that dominates the energy supply by the liver under freely feeding conditions, for example via the effects of absorption on dietary thermogenesis and lipogenesis.

Summarizing then, readily assimilated food calories of any kind probably both satiate temporarily and reinforce a subsequent strengthening of the inclination to eat when the perceived circumstances recur.

This interpretation is rather neatly supported by observations that alcoholic drinks both satiate rats and humans, and condition dietary odor preferences in rats. Such caloric conditioning, alongside the psychoactive effects usually attributed to ethanol, could well play an important role in the development of alcohol abuse and also of bulimic eating disorder involving alcohol.

7.3 SOMATIC CONTEXTUALIZATION

Not only dietary stimuli are conditioned by calories. Somatic and external environmental stimuli can be combined with the taste, aroma, texture and appearance of food in the learned stimulus complex for an appetite or a satiety.

The hunger and stress induced by dieting are therefore likely to become conditioned along with the sensory characteristics of convenience foods if a dieter lapses by eating such items. More daringly, the energy (and protein) contents of snackfoods acting on metabolic lack could condition craving for such foods even when the stomach has been filled by the snack. The learning of such food preferences that depend on a full upper digestive tract has been demonstrated in rats and in people. These appetites cued by a full intestine might be dubbed 'belly bulimia', to distinguish them from the emotion-dependent eating of doom, gloom or 'boom!' (anger) bulimia.

All these conditioned responses will increase the disposition to eat, regardless of rational wishes. Hence, learned appetites will strengthen a sense of being out of control of one's eating, which is characteristic of bulimia nervosa.

7.4 VOMITING-CONTROLLED LOSS OF MEAL-SIZE CONTROL

A mechanism for further loss of control and for increasing size of binges is suggested by the data obtained from experiments on rats and humans. The main effect that was evident during the learning of dietary satiety *Gestalten* was the conditioning of *de*satiation. That is, the major change observed was an enlargement of the meal caused by an increase in liking for a food when already full. This increased liking had been conditioned by effects of dilute carbohydrate. It seems that such a preference, despite ingestion of a considerable amount of food, is conditioned by only a small amount of calories reaching the upper intestine and being absorbed. The same mechanism could explain the further increase in meal size that is observed with repeated withdrawal or drainage of some of the ingested food from the stomach. (The stomach is a pump, not a passive globe. It was found that variable proportions of radio-isotope in a liquid diet drunk with an open gastric fistula rapidly reached the liver and brain via absorption; therefore drainage through a gastric fistula is not at all completely simulation feeding.)

The learning of reduced or at least sustained moderate sizes of meals, on the other hand, seems likely to be reinforced by a bloating effect of duodenal stretch or chemical stimulation after an undigested energy-rich liquid diet has reached the intestine from the stomach. Dilute diets or removal of much of the diet from the stomach will prevent the development of this bloated feeling and so the underlying potential for caloric preference conditioning is not countered. Also, the delay of an energy-rich diet by a mere five minutes in the training meal prevented the conditioning of satiety in rats, as did intravenous infusion of glucose at the rate it is absorbed from the rich diet. In people, a similarly carbohydrate-rich entree conditions dessert satiety. On the other hand, the carbohydrate-rich dessert reaches the intestine too late and so only preference is conditioned. Hence, calories in desserts, unlike calories in entrees, increase the attractions of those desserts when one is already full.

Vomiting after a large meal could be another way to reduce or even prevent the bloating that conditions meal-size moderation. Thus, vomiting after a binge would condition desatiation to the foods eaten toward the end of the binge. At the next binge, this would make those foods all the more attractive, even when the binge is well under way. This mechanism could provide powerful unconscious support for the subjectively obvious physical, emotional and rational relief at getting rid of much of the food that is binged on. The vomiting-conditioned desatiation will also further strengthen the feelings of loss of control. In addition, these feelings may well worsen the general emotional state. Such panic, as well as anxiety and guilt, may create a degree of arousal that itself breaks down learned satiety. Noradrenalin injected into the rat hypothalamus, and so releasing eating from learned control, has the same effect.

8   IMPLICATIONS FOR TREATMENT

On this theory, the bulimic's binge and the anorexic's refusal to eat are largely acquired, involuntary reactions to complex situations that consist of available foods and bodily and socio-affective states, coupled with thoughts about foods and their use to change the shape of the body.

This theory suggests, in addition, a way in which the characteristic loathing for quite normal roundness of the body in the eating disorders might be strengthened. This fear of fatness could partly be a way of making sense of the mysterious compulsion to eat, rather than being an immediate source of the strength of that compulsion. This personal attitude is second nature: our culture links uses of food to control of shape in ways that children learn long before puberty. The individual may well have become deeply involved in the cult of leanness before succumbing to the eating disorder. So the 'lipo-theology' provides a ready rationalization of the emotional turmoil about eating.

8.1   COUNTER-CONDITIONING

Of course, if the traps that are opened up by dieting and relapsing are to be avoided in the longer term, the sufferer must become disentangled from the worst of this web of intolerant attitudes to foods and the body. The present theory, nevertheless, is that the immediate problem is the abnormal control of the behavior toward food that has been acquired by the food stimuli and by the emotional and somatic states to which the foods may have been contextualized.

This implies that exposure to negative or neutral associates for the binge- or fast-evoking stimulus complexes needs somehow to be achieved. Positive associations need to be found for alternative situations that do not evoke disordered eating. Advice on sensible eating, as included in cognitive-behavioral therapy and counseling, might succeed in directing the patient to the appropriate quantities. Nevertheless, more specific advice that suited the person's own daily habits could be more effective.

Allowing a binge but preventing vomiting would restore satiety-conditioning quantities to the bulimic. It remains to be ascertained whether the usefullness that response-prevention therapy may have in bulimia rests mainly on this eating-specific mechanism. It is usually thought to rely on more general processes, such as habituation of anxiety about the fattening effects of a meal or extinction of the desire to vomit (Wilson, 1988).

8.2   DRUG TREATMENT

According to the above-stated theory, there is no reason to expect there to be a bulimia or anorexia 'button' in the head, perhaps with its own transmitter receptor subtype. A drug cannot act on satiety or bingeing as such; it can only act on a pathway in a mechanism contributing to the tendency to stop eating or to binge. Attempts to hit broader systems such as pleasure pathways or addiction mechanisms are likely to have too many side-effects. A drug might in principle reduce the potency of reinforcers of the disordered eating, but it is not clear that this would help the necessary relearning of orderly eating.

Appetite suppressants or mood-altering drugs might reduce the strength of instigators of the disordered eating. That, however, would not weaken the acquired stimulus control in the absence of the drug. At best, a drug would suppress the disordered behavior while some even more urgent therapeutic task was tackled.

Unintended effects of existing pharmaceutical management of bulimia may well provide an opportunity for their integration in dietary and cognitive-behavioral therapy.

The postulate of hepatic energy-supply satiety in the cognitive and physiological theory of appetite implies that the rate of gastric emptying is a powerful modulator of physiological control of appetite. The slowing of the emptying and absorption of a meal should therefore delay the rise of learned and unlearned somatic hunger cues and so reduce the temptations to binge. This aspect of possible modes of action of appetite-suppressant drugs is also supported by mechanistic data.

The suggested benefits of antidepressant drugs in the treatment of bulimia may in fact also arise from slowing of gastric emptying. These drugs are often anticholinergic or have peripheral monoaminergic effects that could contribute to reduced and reversed gastrointestinal motility. Thus, they should prolong the satiating effect of the previous meal and so reduce the instigation of binges. This would reduce the frequency of binges, as observed after antidepressive treatment and as is clinically desirable. Obviously, it would not reduce the size of the binges.

Antidepressants sometimes reduce and sometimes increase appetite or food cravings. In part, this could be because gastro-intestinal stasis induces nausea (and constipation) while a less extreme slowing of transit and absorption induces hunger and stronger caloric conditioning.

Changes in gastric emptying or capacity in adaptation to starvation or to habitual vomiting may contribute to abnormal learned and unlearned cues in anorexia or bulimia nervosa. If so, readaption to normal gastric function will be an important contribution to a normalization of the ingestion in either disorder and to the retention of food early in the treatment of bulimia.

Thus, there is a wide scope for considering the theoretical mechanisms supported by experimental data when seeking to help sufferers from the disturbing problems of disordered eating and body image. Drugs should not be prescribed without planning to exploit their effects on mood and on peripheral physiology to help the sufferer work at improving her condition. Dietary advice, behavior modification and cognitive restructuring should be focussed on, restoring the normal quantities that condition orderly eating.

9   SUMMARY

Physiological and cultural bases for eating disorders are concerned with people's physiological, psychological and social capacities. Moreover the slimming culture, individuals' weight control and the physiology of fat deposition are interlinked causal networks and operate within energy balance across the skin.

Treatment of eating disorders should include dietary advice, behavior modification, and cognitive restructuring, restoring the normal quantities that condition orderly eating.

II

Physiology and endocrinology of food intake regulation

6

Neuroscience of energy substrate homeostasis

Maintenance of the availability of energy substrates is crucial for survival of the organism. Glucose derived from liver and muscle glycogen, free fatty acids (FFA) released by white adipose tissue and - to a minor extent - proteins in the form of aminoacids, may serve as energy sources. Glucose is the predominant metabolic fuel because it is utilized by the brain under almost all physiological conditions. Expenditure of free fatty acids is of increasing importance under energy demanding conditions such as fasting or prolonged exercise. An accurate and fine-tuned regulation of glucose and FFA availability may therefore be considered a prerequisite for an organism with an everchanging balance between substrate supply and the metabolic needs of the body cells. Hormonal and neural factors, often functionally interrelated and acting in a coordinated fashion, form the basis for a sensitive and very specific mechanism regulating the mobilization, storage and utilization of the different energy substrates. The autonomic nervous system plays a key role in the control mechanism that regulates the release of these neural and hormonal factors.

1 THE AUTONOMIC NERVOUS SYSTEM AND ENERGY SUBSTRATE HOMEOSTASIS

1.2 THE AUTONOMIC NERVOUS SYSTEM

The scope of the first part of this chapter is to focus on the autonomic mechanisms involved in the regulation of peripheral energy substrate homeostasis. The autonomic nervous system controls autonomic or visceral mechanisms in the periphery (Guyton, 1986; Kandell et al., 1991; Luiten et al., 1987). This includes a variety of functions such as cardiovascular control (heart rate and blood pressure), regulation of body temperature (vasodilation and sweating), digestion of food (gastric emptying, gastric motility and the release of exocrine factors), and the aforementioned regulation of energy substrate homeostasis. The autonomic nervous system can be divided into three major subdivisions called the parasympathetic, the sympathetic and the enteric nervous system. The first two subdivisions form the neural connection between the brain and the body, are activated under opposite circumstances and act in an antagonistic fashion. Activation of the parasympathetic nervous system can be observed in the non-activated, resting state (sleep, post-prandial) and is characterized by low cardiovascular activity, storage of energy substrates, and increased gastro-intestinal function. Activation of the sympathetic nervous system is required under active, non-baseline conditions such as physical exertion and emotional stress (fight-flight responses), and leads to the mobilization of energy and increases in heart rate, blood pressure and body temperature. The third subdivision, the rather diffuse enteric nervous system, consists of a number of neurons, located in the periphery and connecting the different organs of the gastro-intestinal system.

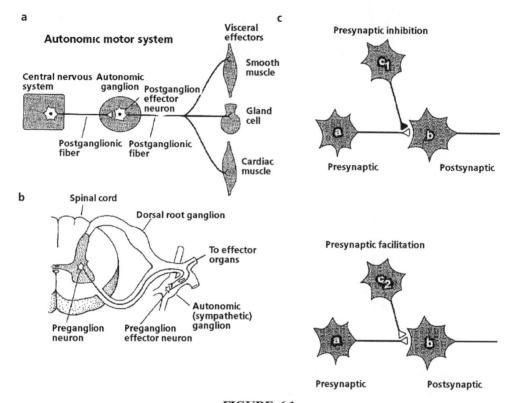

FIGURE 6.1

The autonomic nervous system

a Sympathetic and parasympathetic divisions of the autonomic nervous system

b Preganglionic and postganglionic fibers in the autonomic nervous system

c Presynaptic inhibition occurs when a presynaptic inhibitory neuron (c1) depresses the Ca-current in the terminal of a second presynaptic neuron (a), leading to a reduction in the amount of transmitter released

(From: Kandel, E.R. et al., 1991)

This 'brains of the belly' is involved in gastro-intestinal coordination and seems independent of, but modifiable by, the sympathetic and parasympathetic nervous systems. The enteric nervous system seems to serve as a control and communication unit between the putative autonomic nervous system and the different parts of the gastro-intestinal system, but experimental data are still not available.

The cell bodies of the neurons in the autonomic nervous system that innervate the different target organs in the periphery are all located in small encapsulated clusters of cells (autonomic ganglia) outside the central nervous system. The connections between the central nervous system and the autonomic ganglia are called preganglionic fibers. All autonomic preganglionic fibers are cholinergic, which means that they contain the neurotransmitter acetylcholine. Activation of autonomic preganglionic neurons leads to the release of acetylcholine that stimulates nicotinic receptors on so-called postganglionic fibers, the axons that emerge from a ganglion and project on the target organ. (figure 6.1)

The cell bodies of the preganglionic neurons of the parasympathetic nervous system are located in the brain stem and in the sacral segments of the spinal cord. The vagus nerve, or 10th cranial nerve, carries all the parasympathetic projections of the heart, lungs and the gastro-intestinal tract. The majority (more than 80%) of the fibers in the vagus nerve are travelling from the periphery to the brain, in other words they serve a sensory function. The parasympathetic ganglia are located close to or within the tissue they inner-vate. Acetylcholine is the classical neurotransmitter in the postganglionic neurons of the parasympathetic nervous system. At the target organ, acetylcholine acts on muscarinic re-ceptors. The main parasympathetic connections involved in energy substrate homeostasis are those projecting on the liver (promoting the storage of glucose into glycogen) and the endocrine pancreas (leading to the release of both insulin and glucagon). White adipose tissue and the adrenal glands receive no direct parasympathetic innervation.

The cell bodies of the sympathetic motorneurons are situated in the thoracic and the lumbar levels of the spinal cord. The sympathetic ganglia lie relatively close to the spinal cord, and their postganglionic fibers diverge over large distances to reach the cells in the target organs. Postganglionic sympathetic neurons contain the neurotransmitter nora-drenalin (NA, or norepinephrine, NE) that affects (nor)adrenergic receptors (adrenocep-tors). (figure 6.2) The main sympathetic connections involved in energy substrate homeostasis travel through the preganglionic splanchnic nerve and end in the celiac ganglion. From there, postganglionic fibers innervate important target organs such as the liver (leading to the breakdown of glycogen to glucose) and the pancreas (stimulating glucagon and inhibiting insulin release). White adipose tissue cells are not directly inner-vated by sympathetic neurons. The adrenal medulla is directly innervated by preganglionic sympathetic fibers and can be considered as a modified ganglion/postganglionic sympa-thetic nerve. Stimulation of preganglionic sympathetic nerves projecting on the adrenal medulla leads to the release of the hormone adrenalin (A) into the circulation. Adrenalin differs from noradrenalin in only one CH_3-group but has a different affinity for some adre-nergic receptors leading to essentially different postsynaptic actions on energy metabolism. (figure 6.2) Activation of the sympathetic nervous system is thus accompanied by an increase in the outflow of both noradrenalin and adrenalin. The term sympathoadrenal system is often used to reflect this combined neuronal and hormonal outflow of the two catecholamines.

Neurons synapse with one another not only on the cell body and at dendrites, where they can control impulse activity, but also at their terminals where they can control trans-mitter release. Axo-axonic synapses can either depress or enhance transmitter release through presynaptic inhibition or presynaptic facilitation (see Figure 6.1c).

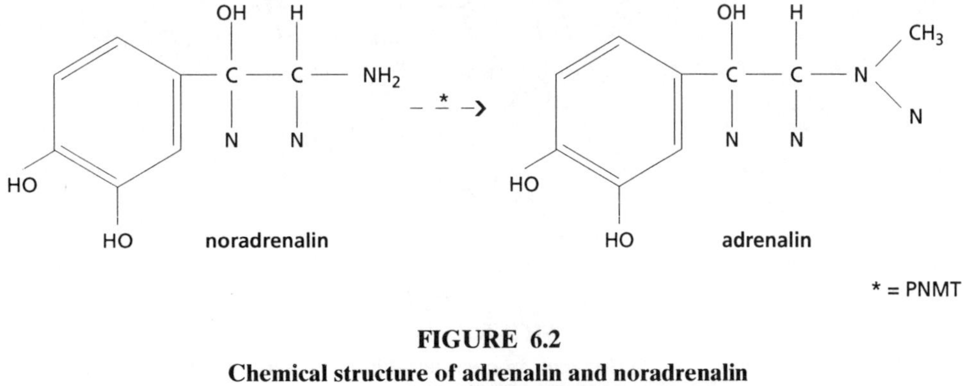

* = PNMT

FIGURE 6.2
Chemical structure of adrenalin and noradrenalin
The enzyme phenylethanolamine-N-methyl transferase (PNMT) converts NA to A.

When the presynaptic terminals of the neurons whose release is modulated contain receptors for the neurotransmitter NA this is called an adrenoceptor-mediated presynaptic regulatory mechanism.

The majority of the autonomic neurons (both pre- and post-ganglionic) contain one or sometimes more cotransmitters, factors that are colocalized with the classical neurotransmitters in the autonomic nerve endings. Cotransmitters are released during stimulation of the autonomic nerves. Typical examples of cotransmitters are VIP (vasoactive intestinal peptide) and galanin that are colocalized with acetylcholine in parasympathetic neurons. Neuropeptide Y, galanin, ATP and adrenalin are identified as sympathetic cotransmitters. Knowledge on the function and release mechanisms of these cotransmitters is still limited.

1.3 AUTONOMIC INFLUENCES ON PERIPHERAL ENERGY METABOLISM

Peripheral energy substrate homeostasis is controlled by the autonomic nervous system (Scheurink & Steffens, 1990; Woods et al., 1984). This is schematically visualized in Figure 6.3. Activation of the parasympathetic branch to the liver directly stimulates the conversion of circulating glucose into glycogen. Stimulation of parasympathetic neurons innervating the pancreas leads to an increased outflow of insulin by pancreatic B-cells. Insulin is the main anabolic hormone. It lowers blood glucose concentrations by suppressing endogenous glucose production and stimulating glucose utilization. More specifically it inhibits hepatic glycogenolysis and gluconeogenesis, and stimulates the storage of glucose in the liver in the form of glycogen. Furthermore, in insulin-sensitive tissues like resting muscle and adipose tissue, insulin stimulates the uptake, storage and utilization of glucose. Insulin is the only major physiological factor involved in a decline in plasma FFA concentrations. Insulin inhibits FFA release, and enhances the re-esterification of free fatty acids by accelerating the transport of glucose into the fat cell.

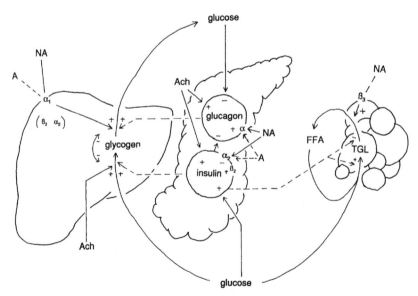

FIGURE 6.3

Neuronal and hormonal influences on peripheral glucose and FFA homeostasis

TGL = triglycerides (stored fatty acids)

Activation of the sympathetic nervous system results in the mobilization of energy substrates from the storage tissues. The effects of adrenalin and noradrenalin include both direct and indirect actions mediated via α- and β-adrenoceptor mechanisms. Catecholamines directly influence hepatic glucose production by stimulating both glycogenolysis and gluconeogenesis in liver. The receptor types involved depend on gender, age and species. Important indirect actions of A and NA on blood glucose are achieved by α-adrenergic inhibition of insulin release and α- and/or β-adrenergic stimulation of glucagon secretion. Glucagon from the A-cells of the pancreatic islets is a potent stimulator of hepatic glycogenolysis and gluconeogenesis. Catecholamines also limit expenditure of glucose via β-adrenoceptor stimulation of muscle glycogenolysis. Lipolysis is predominantly influenced by physiological doses of neuronal NA (and not A) via activation of $β_3$-adrenoceptors on the fat cell. Alpha-adrenoceptor mediated inhibition of FFA release, as reported in studies on human and hamster adipocytes, is of minor importance and is normally masked by the stimulatory β-adrenergic mechanism. The catecholamines are the primary factors influencing FFA release, while other hormones play an additional physiological role (glucagon) or function as factors that maintain or improve the capacity of the fat cell to respond to regulatory influences (corticosterone, growth hormone).

1.4   LEVELS OF ORGANIZATION IN THE AUTONOMIC NERVOUS SYSTEM

Since the classical report of the 'piqûre diabetique' by Claude Bernard in the nineteenth century, the central nervous system has been implicated in the autonomic control of peripheral energy homeostatic mechanisms. The key neuronal mechanisms that control the activity of the autonomic nervous system are located in different levels of organization within the central nervous system. Numerous studies have been performed to identify the neuronal pathways involved in the control of sympathetic functioning. Unfortunately, data on parasympathetic outflow are scarce, mainly because it is almost impossible to obtain reliable measurements of acetylcholine outflow. The underlying theme throughout the following will be the concept of a more or less hierarchical structure in the sympathetic nervous system, consisting of different levels of organization that together determine the ultimate sympathetic outflow at the level of the target organ. It is believed that an analog organization might exist in the parasympathetic branch of the autonomic nervous system.

2   LEVELS OF ORGANIZATION WITHIN THE SYMPATHETIC NERVOUS SYSTEM

2.1   INTRODUCTION

Neuroanatomical and neurophysiological evidence suggests that the organization of an organ-specific activation (or inhibition) of sympathetic activity may take place at different levels in the sympathetic nervous system. (figure 6.4) The hypothalamus can be considered as one of the major levels of organization in the brain coordinating sympathetic output. Extrahypothalamic limbic areas may provide the highest level of organization. Autonomic control mechanisms in various levels of the brain stem and spinal cord may also serve an organ-specific activation of the sympathetic nervous system, but data particularly on the regulation of metabolism are scarce. Finally, presynaptic regulatory mechanisms acting on the pre- and post-ganglionic nerve terminals in the peripheral sympathetic nervous system markedly influence the outflow of NA.

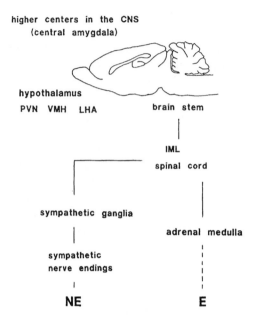

FIGURE 6.4
Levels of organization in the sympathetic nervous system (Scheurink & Steffens, 1990)

The following will focus on the experimental data that provide evidence for the existence of different levels of organization within the sympathetic nervous system. First, information will be given on the type of experiments that are generally used to study fluctuations in sympathetic outflow. Thereafter, experiments will be presented that deal with the actions and the alterations of NA, A and other sympathetic neurotransmitters, followed by studies on the regulatory mechanisms acting at the level of the presynaptic nerve endings and the sympathetic ganglia. Finally, it will be shown that sympathoadrenal outflow can also be regulated at higher levels in the sympathetic nervous system with special emphasis on the role of the hypothalamus.

2.2   EXPERIMENTS

Activation of the sympathetic nervous system leads to an increased mobilization of energy substrates. It is therefore not surprising that the majority of the human and animal studies on the actions and the alterations of the sympathetic nervous system are based on an increased need for glucose and/or FFA. Exercise represents such a physiological state in which the catecholamines have to alter the distribution of glucose and FFA to increase the flow of these nutrients to the exercising muscle, while at the same time an adequate supply of glucose to the brain has to be maintained. Other studies are mainly based on an increase in sympathetic outflow caused by an experimentally induced reduction in the availability of energy substrates by means of administration of hypoglycemic doses of insulin or drugs that block glucose or fatty acid oxidation, such as 2-deoxy-D-glucose (2-DG) or sodium mercaptoacetate (MA). The majority of the following studies are performed in permanently cannulated rats, and deal with the effects of exercise or drug-induced reductions in energy substrate availibility. Exercise consisted of either treadmill running at a speed of 21 m/min, or strenuous swimming against a counter-current (13 m/min) in a pool with water at 33 °C. In all studies, metabolic and hormonal responses were assessed in frequent blood samples taken through permanent catheters before, during and after exercise or drug treatment.

2.3   PERIPHERAL OUTFLOW OF CATECHOLAMINES

As mentioned above, activation of the sympathoadrenal system leads to the outflow of catecholamines from the adrenal medulla and the sympathetic nerve endings. Plasma levels of NA and A increase during all types of sympathetic activation and vary directly with the intensity and the duration of the stimulus (Galbo, 1983). Only few data are available on the individual contribution of the adrenal medulla and the sympathetic nerve endings to the catecholamine alterations during sympathoadrenal activation. In particular the origin of NA in plasma is not well-defined. To illustrate this, adrenodemedullation (surgical removal of the adrenal medulla) was reported to cause a marked reduction in the exercise induced increases in plasma NA concentrations, suggesting that a major portion of NA in plasma originates from the adrenal medulla. Others, however, failed to detect any differences in plasma NA concentrations between intact and adrenodemedullated (Adm) rats, suggesting that either the adrenal medulla does not produce NA or that a compensatory increase in sympathetic nervous activity elsewhere may occur after adrenodemedullation. Experiments were therefore particularly focused on the role of each of the two branches of the sympathoadrenal system on the alterations of plasma catecholamine levels during exercise. Plasma catecholamine concentrations were measured in exercising intact and Adm rats, with and without administration of selective adrenoceptor agonists and antagonists (Scheurink et al., 1989a). The results of these studies are extensively described in Chapter 14 (Figures 14.5 - 14.9). Here it is sufficient to report that a marked reduction in the exercise-induced increase in plasma NA concentrations was observed in both Adm rats as well as in intact rats injected with a β_2-selective adrenoceptor antagonist. Intravenous infusion of a β_2-selective agonist restored the increase in plasma NA in Adm rats. Injection of an α_2-selective antagonist in combination with infusion of a β_2-selective agonist caused an enormous increase in plasma NA. It was concluded that the effect of adrenodemedullation on plasma NA could be entirely explained by a diminished activation of presynaptic β_2-adrenoceptors on the sympathetic nerve terminal. One of the conclusions was that the noradrenergic cells in the adrenal medulla do not essentially contribute to the NA content in plasma, and that practically all NA in plasma originates from the peripheral nerve endings of the sympathetic nervous system. The main message, however, from these experiments is that the data show that adrenoceptor-mediated presynaptic regulatory mechanisms acting on the peripheral nerve endings of the sympathetic nervous system can markedly influence the release of neuronal NA. Data, mainly derived from pharmacological studies *in vitro* and experiments with anesthetized animals, reveal that peripheral sympathetic outflow is also modulated by other factors such as dopamine, acetylcholine, angiotensin, neuropeptide Y, galanin, opioids and serotonin (Langer, 1981; Langer & Arbilla, 1990). The list of these factors is still growing.

In conclusion, the above data reveal that the local outflow of NA depends on the ambient neurotransmitter and hormonal milieu in the synaptic junctional cleft. Variations in the sensitivity and/or the populations of pre- and post-synaptic receptors in the sympathetic nervous system may account for regional differences in sympathetic activation. This means that, in every organ, regulatory mechanisms acting at the pre- and post-synaptic sympathetic nerve terminals may determine the ultimate sympathetic activation of the target cell.

2.4   NORADRENALIN OF ADRENAL ORIGIN

We concluded above that all NA in plasma originates from the peripheral nerve endings of the sympathetic nervous system and that the adrenal medulla does not contribute to circulating NA. However, anatomical evidence reveals that the adrenal medulla contains two types of parenchymal cells with specific granules that store either adrenalin or noradrenalin. In the adult rat, approximately 80-85% of the chromaffin cells in the medulla can be stained on immunoreactivity for the enzyme PNMT (Fig 6.2), indicating that these cells synthesize A. The remaining smaller cells (15-20%) synthesize NA. Adrenalin levels in plasma clearly reflect the adrenal output of this catecholamine. No principal correlation was found between the adrenal secretion of NA and the concentrations in plasma. In addition, the NA/A ratio in the adrenal vein is completely different from the ratio in the general circulation (Bereiter et al., 1986). These data raise an important question on the physiological function of the NA producing cells in the adrenal medulla. It may be that the noradrenergic cells are remnants of a previous, perinatal function of the adrenal medulla. Some evidence for this idea is provided in anatomical studies by Verhofstad (1984), who found that a change in catecholamine content from NA to A appears during the pre- and perinatal development of the adrenal medulla in the rat. Simultaneously the formation of the sympathetic nervous system is completed two or three days after birth. These concomitant stages in the development of the sympathetic nervous system and adrenal medulla seem to suggest that in the pre- and perinatal period adrenal NA may serve as a general substitute for sympathetic activity. After completion of the sympathetic innervation of peripheral tissues NA from the adrenal medulla may be of minor importance in comparison with neuronal NA. This means that in the adult rat, the physiological role of NA of adrenal medullary origin in metabolic and hormonal processes is almost negligible.

2.5   NEURAL AND HORMONAL ACTIONS OF NORADRENALIN

Sympathetic NA may influence peripheral metabolism via two different mechanisms: direct neuronal activation of post-synaptic adrenoceptors in sympathetically innervated tissues such as liver and pancreas, and indirect hormonal stimulation of tissues that are not innervated by the sympathetic nervous system, such as white adipose tissue. During sympathetic stimulation, only a small part of the released neurotransmitter leaks into the blood. Hence, concentrations of neuronal NA in the synaptic cleft will greatly exceed those in plasma. For example, during NA administration in humans the plasma NA concentrations must often exceed 2000 pg/ml in order to produce an increment in blood pressure of 20 mm Hg. Such a pressor response is normally accompanied by an increase in venous NA of a few hundred pg/ml or less (Goldstein et al., 1987). Intravenous infusions of doses of NA that cause a normal increase in plasma NA concentrations, are therefore not sufficient to mimic the concentrations of NA in the synaptic cleft of sympathetically innervated tissues. This means that the effects that can be observed after intravenous administration of a physiological dose of administered NA (for example lipolysis) are typical hormonal actions of neuronal NA.

The leakage or spillover of NA into the blood circulation is not uniform but dependent on the width of the synaptic cleft in the innervated organ (Esler et al., 1984a,b). Noradrenalin spillover rate is relatively high in the lungs and in the endothelial cells in the walls of blood vessels. In contrast, the liver even extracts NA from blood. In addition, recent studies indicated that different patterns of activity may occur within different parts in the sympathetic nervous system. Taken together, these data indicate that one has to be very cautious in using plasma NA levels as an index of sympathetic activation of one single sympathetically innervated organ. On the other hand, plasma NA levels do reflect all hormonal functions of NA, and may also be used as a reliable index for overall sympathetic activation.

2.6   SYMPATHETIC COTRANSMITTERS

Traditionally, NA is thought to mediate all effects of sympathetic nerves. Recent evidence of the localization, release and action of peptides in peripheral sympathetic nerves suggests that NA may not be the sole neurotransmitter in the sympathetic nervous system. Cotransmitters, factors that are colocalized and coreleased with NA might also play an important role. Anatomical and physiological evidence suggest that in particular neuropeptide Y (NPY), galanin, ATP and adrenalin might serve as sympathetic cotransmitters. Knowledge on the actions and the release mechanisms of the cotransmitters in the sympathetic nervous system is still limited. The current hypothesis is that selective outflow of one or more neuropeptides from the peripheral sympathetic nerve terminal allows the organism to differentiate between sympathetic functioning at the level of the target organ.

The following data, that focus on a possible function for galanin and NPY as sympathetic neurotransmitters in the endocrine pancreas, might be used as a typical example for a more general concept of cotransmitter functioning (Dunning and Taborsky 1991, Scheurink et al., 1992). Stimulation of the sympathetic nerves innervating the endocrine pancreas leads to an increase of the plasma levels of galanin and NPY in the pancreatic vein.(figure 6.5)

Administration of an α_2-antagonist markedly enhances the release of galanin and NPY from the sympathetic nerve terminals, indicating that the outflow of sympathetic cotransmitters is also subjected to presynaptic regulatory mechanisms. (figure 6.6)

Anatomical evidence for the presence of galanin and NPY in pancreatic sympathetic nerves has been obtained with immunofluorescent staining. Interestingly, the innervation patterns of the NA, galanin and NPY containing nerves appears to be basically different. Noradrenergic nerves innervate both the endocrine islets and the blood vessels in the pancreas. Noradrenergic nerves also innervate intrapancreatic ganglia, providing an anatomical basis for sympathetic inhibition of parasympathetic neural activity. Galanin-containing nerves preferentially innervate the islets with only a few fibers in the intrapancreatic ganglia and blood vessels.

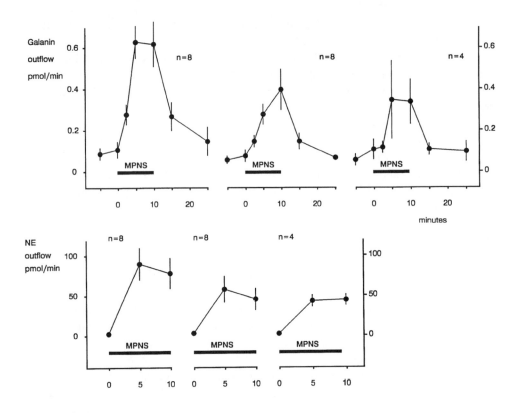

FIGURE 6.5

Outflow of galanin and noradrenalin (NE) into the pancreatic vein during electrical stimulation of the mixed autonomic pancreatic nerve (MPNS) (Scheurink et al., 1992)

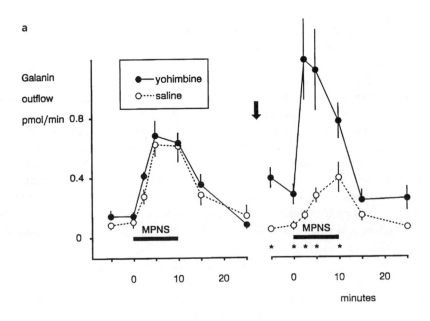

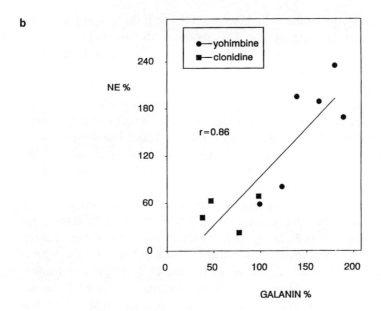

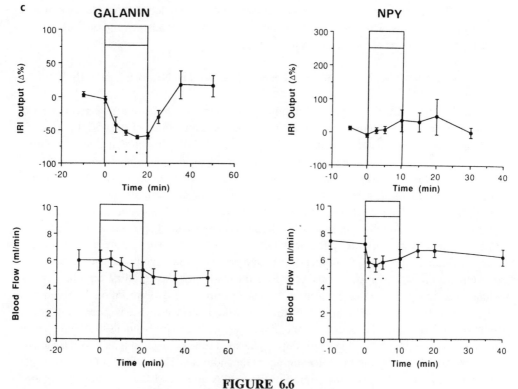

FIGURE 6.6
Galanin and yohimbine, noradrenalin and insulin
a Effect of administration of the α_2-adrenoceptor antagonist yohimbine (indicated by the arrow) on sympathetic outflow of galanin during electrical stimulation of the mixed autonomic pancreatic nerve
b Correlation between galanin and noradrenalin (NE) outflow during electrical stimulation of the mixed autonomic pancreatic nerve (Scheurink et al., 1992)
c Effects of galanin and NPY on pancreatic insulin secretion and blood flow (Dunning & Taborsky, 1991) IRI = immunoreactive insulin

Neuropeptide Y fibers only occasionally innervate the islets; the majority innervate the vasculature and the intrapancreatic ganglia. The anatomical differences (some species variation occurs) between NA, galanin and NPY are reflected in their different functional properties. Noradrenalin is capable of regulating blood flow as well as pancreatic hormone secretion (insulin, glucagon and somatostatin). Galanin exerts a potent and direct action on pancreatic hormone secretion but has no effect on pancreatic blood flow. NPY regulates pancreatic blood flow, analogous to its well-known role in cardiovascular resistance. Neuropeptide Y has no effect on pancreatic hormone release. These data suggest that, in the endocrine pancreas, NA might act as a general sympathetic neurotransmitter, while galanin and NPY are released to potentiate NA's effect on more specific sympathetic functions.

Taken together, the above data reveal that a functional dissociation may occur between the outflow of NA and the different peptidergic cotransmitters at the level of the peripheral sympathetic nerve terminals. Selective activation of one or more cotransmitters will be a useful instrument in determining a fine-tuned sympathetic outflow at the level of the target organ. Future investigations are necessary to unravel the release mechanisms involved in the selective outflow of NA and/or the different cotransmitters. Some investigators suggest that the frequency of the sympathetic stimulus might determine the outflow of the different neurotransmitters: small vesicles containing solely NA would be emptied at lower frequencies, while a high frequency stimulation would also empty the larger vesicles leading to the release of both NA and cotransmitter.

2.7   ADRENAL MEDULLA VERSUS SYMPATHETIC NERVE TERMINALS

As we have discussed, local variations in plasma catecholamine levels can be explained by differences in presynaptic regulatory mechanisms acting at the peripheral nerve endings of the sympathetic nervous system. However, in 1984 Young et al. made an observation that demanded a more centrally located mechanism regulating sympathoadrenal outflow. They demonstrated that, under different experimental conditions, the two branches of the sympathetic nervous system - adrenal medulla and sympathetic nerve endings - are independently activated. This was in sharp contrast to the then generally accepted view of a uniform activation of the sympathoadrenal system. Since then evidence is accumulating that the central nervous system may influence distinctive parts of the body via selective activation of a restricted number of pre- and postganglionic sympathetic nerves.

Our recent study in which rats were forced to swim for the very first time in a swimming pool (emotional stress) forms a typical example for a dissociation between the adrenal and the sympathetic branch of the sympathetic nervous system (Scheurink et al., 1989b). In this study, the catecholamine responses of the naive animals were compared with the plasma catecholamine levels during the sixth time of swimming, e.g. when the animals were well-accustomed to the experimental conditions. The addition of psychological stress to exercise led to an exaggerated outflow of A and a reduction of the exercise-induced increase in plasma NA concentrations (Figure 14.11). In other words, psychological stress (loss of controllability and predictability) shifted the sympathoadrenal response to swimming from the neuronal outflow of NA to a more general adrenomedullary release of A. The two branches of the sympathoadrenal system are not only functionally but also metabolically dissociated. As mentioned before, A and NA have different affinity for some adrenergic receptors leading to very specific alterations in glucose and FFA mobilization. Changes in plasma A levels are always followed by parallel alterations in blood glucose. Changes in plasma NA are generally accompanied by parallel changes in plasma FFA levels. Likewise, in the psychologically stressed animals the shift from NA to A was accompanied by an attenuated FFA response and a highly increased glucose mobilization. (figure 6.7)

Since glucose serves as the most important energy substrate for the brain under all circumstances, it is tempting to speculate that the observed responses to an uncontrollable and unpredictable stressor - a shift from NA to A, and consequently a shift from FFA to glucose - is an appropriate and functional response to guarantee an adequate supply of glucose to the brain under all circumstances.

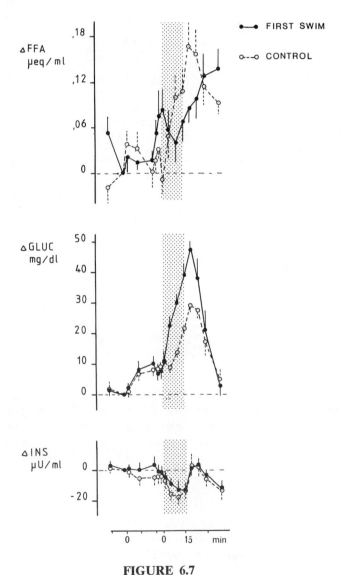

FIGURE 6.7

Blood glucose, plasma FFA and insulin responses to exercise in naive (first swim) and well-accustomed control rats (Scheurink et al., 1989b)

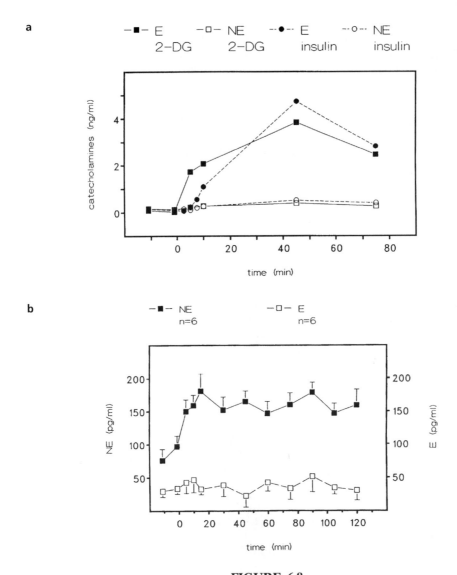

FIGURE 6.8

Adrenalin and noradrenalin responses to glucoprivation and lipoprivation

a Adrenalin (E) and noradrenalin (NE) responses to glucoprivation: effects of insulin and 2-DG (Modified from Scheurink & Ritter, 1993)

b Adrenalin (E) and noradrenalin (NE) responses to lipoprivation: effects of mercaptoacetate (Modified from Scheurink & Ritter, 1993)

Based on the aforementioned A/glucose and NA/FFA correlations, one might hypothesize that a reduction in the availability of glucose or FFA will lead to a selective activation of one of the two branches of the sympathoadrenal system. To test this, rats were subjected to glucoprivation or lipoprivation (decreased glucose or FFA availability), experimentally induced by administrating 2-DG, hypoglycemic doses of insulin, or MA. As can be seen in Figure 6.8, glucoprivation and lipoprivation are counteracted by increased outflow of A and NA, respectively. This demonstrates that the distinct metabolic signals are capable of exerting a selective control of sympathoadrenal outflow.

The finding that decreased glucose availability exerts a powerful and selective control of adrenomedullary secretion is compatible with the evidence that circulating A is an important mediator of hepatic glucose production in rats. Similarly, the selective increase in sympathetic neuronal activity induced by MA is a metabolically appropriate response to blockade of fatty acid oxidation since neuronal NA serves as the physiological mediator of lipolysis in white adipose tissue. These data indicate that a primary function of sympatho-adrenal activation is to increase circulating levels of a specific metabolic fuel.

2.8   SYMPATHETIC GANGLION, SPINAL CORD AND BRAIN STEM

Presynaptic regulatory mechanisms acting at the sympathetic nerve terminals are not sufficient to explain the differences in sympathoadrenal outflow seen in the experiments described in the previous paragraph. Therefore, we hypothesize that higher levels of organization in the sympathetic nervous system must be involved in the regulation of neuronal activity in pre- and postganglionic sympathetic nerves. For example, sympathetic outflow might be affected at the level of the sympathetic ganglia. As shown in Figure 6.9, subcutaneous administration of vasopressin (AVP) markedly reduced baseline and stimulated neuronal outflow of NA (Buwalda et al.). Evidence from pharmacological and anatomical studies suggests that this effect of AVP on neuronal outflow is probably mediated by vasopressinergic receptors located at the cell bodies of the postganglionic sympathetic nerves rather than at the level of the presynaptic nerve endings. In other words AVP seems to modulate sympathetic outflow at the level of the sympathetic ganglion.

Higher levels of organization in the sympathetic nervous system include the cell bodies of the motorneurons in the intermediolateral column in the spinal cord and the autonomic control mechanisms in various levels of the brain stem. Bereiter and his colleagues (1986) were the first to show that electrical stimulation of certain areas in the brain stem resulted in very specific changes in plasma A and NA concentrations. Their data and the results from many other experiments (in particular studies investigating the role of the sympathetic nervous system in cardiovascular functioning) revealed that the sympathetic motorneurons in the brain stem and spinal cord play an important role in the organ-specific activation of the sympathetic nervous system. Unfortunately, data that combine sympathoadrenal outflow with the regulation of energy substrate metabolism are still scarce.

At this point, it is important to note that several areas in the brain stem contain a wide variety of receptors that inform the central nervous system on changes that occur in the periphery. Glucose-sensitive neurons located in the area postrema (AP) and the nucleus of the solitary tract (NTS) fulfill an important role in the regulation of energy substrate homeostasis (Ritter et al., 1983). Stimulation of these glucose-sensitive neurons directly affects the activity of central noradrenergic pathways that project on glucose and FFA regulating areas in the hypothalamus (Oomura, 1983). This implies that it is almost impossible to distinguish between a sensory or a motor function of a specific neuron, since experimentally induced changes in brain stem activity might influence both efferent (brain→periphery) and afferent (periphery→brain) pathways. In other words, changes in sympathoadrenal outflow that occur after electrical stimulation of the brain stem may be interpreted as the result of direct stimulation of motorneurons innervating the peripheral sympathetic nervous system, but may also be secondary to the activation of a sensory mechanism in the brain stem projecting on glucoregulatory areas in the hypothalamus.

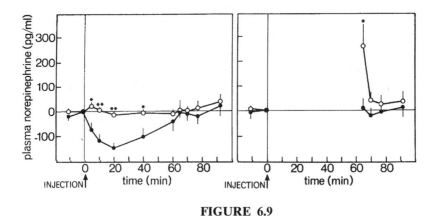

FIGURE 6.9

Effect of subcutaneous injection of vasopressin on baseline and stimulated noradrenalin responses in conscious rats (Buwalda et al., 1991)

2.9   HYPOTHALAMUS AND HIGHER LEVELS

The hypothalamus is regarded as an important integrative station for the neural and hormonal control of peripheral metabolic processes. Physiological evidence for the hypothalamic control of peripheral energy substrate homeostasis has emerged mainly from studies on changes in peripheral glucose and FFA metabolism in response to direct chemical and electrical stimulation or lesioning at various locations within the hypothalamus, and from single unit recording studies of hypothalamic neurons in response to changes in hypothalamic and peripheral concentrations of glucose and other nutrients (Oomura, 1983; Shimazu, 1986; Steffens & Strubbe, 1983). The ventromedial (VMH), lateral (LHA), and paraventricular (PVN) hypothalamic areas are the predominant regions within the hypothalamus involved in the control of peripheral metabolism. In detail, more principally catecholamine and serotonin sensitive neurons within the VMH, LHA and PVN seem to be involved, since administration of NA, A and serotonin (5-HT) into these hypothalamic areas markedly influences glycolysis, glycogenolysis as well as the peripheral concentrations of glucose, FFA and their regulating hormones (Scheurink et al., 1989, 1990b, 1993).

The changes in peripheral energy homeostasis observed after stimulation - lesioning the hypothalamus - are probably secondary to the alterations in autonomic activity. Lesioning the VMH leads to an increase in the neural activity of the parasympathetic nerves and a reduction in the activity of the sympathetic nerves innervating the endocrine pancreas. These studies, and many other experiments dealing with the effects of hypothalamic manipulations on the firing rate of several peripheral autonomic neurons, are extensively described by Oomura (1983) and Shimazu (1986). We shall focus on the effects of chemical stimulation of the hypothalamus on sympathoadrenal outflow. For example, infusion of selective adrenoceptor agonists or antagonists in the hypothalamus selectively influences sympathoadrenal output in resting or exercising rats. Administration of an α-adrenoceptor blocker to the VMH markedly reduced the exercise-induced increase in plasma A concentrations without changes in neuronal outflow of NA. The opposite, a clear reduction in plasma NA concentrations without any effect on adrenal A outflow was observed after administration of a β-adrenoceptor antagonist into the PVN (Scheurink et al., 1990a,b; Scheurink & Steffens, 1990) (figure 6.10).

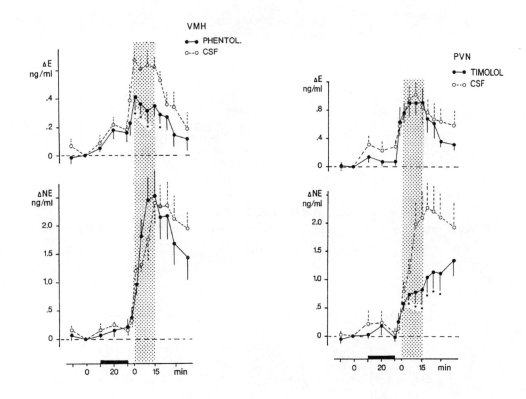

FIGURE 6.10

Effects of intrahypothalamic administration of adrenoceptor antagonists on the exercise-induced changes in plasma adrenalin (E) and noradrenalin (NE)

a effect of phentolamine in the VMH

b effect of timolol into the PVN (Scheurink et al., 1990a,b)

Peptidergic mechanisms in the hypothalamus are also involved in the regulation of peripheral catecholamine outflow. We recently observed that injection of the peptidergic cotransmitters NPY and galanin into the PVN significantly changed baseline catecholamine levels and cardiovascular activity in resting animals (Van Dijk et al.). Taken together, these data indicate that the hypothalamus serves as an important area in the brain controlling peripheral sympathoadrenal outflow. Data on a possible role of the hypothalamus in the peripheral outflow of peptidergic cotransmitters such as galanin and NPY are not yet available. The hypothalamic influence on peripheral catecholamine outflow probably represents one part of a more general central nervous system controlled activity that includes metabolic, cardiovascular as well as locomotory alterations. The central noradrenergic pathways, arising in the brain stem and projecting to the hypothalamus through the ventral noradrenergic bundle, seem to be specifically involved in the transmission of signals concerning peripheral glucose concentrations to the autonomic regulatory regions in the hypothalamus.

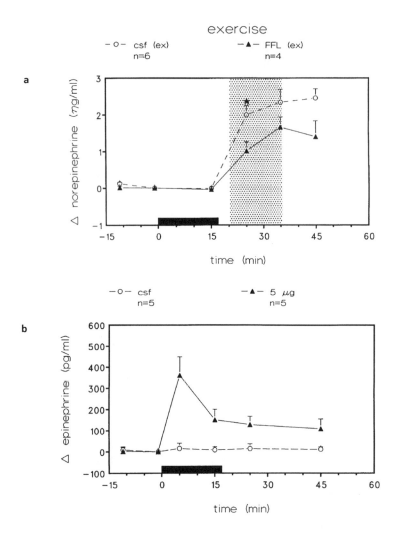

FIGURE 6.11

Effects of intrahypothalamic administration of the serotonergic reuptake blocker fenfluramine on baseline plasma adrenalin (E) (b) and exercise noradrenalin (NE) levels (a) (Scheurink et al., 1993)

Intrahypothalamic (PVN) infusion of a serotonergic 5-HT_{1a}-agonist and dex-Fenfluramine, a serotonergic reuptake blocker, markedly increased blood glucose and plasma concentrations of A and corticosterone, while plasma NA levels are significantly reduced (Korte et al., 1992; Scheurink et al., 1993) (figure 6.11). The effects of activating serotonergic receptors in the PVN - a shift from NA to A, increased blood glucose, decreased plasma FFAs and increased corticosterone - are remarkably identical to the (neuro)hormonal and metabolic responses that can be observed in psychologically stressed animals. It may be hypothesized that the addition of emotional stress to exercise leads to an increased outflow of endogenous serotonin in the PVN, finally leading to the activation of adrenal A release and a reduction in the neuronal outflow of NA. This hypothesis is in agreement with current opinions in the stress literature that hypothalamic serotonergic mechanisms are activated during anxiety, stress and depression, and that hypothalamic 5-HT might mediate the hormonal and behavioral responses to these pathophysiological conditions. Likewise, the metabolic and cardiovascular changes that can be observed during stress and after administration of 5-HT - switches from NA toward A, from FFA to glucose mobilization, increased blood pressure - makes it very tempting to speculate that hypothalamic serotonergic mechanisms might be involved in the causal relationship between stress and the occurrence of obesity and hypertension, as seen in human pathological and epidemiological studies.

In conclusion, the hypothalamus can be considered an important integrative station in the brain coordinating sympathoadrenal outflow. Extrahypothalamic limbic areas may provide the higher levels of organization. In particular, the central amygdala, an area with monosynaptic connections to both sympathetic and parasympathetic medullary nuclei, can be considered an important limbic autonomic nucleus. The limbic organization may take place via the basal hypothalamic nuclei, but also via a direct projection on lower brain stem autonomic nuclei. Future experiments are needed to identify the pathways by which the higher centers in the central nervous system exert their influence on the regulation of sympathoadrenal activity.

3   SUMMARY

Neuroanatomical and neurophysiological evidence suggests that the organization of an organ-specific activation (or inhibition) of sympathetic activity may take place at different levels in the sympathetic nervous system (figure 6.12).

Extrahypothalamic limbic areas may provide the highest level of organization, but firm data concerning their role in peripheral glucose and FFA regulation are not available.

The hypothalamus can be considered as one of the major levels of organization in the brain coordinating sympathetic output. The hypothalamic influences on the peripheral autonomic nervous system may be achieved via multisynaptic connections with the autonomic brain stem mechanisms. Additionally, single neuronal connections from the PVN to the intermediolateral column of the spinal cord may be involved.

Autonomic control mechanisms in various levels of the brain stem and spinal cord may also serve the organ-specific activation of the sympathetic nervous system, but data particularly on the regulation of metabolism are scarce.

Presynaptic regulatory mechanisms acting on the pre- and postganglionic nerve terminals in the peripheral sympathetic nervous system markedly influence the outflow of NA and the sympathetic cotransmitters. The ambient neurotransmitter and hormonal milieu in the synaptic junctional cleft may therefore be considered an important factor that influences the outflow of NE and the sympathetic cotransmitters from individual sympathetic nerve terminals.

Regional variations in the sensitivity and/or the populations of pre- and postsynaptic receptors in the peripheral sympathetic nervous system may determine the ultimate effect of sympathoadrenal activation. An analog hierarchical organization might exist in the parasympathetic branch of the autonomic nervous system.

CENTRAL NERVOUS SYSTEM
- higher centers in the CNS
- hypothalamus
- brain stem

PERIPHERY
- IML (spinal cord)
- sympathetic ganglia and adrenal medulla
- sympathetic nerve endings
    - presynaptic regulatory mechanisms
    - neurotransmitters and co-peptides

POSTSYNAPTIC MECHANISMS
- receptor
    - subtypes
    - # / sensitivity / internalization
- postreceptor
    - G proteins
    - 2nd messengers
    - mRNA synthesis

FIGURE 6.12

Levels of organization in the sympathetic nervous system

7

Feeding, digestion, absorption and thermic effect of food

The functioning of living cells depends on a continuous flow of nutrients to cover their metabolic needs. Cells utilize nutrients, firstly as fuel for energy metabolism (catabolism) and secondly as basal elements for growth and maintenance of body cells (anabolism). Carbohydrates, fats, and proteins are the main fuels. The other nutrients function as anabolic elements, e.g. calcium, phosphates, etc., or as co-factors, e.g. trace elements and vitamins, for the optimal functioning of physiological processes. Some of these materials are released in the extracellular fluid from stores in the body, e.g. glucose from the liver, calcium from bones. When the stores are depleted, the animal needs to replenish them by feeding, digestion and absorption. To be absorbed, nutrients must pass the epithelial cells in the intestines. To that end, the ingested nutrients have to be converted into substances of small molecular size. This starts already in the oral cavity where food is sensed by taste receptors, and mechanically broken down by chewing. Subsequently, it is transported via the esophagus to the stomach where it stays for a while, and where pathogenic organisms are destroyed by the low pH so that disturbances of the subsequent digestive processes in the small intestine are prevented. Several enzymes are already released in the oral cavity and in the stomach. The stomach very accurately regulates the subsequent supply of nutrients to the intestinal tract. The food is rapidly transported through the duodenum and jejunum and more slowly through the ileum. The final digestion and absorption occurs mainly in the small intestine after the action of several digestive enzymes released from the pancreas and duodenal wall. The intestinal tract is a very specialized organ and every part of it has its own function in the processing of food. In plant eaters the caecum has a function to destroy and convert cellulose to digestible and absorbable carbohydrates. What is left over after absorption of the nutrients leaves the body via colon and rectum.

This chapter provides an overview of the various aspects of the processes our food undergoes from intake to defecation, and ends with a brief look at the energy provided by and the thermic effect of food.

1   TRANSPORT OF NUTRIENTS THROUGH THE INTESTINAL TRACT

1.1   CHEWING

Chewing is regulated by a chewing centre in the medulla oblongata (Figure 7.1). The movements consist of rhythmic contractions of the muscles moving the underjaw upwards and downwards in combination with the lips and tongue. Food in the oral cavity activates mechanoreceptors, and via afferent pathways of the trigeminus nerve (V) the motorneurons of the hypoglossus nerve (VII) are activated, so that the cavity is opened. At the same time the motorneurons of the trigeminus that cause the closure of the cavity, are inhibited.

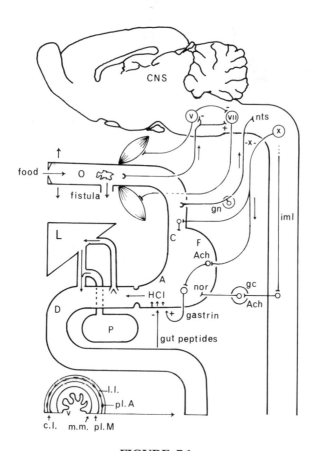

FIGURE 7.1

Schematic presentation of the physiological processes involved in chewing, swallowing, gastro-intestinal motility and secretion

For abbreviations see text and glossary.

At that time the mechanoreceptors do not sense the presence of food since the mouth is open. However, it is the stretch in the closing muscles that activates their muscle spindles. Their signals reach the closing muscles via the afferent and efferent pathways of the trigeminal nerve, and activate them to contract. The underjaw is raised and the cycle starts again. Although the whole cycle of these reflexes is rather autonomic there is also an important input from the cerebral cortex telling the animal to stop or start the cycle.

1.2 SWALLOWING

The entrance and exit of the esophagus are mostly closed by tonic contraction of the sphincters. The upper rostral part of the esophagus contains striated and the lower part smooth muscles. The movements of the tongue bring the food to the entrance of the esophagus. Receptors connected with the afferent vagus nerve (X) and glossopharyngeal nerve (IX) activate a swallowing centre in the medulla oblongata (Figure 7.1). Via cranial nerves IX, X, and XII, this centre regulates the closure of the nasal cavity, the interruption of breathing movements and contraction of the muscles at the entrance of the esophagus. This enables food to be transported from the oral cavity to the esophagus. Subsequently, peristaltic movements transport the food to the stomach. These peristaltic movements are local muscle relaxations followed at the rostral side of the relaxation by muscle contractions so that the food is pushed caudally. This pattern of relaxation and constriction is repeated, resulting in a pattern of peristaltic movements. The swallowing center regulates the peristaltic movement along the entire length of the esophagus via efferent vagal nerves (Lundgren, 1983). Afferent vagal branches inform this centre about the position of the food in the esophagus so that proper action can be taken. Therefore, cutting the esophagus has no influence on the peristaltic movement, which is in contrast to peristalsis in other regions of the gastro-intestinal tract. Cutting the vagus nerve, however, causes total paralysis of the esophagus, except in the caudal part of the esophagus, where some peristaltic movement is still possible because of the presence of smooth muscles and local intrinsic neural plexuses.

Finally, the food reaches the stomach via the sphincter of the cardia.

1.3   MOTILITY OF THE STOMACH

The following anatomical structures deserve mentioning in the context of this book (see Figure 7.1). The part of the stomach where the esophagus ends is called the cardia (C). The fundus (F) and corpus contain most of the stored food. The antrum (A) has a thick wall of muscles and is connected to the pylorus part, i.e. the exit to the duodenum (D). In humans, an empty stomach performs a series of strong contractions every ten minutes and these are associated with feelings of hunger. This may be one of the many factors regulating food intake but it probably has no influence on the regulation of caloric homeostasis. Under fasting conditions rats also have bursts of contractions of the stomach, as shown in an experiment in which a rat was provided with a permanently implanted water-filled balloon in its stomach. A long tube connected the balloon with a membrane from which the up and down movements could be registered. When food enters the stomach the relative increase in pressure is only small, since the smooth muscles in the wall will relax. Only when large amounts of food enter the stomach, further relaxation is not possible and the pressure suddenly increases (Figure 7.2a). Therefore, the stomach has a defined capacity which can be used as a regulating factor in the control of feeding behavior (see Chapter 9).

The stomach performs peristaltic contractions that push the contents in the direction of the pylorus. The latter does not function as a sphincter, which is normally closed; the peristaltic wave is not interrupted by it.

Peristalsis is autonomic, i.e. it continues after the extrinsic innervation of the stomach is cut. The contractions are coordinated by the nervous plexuses situated in between the muscle layers in the stomach wall. If the plexuses are anesthetized, peristalsis is stopped. The frequency of the contractions is governed by the electric activity of pacemaker cells in the cardiac region of the stomach. The membrane potential of these cells changes rhythmically which generates slow depolarization waves that are conducted over the entire stomach wall. The depolarizations themselves do not cause action potentials resulting in contractions, but provide the 'electrical substrate' on which external triggers, such as muscle stretch and acetylcholine, can act.

Action potentials are elicited when the lowest point of the slow waves reaches the threshold potential which is facilitated by muscle stretch and acetylcholine. In fact, the slow waves determine when the contractions occur, whereas the plexuses coordinate the movements so that a real peristaltic wave develops. The external innervation is important in modifying the rate and strength of the contractions and will be discussed in the next section.

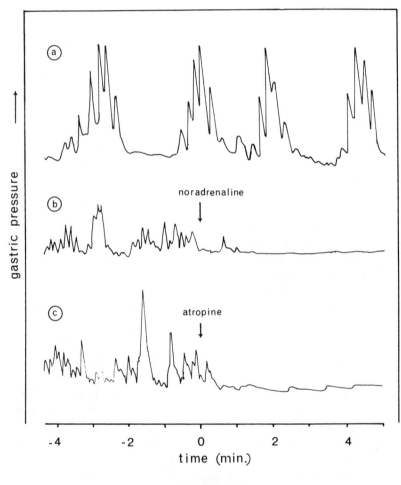

FIGURE 7.2
Gastric contractions in the rat

a Control, contractions in untreated rat
b Gastric contractions in the rat during treatment with noradrenalin (3 mg/kg)
c Gastric contractions in the rat during treatment with atropine (0.75 mg/kg)
(From: Strubbe, 1990)

1.4   GASTRIC EMPTYING RATE

Already in the 19th and at the beginning of this century, the regulation of gastric emptying was the subject of investigation. During the last decade, however, the progress in this field has increased markedly due to new techniques and renewed interest in peripheral regulatory physiological mechanisms. As pointed out already, the function of the stomach is the temporary storage of food, mixing this with enzymes and gastric acid to a watery phase, and transporting it to the entrance of the duodenum. The emptying rate of the stomach should not surpass the uptake capacity of the duodenum. Therefore, a major control of the emptying rate is performed by the duodenum which is able to activate and inhibit gastric motility (McHugh, 1983). It can do so via various routes:

1   Nervous feedback from chemo- and mechanoreceptors in the duodenum on the central nervous system (CNS) can act on the motor neurons of the autonomic nervous system and modify the strength and rate of stomach contractions. The autonomic nervous system consists of two major parts, the parasympathetic nervous system and the sympathetic nervous system, as dealt with in Chapter 6. They differ from the somatic nervous system, that innervates the striated muscles, by a more peripheral location and a ganglion connection between the motor neurons in the CNS and the target organ, so that there is a presynaptic and a postsynaptic part. For the *sympathetic* nervous system these ganglia are located in a chain close to the vertebral column. These so-called *paravertebral* ganglia innervate the target organs. There are also motor neurons that do not have their synapses in the paravertebral ganglia but in separate *prevertebral* ganglia. The prevertebral ganglion *ganglion coeliacum (GC)* is very important for the innervation of the intestinal organs. These ganglia are situated *outside* the target organs. The presynaptic neurons release acetylcholine which acts on nicotinic receptors. These receptors can be activated by low doses of nicotine and blocked with substances like curare and hexamethonium, but also with high doses of nicotine.

The final transmission to the target is by the catecholamine noradrenalin secreted by sympathetic nerve ending. Noradrenalin acts on α- and β-receptors of the target. These receptors can be blocked by specific blocking drugs such as phentolamine and propranolol respectively. Another source of catecholamines is the adrenal medulla which can be considered as a modified sympathetic ganglion. The products of the adrenal medulla are released in the blood circulation. There are also sympathetic afferent pathways transferring messages from the gastro-intestinal tract to the central nervous system.

2   The main *parasympathetic* neuron innervating the intestinal organs is the *vagus nerve*, or tenth cranial nerve (X). The motor neurons are located in the medulla oblongata and the efferent ganglion connection is located in the target organ. Again, the presynaptic neuron releases acetylcholine which acts on the postsynaptic nicotinic receptors. The transmission of the parasympathetic nervous system in the postsynaptical target area is performed by acetylcholine acting on muscarinic receptors. These receptors can be activated by muscarine and blocked by the drug atropine. The many feedback messages from the gastro-intestinal system are transferred via the afferent branches of the vagus nerve to the *nucleus tractus solitarius (NTS)*. The neuronal cell bodies are located in the *ganglion nodosum (GN)* (Figure 7.1). Although the contents of these messages are not yet clear, they are probably very important for regulation, since 80 per cent of the vagus neurons are afferent. The autonomic nervous system is not only important for gastro-intestinal processes but also for many other physiological processes, such as the regulation of heart rate, blood pressure, blood flow through different organs, thermoregulation etc.

3   Experiments on stomach contraction and pressure in the lumen show the regulatory function of the autonomic nervous system. Infusion of noradrenalin causes inhibition of stomach contractions (Figure 7.2b). A similar effect occurs after treatment with pharmacological blockade of the muscarinic receptors with atropine (Figure 7.2c) Thus, parasympathetic activity stimulates motility and sympathetic activity inhibits it. Parasympathetic activity can also induce the stomach to release the hormone gastrin, which stimulates

contraction. On the other hand, the presence of food in the stomach also stimulates the release of gastrin. Other transmitters involved in modulating the contractions, are dopamine and vaso-inhibitory peptide (VIP). When food enters the duodenum, feedback actions start operating on the gastric emptying rate from the duodenum. Nervous feedbacks from the duodenum reporting the osmolarity and pH to the CNS, may exert their influence via the autonomic nervous system. However, there may also be a direct feedback via the plexus. Moreover, there is a feedback from intestinal hormones such as enteroglucagon, secretin and cholecystokinin (CCK), released upon the presence of H^+, fats and carbohydrates in the duodenum, which inhibit the motility of the stomach and the release of gastrin. Even the presence of food in the ileum (distal part of the small intestine) can modulate the contractions of the stomach via nervous pathways and via the release of hormones such as enteroglucagon and motilin, neurotensin and neuropeptide YY (Figure 7.1). At regular times in the intervals between meals the so-called *migrating motor complex (MMC)* appears, i.e. a rhythm of strong contractions separated by silent periods as can be seen in Figure 7.2a. There is evidence that the MMC is caused by cyclic variations of plasma motilin.

4 If the content of the stomach checked by chemoreceptors is not right for its health, the animal sometimes has the possibility of vomiting. The stomach wall distends, the pylorus closes and after a few deep breathing movements the nasal cavity and trachea are closed. Subsequently, the diaphragm and abdominal wall contract and press on the stomach so that it is emptied. This complicated reflex is coordinated by a centre in the medulla oblongata. This centre can be activated by mechanical and chemical receptors in the stomach reporting, via vagal pathways, on the stomach content. Moreover, vomiting can also be caused by continuous stimulation of the labyrinth (e.g. sea sickness). Several species (among others the rat) do not possess the ability to vomit. That is why these animals need to taste novel food very carefully, in small quantities.

1.5   MOTILITY OF THE SMALL INTESTINE

The general structure of duodenum, jejunum and ileum is almost the same as described for the stomach. From the serosal side (outside) to the mucosal side (inside) the following structures can be seen: a longitudinal oriented muscle layer, the plexus of Auerbach, a circular muscle layer with the plexus of Meissner, the submucosa with glands and muscularis mucosa, and finally the mucosa with many villi covered with epithelial cells. Two kinds of motility can be distinguished:

1 Segment movements mix food and enzymes by subsequent contractions of different segments. The movements continue after nervous blockade and are therefore myogenic. The muscle cells possess an instable membrane potential generating slow electric waves comparable to those of the stomach.

2 Peristaltic movements are slow and extinct after a few centimeters. Coordination is achieved by the nervous plexuses in the intestinal wall. The contractions are non-myogenic. Peristalsis is activated by stretch and chemical receptors. Antiperistalsis is not possible.

Although part of the muscular activity of the small intestine is of myogenic origin, muscular activity can be stimulated by parasympathetic activity and inhibited by an increase in sympathetic activity.

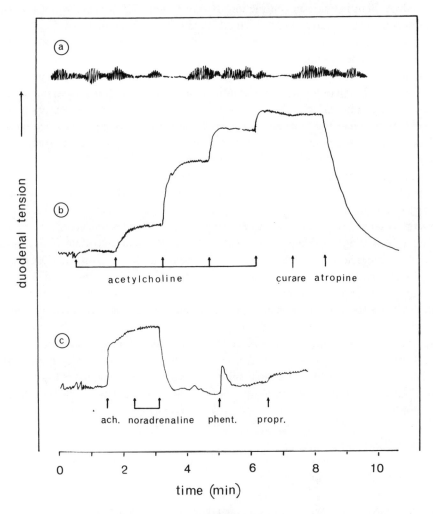

FIGURE 7.3
Motility of the isolated small intestine of a rat

a   Control, untreated. Notice the interference between the different segmental elements resulting in silent periods and periods of strong contractions.

b   After treatment with graded doses of acetylcholine. Curare and atropine are used as cholinergic blocking agents.

c   After treatment with noradrenalin, phentolamine and propranolol are used to block a- and b-receptors respectively.

(From: Strubbe, 1990)

This can be demonstrated by an experiment with an isolated piece of rat duodenum, kept in a beaker filled with a physiological solution of the right pH, oxygen tension and temperature. One side of the piece of duodenum is fixed to the bottom of the beaker and the other side is stretched and connected to a recording system. In this way the contractions can be monitored under control conditions (Figure 7.3a) and after treatment (injection in the beaker) with acetylcholine in varying concentrations. After raising the dosage a clear dose-dependent increase in contractions is seen (Figure 7.3b). Blockage of the nicotinic receptors with curare appeared to be ineffective in contrast to the blockade of muscarinic receptors with atropine (Figure 7.3b). The reason why the nicotinic receptor blockade has no effect is that acetylcholine acts directly on the muscarinic receptors of the muscle cells. The sympathetic influence is mimicked by injecting adrenalin after increasing the general tension with acetylcholine. A clear inhibition is observed after raising the concentration of noradrenalin (Figure 7.3c). When blocking the α-adrenergic receptors a small increase is seen, whereas the blockage of β-receptors by propranolol proves to be ineffective (Figure 7.3c). This means that in the investigated part of the duodenum sympathetic inhibition of intestinal motility occurs via α-adrenergic receptors. The inhibiting influence of the sympathetic nervous system on intestinal activity is very functional during exercise and stress when blood and energy are needed in the brain and striated muscles, and in the periphery to dissipate heat.

For completeness' sake, a few words on the motility of the caecum, colon and rectum. The colon and caecum have the same characteristics as the small intestine. The contractions are also autonomous but much slower than those of the small intestine. They are potentiated by the parasympathetic nervous system. There is a sphincter between the ileum and caecum which prevents movement of the intestinal contents from the caecum and colon to the ileum. This sphincter is only opened when a peristaltic wave arrives from the ileum. Filling of the rectum activates pressor receptors which influence a defecation centre in the sacral cord. This is under the influence of the medulla oblongata and hypothalamus. Via sacral parasympathetic efferents the pressure in the distal colon is increased and, at the same time, the distention of the anal sphincter. Not only this sphincter has to be relaxed, but also an external striated muscle sphincter which is under the control of the motor cortex of the CNS.

2   DIGESTION

2.1   SALIVARY GLANDS

Mammals have three pairs of salivary glands with two types of cells. One type produces a serous product (water), and the other produces mucus. In several species the glands produce the carbohydrate splitting enzyme amylase. Amylase breaks down starch, forming the disaccharide maltose. The optimum pH for amylase is 6.5. Although the stomach has a much lower pH, the enzyme can continue its action for a while, since the food is stored in the stomach in concentric layers which prevents the immediate penetration of the acid. The glands are only controlled by parasympathetic and sympathetic activity on command of the nucleus salivatorius which is located in the medulla oblongata.

2.2   GLANDS IN THE STOMACH WALL

The entire stomach wall is covered with mucous-producing epithelial cells in which many tubular glands, of which the openings are known as gastric foveolae, are located. In the region near the pylorus and in the cardia the gland cells only secrete mucus. Apart from mucus, the glands in the corpus and fundus produce hydrochloric acid and enzymes as well: the parietal cells secrete HCl and the chief cells enzymes such as pepsinogen. Cells of the upper part of the tubular glands also produce the intrinsic factor that enables absorption in the small intestine of vitamin B_{12}, an important factor in the synthesis of red blood cells.

Although the pH of the contents of the stomach is very low, the stomach wall is, to a certain extent, protected by an alkalic mucous layer of about 1 mm. The mucus neutralizes the penetrating HCl. There is a continuous production of new mucus stimulated by the local action of food in the stomach and by vagal action. The wall is further protected by tight junctions which connect the epithelial cells to each other so that the acid cannot penetrate. Moreover, the epithelial wall is regenerated every 3 days.

The enzymes secreted by the stomach wall are the protein-digesting pepsins and the fat-digesting lipase which only hydrolyses tributyrates and only occurs in small quantities. *Pepsins* split proteins to polypeptides. Their optimum pH is 2. They are released in the inactive form as pepsinogen which is changed to active pepsin in the stomach by splitting off a small peptide. The *lipase* is not very active in the stomach since the optimum pH is close to 7. The stomach of young mammals also contains small amounts of *rennin* which plays a role in the curdling of milk.

Hydrochloric acid secretion by the parietal cells occurs as follows. Cl⁻ is actively secreted across the luminal membrane and K⁺ follows passively.

Then K⁺ is actively exchanged for H⁺ which originates from the dissociation of H_2CO_3. The carbonic acid originates from the coupling of CO_2 and H_2O by carbonic anhydrase. HCO_3^- diffuses to the plasma and is exchanged for Cl⁻ from the blood. Finally it is released at the apical pole of the cell. The physiological control of acid secretion is one of the most extensively investigated physiological processes. A cephalic, a gastric and an intestinal phase can be distinguished.

Cephalic phase

If an animal is provided with a fistula (connection between esophagus and the outer world) in the esophagus, so that the ingested food does not reach the stomach but is removed via the fistula, gastric secretion occurs when the animal is eating (Figure 7.1). The secretion stops after cutting the vagus nerve which indicates that this nerve is important in triggering the response. Apparantly, afferent signals from the oral cavity are involved. This method of so-called sham feeding is one of the approaches to investigate the influence of oropharyngeal stimulation on the nervous regulation of intestinal physiology. Another approach that is often used in humans, is chewing on food that has no nutritional value, and subsequently removing the 'food' by spitting it out. There is some evidence that the vagal action of releasing acetylcholine, which acts on muscarinic receptors, causes the release of the hormonal factor histamine. Histamine has a more locally stimulating effect on the acid production by parietal cells. Moreover, as was mentioned in Section 1.4 parasympathetic activity induces the release of gastrin which in its turn can also promote the release of gastric acid (see below). The sympathetic nervous system can inhibit these processes. When food reaches the stomach other factors play a role in the stimulation of gastric acid release.

Gastric phase

If an animal is provided with an extra, transplanted and therefore denervated gastric pouch (Pavlov pouch) from which the acid secretion can be measured but though which no food passes, there is still acid secretion from the pouch as soon as food is put into the stomach. It is suggested therefore that the contents of the stomach induces, by chemical stimulation, the release of a hormone which can reach the parietal cells via the circulation. This hormone, *gastrin*, is released by the antrum. Indeed, antrum extracts proved capable of inducing gastric secretion. Electrical stimulation of the vagus nerve can also stimulate the release of gastrin. However, this is not the only route of electrical stimulation, since vagal efferents also stimulate the parietal cells directly.

Ulcer formation is caused by excessive stimulation of acid secretion. This can be remedied by a pharmacological blockade with ganglion blocking agents, especially with histamine antagonists. If this is not successful, cutting of vagus branches leading to the stomach is the next choice of treatment. When food mixed with acid is transported to the duodenum, the pH is sensed there and feedback signals arise to modulate acid secretion. The processes involved in this mechanism are called the intestinal phase.

Intestinal phase

If the stomach is completely isolated from the intestinal tract the presence of food in the upper part of the duodenum can still induce acid secretion in the stomach albeit at a low rate. This secretion is caused by the release of intestinal gastrin which is identical to gastric gastrin. When the pH is too low, another hormonal action comes into play, namely the release of enteroglucagon which inhibits gastric secretion.

2.3   EXOCRINE GLANDS IN THE INTESTINAL WALL AND PANCREAS

The products of pancreas and liver reach the duodenum via the pancreatic and bile ducts, which join together to form one common duct. The entrance of this duct into the duodenum is closed by the *sphincter of Oddi*. Apart from these products there are also secretions from glands in the wall of the intestine, the *glands of Brünner* in the duodenum and the *glands of Lieberkühn* along the whole of the small intestine. Moreover, epithelial cells in the villi produce mucus. Several enzymes are produced by these intestinal glands such as peptidases, lipase, and maltases (see Figure 7.4). Most of these enzymes are located in the epithelium cells themselves and are released after shedding of the cells. Depending on the species and strain the microvilli also contain lactases, separating lactose from milk. If no lactase is present, the consumption of milk and milk products will lead to gastro-intestinal problems.

The *pancreas* has both endocrine and exocrine functions. The endocrine functions will be discussed in Chapter 8. Here, we shall concentrate on the exocrine function, and only the main enzymes secreted by the pancreas will be reviewed.

Trypsin splits denatured proteins and polypeptides. It is secreted in the inactive form as trypsinogen. In the intestinal tract this is activated by enterokinase, an intestinal enzyme. The process is autocatalytic, i.e. trypsine in its turn can activate trypsinogen by splitting off a small peptide. This activation in the intestines prevents the pancreas from lysis by its own enzymes and is therefore an important physiological adaptation.

Other products secreted by the pancreas are *chymotrypsin* which is synthesized as its precursor chymotrypsinogen, and acts similar to trypsin, several peptidases, lipases (split fatty acids from triglyceride), amylase (splits glycogen and amylopectin) and nucleotidases (break down nucleotides). The pancreas also produces bicarbonate which neutralizes the acids secreted by the stomach.

The optimum pH for most enzymes in the small intestine is 7.

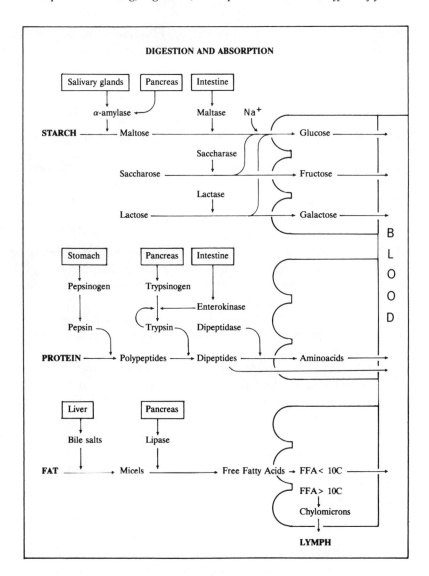

FIGURE 7.4
Schematic presentation of the physiological processes underlying digestion and absorption
of carbohydrates, proteins and fats

Comparable to the acid secretion of the stomach, the regulation of the secretion of the exocrine products of the pancreas is based on three phases, a cephalic, a gastric and an intestinal phase. Sham feeding, as described above, can induce enzymatic secretion if the vagus nerve is intact. This indicates a *cephalic* contribution. The *gastric phase* consists of the release of gastrin upon food stimulation of the gastric wall. Gastrin stimulates the release of bicarbonate and certain enzymes by the pancreas. When food passes the duodenum, other hormones are secreted by the intestines as well. In this respect the intestinal hormones secretin and cholecystokinin (CCK) can be mentioned. Formerly it was thought that yet another hormone, pancreozymin, was released. Later it was found that pancreozymin and CCK have the same chemical structure. CCK triggers the release of enzymes from the pancreas. Secretion of CCK is caused by the entrance of digestive products into the duodenum. Secretin is released after acid infusion into the duodenum and induces the release of bicarbonate to neutralize the pH.

2.4   LIVER AND BILE

Since the liver plays a major role in energy metabolism, some of the processes concerned will be briefly summarized.

The liver is a large organ and receives 20 per cent of the blood of the aorta i.e. 5 per cent via the hepatic artery and 15 per cent via the portal vein of the liver. All absorbed nutrients reach the liver via the portal circuit where they are absorbed or transmitted to the general circulation. Some of the liver's metabolic functions are:

Carbohydrate metabolism: glycogenesis (formation of glycogen from blood glucose), glycogenolysis (cleavage of glycogen to glucose which is released to the blood), gluconeogenesis (synthesis of glucose obtained from lactate, glycerol and deaminated amino acids).

Lipid metabolism: synthesis and splitting of free fatty acids.

Protein metabolism: amino acid synthesis from keto acids by transamination, synthesis of many proteins circulating in the blood, deassimilation of amino acids and formation of ureum which is secreted in the urine.

Cholesterol metabolism: the liver determines the circulating levels of cholesterol.

Bile production. One of the many functions of the liver is also the production of bile which is secreted into small canaliculi and finally led to the main bile ducts. Most animals have a storage place in the form of the gall bladder where the bile is concentrated. Bile consists of a watery solution of bile salts and bile pigments which are rest products of haemoglobin leaving the body via the intestinal tract. The bile salts are synthesized in the liver from cholesterol. These bile salts or *chelates* are important for the digestion of lipids in the intestine. They render the lipids more soluble in water because of their polar groups which are situated in the border area between water and fat. In this way, fat is emulsified and can then be effectively attacked by the lipases. About 90 per cent of the bile salts are absorbed by the intestines and transported via the portal vein back to the liver. This cycle is known as the *enterohepatic circulation.* Bile secretion is stimulated by the presence of chelates in the blood and by secretin. The vagus nerve has a stimulating and the sympathetic system an inhibiting effect on bile secretion. The hormone cholecystokinin plays a role in emptying the gall bladder by stimulating the smooth muscles in the wall to contract.

3   ABSORPTION THROUGH THE INTESTINAL WALL

The small intestine is the main site of absorption for nutritional compounds. The wall is adapted to this function by surface enlargement. This is achieved by many villi that are lined with epithelial cells. These again, are covered with microvilli. Only the very end products of enzymatic digestion can be absorbed by the intestinal wall. In the following, these processes will be described in brief for the main nutrients only (see Figure 7.4).

Carbohydrates

To be absorbed carbohydrates need to be degraded to monosaccharides first. The disaccharide saccharose is split by a saccharase to fructose and glucose. Another disaccharide already mentioned above, lactose, is split by lactase to galactose. Starch, the main carbo-hydrate in food, is split by amylase to maltose which is subsequently split by maltase into two glucose molecules. The monosaccharide glucose is actively absorbed by coupling to a carrier. In fact, this glucose carrier has two binding sites, one for glucose and one for a sodium ion. When no sodium is available at the brush border there can be no transport. The electrochemical gradient is maintained by an energy-requiring sodium pump which is located at the basal lateral membrane of the cell and causes the release of sodium to the blood. The facilitated uptake of glucose, which takes place in the duodenum and ileum, is therefore coupled to an energy demanding transport system.

Proteins

To be absorbed proteins must be broken down to free amino acids. The main enzymes splitting proteins are pepsin, carboxypeptidase, trypsin and chymotrypsin. The main prod-ucts are polypeptides and finally dipeptides. Some of the dipeptides can be absorbed, most of them are split by aminopeptidase and dipeptidase to amino acids. The amino acids are absorbed in the proximal part of the jejunum. There is also an active transport system for amino acids by a carrier system which shows some similarity to that for glucose absorption.

Fats

Fat absorption involves a long chain of processes. Firstly, the already mentioned emul-sification by bile. Secondly, the attack by the lipases and micel formation, and thirdly the uptake of free fatty acids in the epithelial cells of the villi. Inside these cells, which are mainly located in the villi of the jejunum, re-esterification occurs in the smooth endoplas-matic reticulum. In the Golgi apparatus a lipoprotein coat is built around the fat droplet to form the so-called *chylomicrons*. These chylomicrons are released into the lymph vessels so that they do not travel immediately through the liver. Short-chained fatty acids are not re-esterified and are transported by the blood to the liver. In the liver, they are lengthened and esterified. These lipids are then released from the liver as very low density lipopro-teins (VLDL). These are formed in the liver cells in a way similar to that of the chylomi-crons in the intestinal wall.

Water

Water is the carrier for all digestive juices containing the enzymes secreted by the dif-ferent organs along the intestinal tract. In humans, this can amount to 8 l/day. If this vol-ume were to leave the body via the faeces, it would disturb the whole physiology, as happens in the case of cholera infection. To avoid this, water is reabsorbed by the lower part of the intestine. The absorption of water is a passive process which is regulated by osmotic pressure. Water and electrolyte absorption occur mainly in the tip of the villi. This is regulated by a concentration gradient over the villus, with the highest sodium concentra-tion at the tip. This concentration gradient is actively maintained by the way the circula-tion in the villus is organized and is achieved by the uptake of sodium chloride from the venous circulation in the villus by the entering arterial circulation. After a meal many nutrients are absorbed. However, they cannot stay in the circulation until they are utilized, as the kidneys might then become flooded with them, with subsequent loss in the urine. Therefore, the absorbed nutrient must be kept in storage tissues such as the liver and adipose tissue. The physiological mechanisms involved in storage and release of energy in the form of carbohydrate and fat in liver and adipose tissue will be discussed in Chapter 8.

4   SUMMARY

The autonomic nervous system is important for the transport of nutrients through the intestinal tract : chewing, swallowing, motility of the stomach and the small intestine, gastric emptying rate.

External innervation is important for the motility of the stomach in modifying the rate and strength of the contractions, while the autonomic innervation determines on which times the contractions occur and coordinates the peristaltic contractions.

A major control of the gastric emptying rate is performed by the duodenum which is able to activate - parasympathetic - and inhibit - sympathetic - gastric motility; besides this nervous feedback, there is a feedback from gut hormones.

Two kinds of motility of the small intestine can be distinguished: segment movements - which are myogenic - mixing food and enzymes by subsequent contractions of different segments, and peristaltic movements - which are non-myogenic - which move slowly and extinct after a few centimeters. Although part of the muscular activity of the small intestine is of myogenic origin muscular activity can be stimulated by the parasympathetic activity and inhibited by an increase in sympathetic activity.

The contractions of caecum, colon and rectum are autonomous and potentiated by the parasympathetic nervous system.

The salivary glands are only controlled by parasympathetic and sympathetic activity.

Regulation of the secretion of the exocrine product of the pancreas is based on three phases (comparable to the acid secretion of the stomach): a cephalic, a gastric and an intestinal phase.

Bile secretion is stimulated by the nervus vagus and inhibited by the sympathetic nervous system.

Only the very end products of digestion by the enzymes can be absorbed by the intestinal wall.

8

Endocrine regulation of metabolism

Cells require energy and the fuel for the production of this energy is taken up as food. This is subsequently digested, absorbed and transported to the cells. In mammals the most important fuels are glucose, free fatty acids, and amino acids.

The absorbed nutrients can be used directly, but they can also be stored in the reserve tissues - liver, and adipose tissues - from which they are released when needed. It is in particular the central nervous system (CNS) that is in command of these processes, not only by controlling feeding behavior but also by regulating the release of several hormones involved in storage, release and metabolism of fuels. The anabolic and catabolic functions are controlled by hormones from the pituitary, thyroid, gonads, adrenals and pancreas. Especially the hormones insulin and glucagon produced by the islets of Langerhans in the pancreas play an important role in the control of anabolic and catabolic processes respectively. Insulin is the main factor in the storage of carbohydrates and lipids in adipose tissue. It is therefore one of the key factors in the causation of obesity.

The secretion of insulin and glucagon is also controlled by fuels, the main controller being glucose which stimulates insulin secretion and inhibits the secretion of glucagon. The other fuels, amino acids and, to a minor extent, free fatty acids, act in a more synergistic way. By monitoring the plasma levels after food ingestion, the islets of Langerhans secrete enough insulin to store the absorbed fuels as glycogen and/or fat. The endocrine pancreas derives additional information on the incoming nutrients from the numerous gastro-intestinal (GI) hormones released by different parts of the intestines in anticipation of or when food passes. Several of these GI hormones are capable of influencing the glucagon and insulin release from the endocrine pancreas. This peripheral control of insulin and glucagon secretion satisfies most metabolic needs. However, an animal also needs a finely tuned regulation of the secretion of these hormones in order to cover long-term metabolic demands and emergency conditions. In this respect it is sometimes important to anticipate or adapt to external influences. Such a main controlling role can be fulfilled by the CNS. Not only does the endocrine pancreas play a key role in the hormonal control of nutrient metabolism but, as will appear from the following chapters, it may also function in the regulation of food intake.

This chapter provides an overview of the hormonal control of nutrient metabolism by the endocrine pancreas and its interaction with the central nervous system.

1   SIGNIFICANCE OF THE ENDOCRINE PANCREAS: SOME HISTORICAL FACTS

In 1862 Georg Ebers dug up a 20 meter long papyrus from a grave in Thebes, Egypt, the so-called Ebers papyrus, dating from 1500 BC. The papyrus contained a complete de-scription of what was known about the diseases of the times in which the Pharaohs lived.

Among the various recipes and prescriptions the treatment of diabetes was mentioned, describing the condition as a high level of urine production.

The name 'diabetes' was already in the first century in use. It literally means non-stop passage, and indicates that in those times diabetes was considered to be a kidney disease. It was described as a characteristic flow of urine and concomitant polydipsia (i.e. excessive fluid intake). There was a reduction of body weight despite a high level of food intake. In England the famous physician Thomas Willis (1621-1675) discovered that the urine of patients with diabetes was sweet. This would have been an enormous step forward in the discovery of the cause of the disease, if Conrad Brunner (1653-1727) had not misinterpreted Willis' results. Brunner excised the pancreas of a dog which showed the typical signs of the disease. However, he left a small part of the duodenal pancreas intact. The pancreas regenerated to a certain extent and the animal turned healthy again. However, Brunner did not see the significance of his results as we see them know and published his conclusion that the pancreas was not of vital importance for life. In this way, he blocked the progress of diabetes research for over 150 years.

In 1869, in his thesis Paul Langerhans described the cell groups (islets) that were later named after him. These cell groups showed a structure that was quite different from the surrounding tissue. The islets possessed a very rich vascularization with typical sinuses which were not present in the surrounding tissue. Langerhans did not make any suggestions as to the possible function of these structures though.

After correct excision of the pancreas in dogs as carried out by Oskar Minkowsky (1858-1931), polyuria developed and high levels of glucose appeared in the urine as well as in the blood. From then onwards, research was focused on the relationship between the pancreas and diabetes. However, extensive research using extracts of the pancreas did not lead to an active substance. When the pancreatic duct was ligated, the exocrine part of the pancreas degenerated, but the islets of Langerhans remained intact. Since this surgical procedure did not result in hyperglycemia, it was concluded that the islets produced a hypoglycemic substance, which was named 'isletin'. However, extractions were without result, until Banting and Best used alcoholic extraction of a ligated pancreas under cool conditions, thereby preventing proteolytic action of the exocrine tissue of the pancreas. In 1921 these researchers succeeded in obtaining an active hypoglycemic substance, which suppressed the blood sugar level of diabetic dogs from which the pancreas was excised. *Insulin* was discovered and within five years the hormone was purified to a high enough degree to be used for daily injection in humans. From then onwards, insulin has saved millions of human lives. It will be obvious that since the discovery of insulin, research has focused on the islets of Langerhans, which contain the endocrine cells of the pancreas but take up only 1 per cent of the whole pancreas. (For a more extensive review of the history of insulin see Engelhardt, 1989.)

Long-term insulin treatment, however, leads to several complications such as retinopathy, neuropathy and nephropathy. It is suggested that the main cause for these disorders is the inability to mimic natural insulin release patterns.Treatment with insulin injections results in fluctuating blood glucose levels with unnaturally high amplitudes. Under normal physiological conditions the islets of Langerhans exert a fine regulation of the release of insulin by influences from hormones and the CNS. This fine regulation and the adaptation to changes in food availability will be discussed below, but first the anatomy of the endocrine pancreas and the islets of Langerhans will be described, in order to gain a better understanding of the events taking place there.

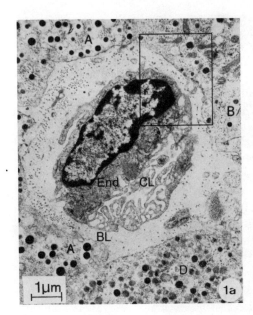
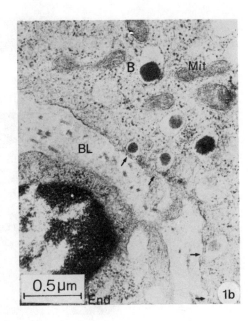

FIGURE 8.1

Cells in the islets of Langerhans

a Thin section of a part of the islet of Langerhans in the rat.Three types of endocrine cells can be seen grouped around a blood vessel. a-cell (A), b-cell (B) and d-cell (D), capillary lumen (CL), basal lamina (BL), endothelial cell (End). x4,500. The oulined area is shown in detail in figure b.

b Mitochondria (Mit). Signs of exocytosis are indicated by arrows. x54,000.

(Source: Strubbe & Van Wachem, 1981)

2   ANATOMY OF THE ENDOCRINE PANCREAS

2.1   CELLS AND MICROCIRCULATION

Although for a long time it was believed that the various islets of Langerhans do not differ from each other, in the last decades it has been shown that they do differ in composition depending on their location in the pancreas. In the so-called 'head' of the pancreas there are relatively more cells secreting pancreatic polypeptide and somatostatin (D or δ-cells), whereas the 'tail', i.e. the part close to the spleen, contains relatively more glucagon-secreting cells (A or α-cells). In all parts the insulin secreting B or β-cells are dominant in number. The secretion products of the various cell types can influence each other's secretion. There is evidence that somatostatin inhibits the secretion rate of glucagon and insulin. Insulin in its turn inhibits glucagon secretion and glucagon stimulates insulin secretion. The secretion products can reach the endocrine cells via the general circulation, but they can also arrive at neighbouring cells via the intercellular space, in which case in a much higher concentration than via the general circulation. Such influences are known as *paracrine*, meaning the influence of one cell on a neighbouring cell via the intercellular space or even via small cell-to-cell contacts, called gap junctions. It has been suggested that ions and small molecules can cross from one cell to another via these gap junctions. The close association between the different cell types clearly shows in Figure 8.1 where α-, β- and δ-cells are located around a small blood vessel.

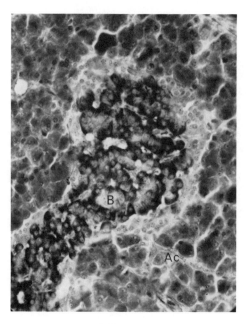

FIGURE 8.2
Histological section of an islet of Langerhans
Dark aldehyde fuchsin positive, β-cells predominate (B). The light border of the islets contains other endocrine α- and δ-cells. Acinar tissue surrounds the islet (Ac). Aldehyde fuchsin, x 580.
(Source: Strubbe & Van Wachem, 1981)

These cells are characterized by the degree of electron density and the shape of the granules containing the various secretory products. The micrograph also shows another possible route of influence on neighbouring cells, namely via the blood stream within the islets. Magnification of the outlined area (Figure 8.1b) shows signs of exocytosis, i.e. opening of granulated vesicles close to the cell border. In this way, insulin is delivered into the extracellular space. In most species abundant β-cells are situated in the center, whereas the other cell types are located at the periphery of the islet (see Figure 8.2). So far, the function of this anatomical arrangement is not known, but if it does have a function it presumably has to do with the microcirculation within the islet. There are indications in the literature that the blood flow is from the periphery to the centre of the islet. In that case insulin-secreting β-cells may be affected by glucagon and somatostatin in very high concentrations.

2.2   INNERVATION OF THE ISLETS OF LANGERHANS

The pancreas is innervated by parasympathetic fibers from the vagal and sympathetic fibers from the splanchnic nerve. These nerves enter the pancreas via the wall of the blood vessels. Especially the islets of Langerhans have abundant *cholinergic* innervation. Cholinergic terminals from the parasympathetic nervous system are found in the neighbourhood of all cell types but particularly in the periphery of the islet. The terminals are also in close connection with the blood vessels and are enclosed in the space between the two basement membranes. In the past few years evidence has accumulated that besides this classical autonomic innervation a *peptidergic* innervation system exists. This system is mainly concerned with the innervation of the gastropancreatic regions.

Immunohistochemical techniques have shown intestinal hormones such as gastrin, somatostatin, substance P and cholecystokinine (CCK) to be closely linked to parasympathetic nerve endings. These peptides are probably transported through the nerves to the target cells. Apart from this innervation of the gastro-intestinal region, there is also evidence of peptidergic innervation. Nerves containing vaso-inhibitory peptide (VIP) are distributed in the smooth muscles of the blood vessels where vasodilatation can be stimulated. Many CCK nerves are found in the pancreatic islets. The predominant molecular forms are CCK-4 and CCK-8. These molecules have been shown to stimulate the secretion of islet peptides (Balkan *et al.*, 1990).

3   FUNCTION OF THE ISLETS OF LANGERHANS

The function of insulin will be clear from the experiments described above. It appears that diabetics have either a relative or an absolute deficiency in insulin production. In the overweight type 2 diabetic a high insulin level is sometimes found in moderate *hyperglycemia*. This indicates that in those cases insulin is less effective, resulting in a rise in blood glucose level. However, when the glucose level becomes too high (above 180 mg per dl) the kidney can no longer reabsorb all glucose and this then appears in the urine. As a result of this *glucosurea*, water filtered in the glomeruli cannot be reabsorbed in sufficient quantities from the tubuli because of the high osmolarity of the filtrate (polyuria). Therefore, the diabetic has to drink more (polydipsia). As significant amounts of glucose leave the body, there is a continuous feeling of hunger so that the patient eats more *(hyperphagia)*. As mentioned earlier, the lack of insulin causes the uptake of glucose from the blood to be reduced. Therefore, the tissues utilize free fatty acids (FFA) and amino acids as energy sources. Since the key factor for fatty acid storage (insulin) is missing, extensive mobilization of depot fat sets in. The liver then needs to oxidize far more fat than it can metabolize. If the glucose oxidation rate is too low, as is the case in the diabetic liver, the supply of oxaloacetate is low and therefore the rate of the citric acid cycle, in which acetyl-CoA is metabolized, decreases. Subsequently, a surplus of acetyl-CoA develops. This leads to aggregation of acetyl-CoA with acetoacetic acid and β-hydroxy-butyric acids. The aggregates are released into the general circulation where they cause increased ketone levels. This *ketonemia* leads to progressive metabolic acidosis which eventually causes hyperglycemic coma and shock, finally resulting in death if no insulin treatment is applied.

The physiological effects of insulin administration are the opposite of what is described above. The most striking effect is the lowering of the blood glucose level after insulin treatment (see Figure 8.3). An insulin injection causes an immediate decrease in blood glucose *(hypoglycemia)*. Insulin is eliminated very rapidly with very short half-life values (1-3 minutes). As shown in Figure 8.3, the linear decrease in the semilog plot indicates an exponential removal rate. The short half-lives are probably of adaptive value to prevent long-term hypoglycemia which can lead to hypoglycemic coma. When insulin is removed from the blood, the rapid return of glucose to the original level is facilitated. Apart from this, there is also increased glycogenolysis (i.e. glucose release from glycogen in the liver), which is caused by an increased contraregulation by catecholamines and glucagon. It is very important to prevent hypoglycemia since it can lead to coma and eventually to death.

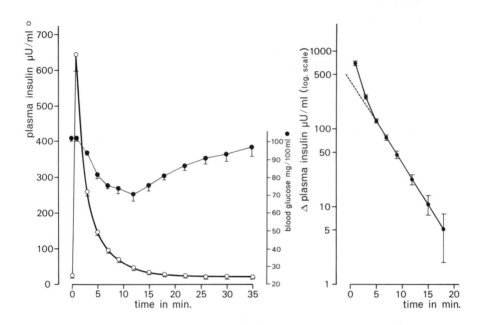

FIGURE 8.3

Effect of insulin injection (32 mU) on plasma insulin and blood glucose levels

The right panel shows the exponential disappearance rate of insulin from the circulation.

The decrease in blood glucose is caused by an increased permeability of the cell membranes of most tissues, and in particular of the resting muscle and adipose tissue. After uptake, glucose can be used intracellularly for metabolism, glycogenesis, protein synthesis and lipogenesis. Nerve cells, active muscle cells and liver cells can utilize glucose without the influence of insulin. The liver cells, however, need insulin to activate glycogen synthetase to activate glycogenesis. (For a review of these processes, see Engelhardt, 1989.)

4   SECRETION OF INSULIN

Carbohydrate, fat and protein metabolism are strongly dependent on the concentration of pancreatic hormones in the blood. This section gives an overview of the processes involved in the release of insulin.

Food intake is commonly regarded as the main stimulus for insulin secretion. After the start of a meal, glucose appears in the general circulation relatively quickly, depending on the species. So, more glucose enters the blood than is utilized by the tissues. This means that the blood glucose level increases which is one of the main triggers for insulin secretion. This can be illustrated by the following experiment. A rat, provided with permanently implanted heart catheters, was given an intravenous glucose infusion. The catheters allow blood sampling as well as infusion without disturbing the animal. Immediately after the β-cells had measured an increase in blood glucose, the release mechanisms were activated and the blood insulin level was raised (Figure 8.4a; Strubbe & Bouman, 1978).

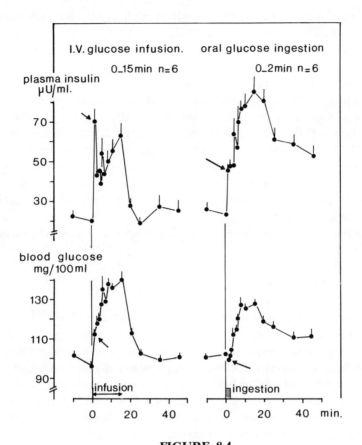

FIGURE 8.4

Plasma insulin and blood glucose levels after intravenous infusion and after oral ingestion of glucose

(Source: Strubbe & Bouman, 1978)

The speed of secretion was remarkable, i.e. within a minute, and an additional experiment showed that secretion can even occur within 10 seconds after stimulation. Despite the rise in blood glucose level there was a drop in insulin secretion after the first 2 minutes. In fact, the β-cells did not react to further stimulation. After about 5 minutes the insulin level increased again but at a slower rate. The latter was proportional to the level of glycemic stimulation. This means that the β-cells have a compartment that is sensitive to glucose and one that is less sensitive (Strubbe & Bouman, 1978). The reason for this biphasic response is explained below.

Figure 8.1b shows that there are granules near the cell border that can be released immediately upon stimulation. Other granules, however, need to be transported to the cell membrane by means of contractile microtubuli. The activity of this transport mechanism can be inhibited by pretreatment with colchicine or vincristine. In that case only the first phase is released upon glycemic stimulation. Thus, the first phase consists of stored insulin granules close to the cell border and the second phase consists of stored granules transported over a longer distance.

The second phase cannot be the result of the newly synthetized insulin, since it does not appear in the blood until 1 hour after stimulation with glucose. This was shown by administration of radioactive leucine to an isolated islet preparation. The labeled amino acid appeared after 1 hour in the released insulin molecules. These results show that intravenous glucose is an important stimulant for insulin secretion and synthesis.

For obvious reasons the most important route by which glucose enters the body is by ingestion of food. Three minutes after the ingestion of carbohydrate-rich food, an increase in glucose level is seen in the peripheral circulation of rats. Ingestion of radioactively labeled glucose showed that the rise in glucose after 3 minutes derived from the just ingested food. The insulin level, however, had already increased at the first minute, i.e. 2 minutes prior to the first noticeable rise in glucose level (Figure 8.4b). This rapid pre-absorptive insulin release (PIR) was also observed in rats eating a 'dummy' meal without any digestible nutrients. Since this response is extremely rapid, mediation by the central nervous system is very likely (Strubbe & Steffens, 1975).

The following sections describe how the central nervous system influences the endocrine pancreas.

5   PARASYMPATHETIC NERVOUS SYSTEM

In the middle of the 19th century the famous French physiologist Claude Bernard suggested that several metabolic processes concerned with energy regulation, and in particular with blood glucose homeostasis, are controlled by the central nervous system. He suggested that there are nerves that cause hyperglycemia and other nerves that stimulate glucose uptake by the tissues. As we have discussed, these theories changed half a century later. The main regulatory region for blood glucose became the pancreas instead of the brain. In the last decades, however, there is increasing evidence for participation of the central nervous system in blood glucose homeostasis via direct innervation of the liver or via innervation of the pancreas and adrenals.

In 1967 it was shown that electrical stimulation of vagal nerves to the pancreas resulted in *in vitro* insulin secretion (Kaneto *et al.*, 1967). Under more physiological conditions the contribution of the vagus nerve was demonstrated by cutting the nerve and offering the animal glucose. After spontaneous ingestion of glucose there was no PIR but there was a simultaneous exaggerated glucose response (Figure 8.5; Strubbe, 1989). Therefore, it is suggested that the vagus nerve plays an important role in the causation of PIR. Failure of this response could be responsible for wider fluctuations in blood glucose as will be shown in the following experiment on increased insulin responses.

Since the efferent parasympathetic influences originate from the vagal motor nucleus in the brainstem, it is also of interest to investigate the pathways to this motor nucleus. In this respect olfactory, visual and oropharyngeal mechanical signals might play a role via their afferent nerves in the causation of PIR. Figure 8.6 compares the insulin response to an oral glucose load with that to the same load infused in the same time directly into the stomach via a permanently implanted catheter. PIR failed to occur when glucose was infused directly into the stomach. This strongly suggests that the origin of the reflex is located in the oropharyngeal region (Strubbe, 1989). The information can be transmitted to the nucleus solitarius via the trigeminal nerve, facial nerve, glossopharyngeal nerve and the vagus nerve. Finally, the information can be transmitted to the hypothalamic areas. Some of these areas show changes in activity during feeding. For instance, it has been shown that during food intake the nerve terminals in the lateral hypothalamus (LHA) secrete increased amounts of the transmitter noradrenalin in the synapses. Therefore, it could be that the hypothalamus functions as a relay area to connect the afferent information to the efferent output, which is directed at the motor nucleus of the vagus nerve. To investigate this, transmitter activity was mimicked by local injection of noradrenalin into the LHA. During the noradrenalin treatment the insulin level increased. This response was absent in rats in which diabetes was chemically induced by alloxan (Figure 8.7; Strubbe, 1990). Alloxan destroys the insulin secreting β-cells. When rats were premedicated with atropine, a belladona alkaloid that blocks the parasympathetic transmission of acetylcholine to muscarinic receptors, this response was also absent. Thus, the lateral hypothalamus seems to control noradrenalin-induced insulin secretion via the vagus nerve and its efferents (De Jong *et al.*, 1977; Strubbe & Steffens, 1988).

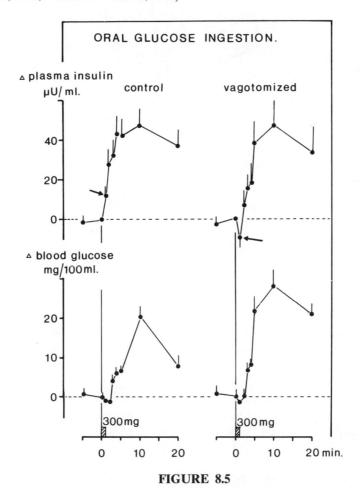

FIGURE 8.5

Effect of subdiaphragmatic vagotomy on plasma insulin and blood glucose responses during spontaneous ingestion of glucose

The arrow indicates the insulin level 1 min after the start of ingestion.

(Source: Strubbe, 1989)

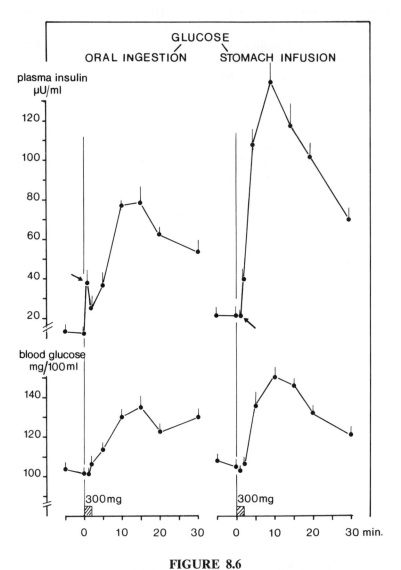

FIGURE 8.6
**Blood glucose and plasma insulin patterns during oral ingestion of glucose (left panel) and
during glucose infusion into the stomach (right panel)**
The arrow indicates the insulin level 1 minute after the start of ingestion of glucose
(Source: Strubbe, 1989)

The following question merits attention now. What is the role of these neural influences on blood glucose homeostasis after the ingestion of food? The classical approach to
find out is to eliminate the neural influences and to observe what happens. The experiment
described above showed that after oral glucose intake, cutting of the vagus nerve resulted
in abstinence of PIR while the blood glucose level increased slightly above control values.
The second approach, infusion of glucose into the stomach, showed the same phenomenon.
The third approach not yet discussed here, is the transplantation of denervated islet tissue
into diabetic recipients. Research on this technique is performed worldwide in order to be

able to treat diabetic patients. There is some evidence that the artificial administration of insulin by injection may be the cause of many side effects. It is suggested that the concomitant irregularly occurring hyperglycemia is the cause of these complications. The idea is that a transplanted pancreas can secrete insulin in a more physiological way, thereby preventing severe hyperglycemia or hypoglycemia, as often occurs during the injection treatment. However, as pointed out above, the transplanted tissue may lack innervation and this could result in inadequate insulin secretion and, subsequently, in glucose intolerance. When neonatal pancreatic tissue taken from rats is transplanted under the kidney capsule of diabetic rats, there is no PIR either. Meal ingestion results in a higher glucose response and, as expected, in a delayed insulin response but with a higher amplitude (Figure 8.8). Thus, as an answer to the question, the absence of PIR results in an enhanced blood glucose level and a concomitant insulin response (Strubbe & Van Wachem, 1981).

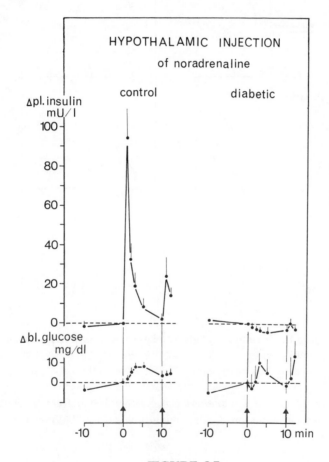

FIGURE 8.7
Effect of noradrenalin injection into the hypothalamus on plasma insulin and blood glucose in normal healthy and diabetic rats
(Source: Strubbe, 1990)

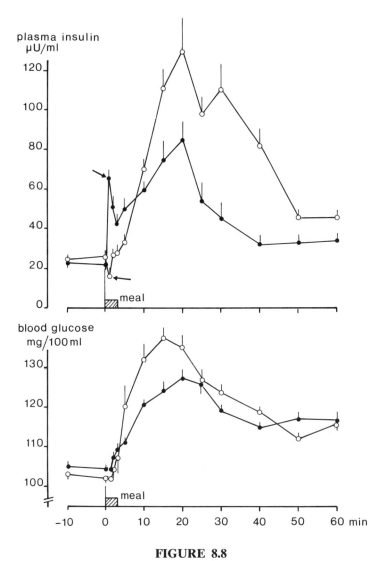

FIGURE 8.8

Comparison of plasma insulin and blood glucose between control and transplanted rats
Control (•—•), transplanted (o—o)
(Source: Strubbe, 1990)

Summarizing, the results show that interference with either efferent or afferent central nervous signals impairs insulin secretion, resulting in wider oscillations of blood glucose levels following a meal.

6   SYMPATHETIC NERVOUS SYSTEM

Catecholamines from the adrenal medulla and the nerve terminals of the sympathetic nervous system may act on both α- and β-adrenergic receptors. α-Adrenergic action inhibits insulin secretion whereas β-adrenergic action activates it. Because of these opposing effects an increase in sympathetic tone may cause a complex dual effect.

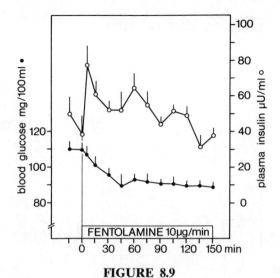

FIGURE 8.9
Effect of long-term infusion of phentolamine on plasma insulin and blood glucose
(Source: Strubbe, 1989)

The existence of a tonic inhibition by sympathetic activity is demonstrated in Figure 8.9, showing the results of an experiment in which α-adrenergic receptors are blocked by phentolamine: plasma insulin increases while blood glucose falls. On p. 124 it was mentioned that the plasma insulin level increases during the first minute after oral ingestion before the glucose level rises. Figure 8.10 (left part) shows this for control rats ingesting 150 mg of glucose. The right hand part of the figure shows that the immediate insulin secretion is enhanced during α-adrenergic blockade with a phentolamine infusion, whereas blood glucose levels decrease. These results show therefore that the sympathetic nervous system can modulate the sensitivity of the β-cells to influences of the parasympathetic nervous system and probably to several other stimulating factors.

Blood catecholamines from the adrenals and glucagon from the pancreas are also released during the beginning of a meal (Figure 8.11; De Jong et al., 1977; Strubbe, 1990). This may indicate that not only parasympathetic activity is increased at the beginning of a meal but also sympathetic activity. Glucagon secretion can be stimulated by both autonomic nervous systems.

During severe stress or during physical activity, the sympathetic nervous system plays an important role in stimulating glycogenolysis in the liver and lipolysis in the adipose tissues. The release of glucose and FFA respectively causes an increased availability of these fuels for muscular activity. However, when the insulin level is too high in these situations, the mobilization of fuels is strongly inhibited. Therefore, during physical activity insulin secretion is inhibited so that glycogenolysis and lipolysis can exert their effects. The sympathetic nervous system plays a role in this inhibition, since during an α-adrenergic blockade with phentolamine the animals do not mobilize glucose and FFA sufficiently and become tired sooner.

These experiments show therefore that during exercise and probably also during stress, α-adrenergic action is dominant over β-adrenergic action so that insulin secretion can be suppressed.

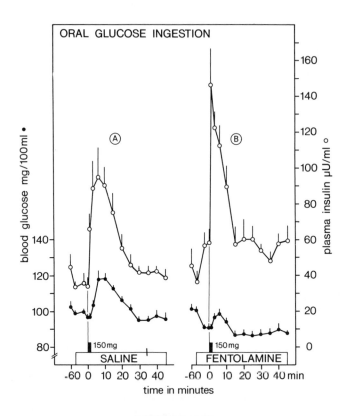

FIGURE 8.10
Plasma insulin, and blood glucose patterns following rapid spontaneous ingestion of 150 mg of glucose during saline infusion (A) and during infusion of phentolamine (B)
(Source: Strubbe, 1989)

Although the CNS plays an important role in the fine regulation of hormone secretion from the islets of Langerhans, there are also many hormonal influences involved in it. These processes will be discussed in the next sections.

7   HORMONAL CONTROL OF THE ENDOCRINE PANCREAS

7.1   GASTRO-INTESTINAL HORMONES AND THE CONTROL OF INSULIN
         SECRETION

The experiments described on p. 123 (Figure 8.4) showed that during glucose infusion insulin secretion can be strongly stimulated by the enhanced glucose levels. The normal physiological route, however, is oral ingestion. In rats this can be performed by spontaneous licking of the sweet solution. Figure 8.4 shows that in comparison with the effects of intravenous infusion, oral glucose ingestion causes a lower glucose response and an increased insulin response. This means that oral glucose is more effective in stimulating insulin secretion than intravenous glucose. Apart from the cephalic influences described in the previous sections, this effect is attributed to an insulinogenic hormonal factor released by the intestines. As described in Chapter 7, the intestinal tract is the largest endocrine organ in the body and secretes many hormones.

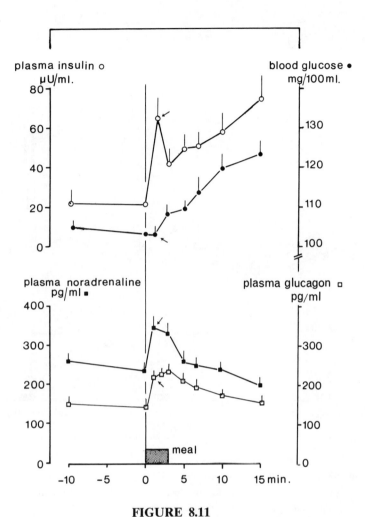

FIGURE 8.11
Plasma insulin blood glucose, plasma noradrenalin and plasma glucagon levels during food intake

(Source: Strubbe, 1990)

Since the hormone-producing cells are dispersed over the intestinal tract it is difficult to perform research on them by removing parts of the intestines. Moreover, the different cell types show some overlap in different areas. Several of these intestinal hormones are involved in controlling the digestive processes and the intestinal motility. Some of them, e.g. cholecystokinin (CCK) and gastro-inhibitory polypeptide (GIP), can stimulate and potentiate insulin secretion. The octapeptide of CCK proved to be capable of stimulating insulin secretion immediately after intravenous infusion (Figure 8.12; Balkan *et al.*, 1990). However, the intraportal route was ineffective. Therefore it is suggested that the liver removes CCK-8 from the circulation. From this it can be concluded that CCK may act in a more paracrine way. We have already discussed the possibility of peptidergic innervation via efferent pathways of the vagus nerve, where CCK-8 and CCK-4 are released at the terminal ends in the islets of Langerhans.

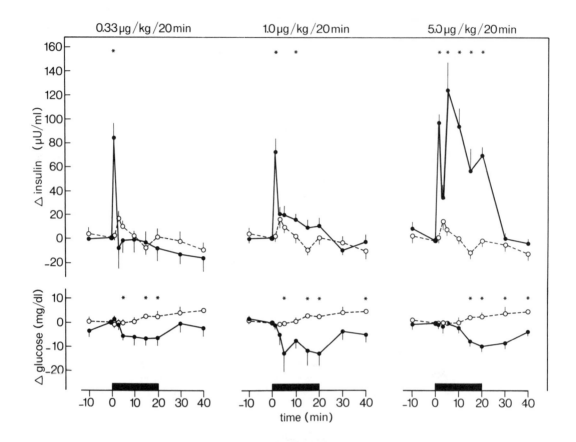

FIGURE 8.12

Plasma insulin and blood glucose responses during and after 20 min intravenous infusion of different amounts of cholecystokinin and saline infusion in rats

CCK-8 (• •), saline (o o). Solid bars indicate the infusion periods.

* represents significant differences between CCK-8 and saline infusions.

(Source: Balkan *et al.*, 1990)

The release of GIP is not controlled by the CNS but the polypeptide is released upon food stimuli in the intestines.

Together, the various interactions between intestines and islets of Langerhans are named the 'entero-insular axis'. The physiological importance of these potentiating actions which occur 'in concert' after the ingestion of a meal, is probably to maintain the blood glucose level between fixed limits for all meal sizes (see Figure 8.13; Strubbe & Bouman, 1978). In this way, hyperglycemia, which is probably a causal factor in the development of the complications seen in diabetics, is prevented.

Summarizing, the insulin response to food intake results from the successive and cumulative operation of anticipatory, nervously triggered insulin secretion and the anticipatory load-dependent potentiation of secretory stimulation by intestinal hormones that raises the blood glucose level.

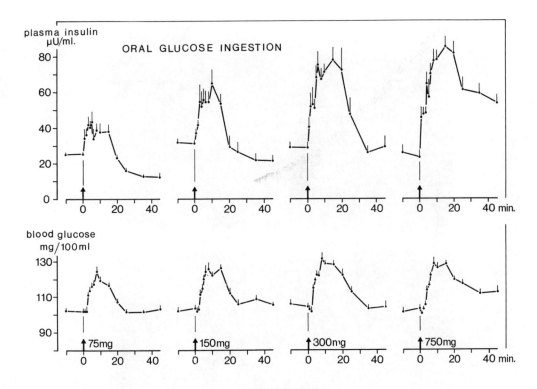

FIGURE 8.13

Plasma insulin and blood glucose patterns following rapid spontaneous ingestion of 75, 150, 300 and 750 mg glucose

(Source: Strubbe & Bouman, 1978)

7.2   PHYSIOLOGICAL ROLE OF SOMATOTROPIC HORMONE, SOMATOMEDINS AND SOMATOSTATIN

In addition to the catecholamines from the sympathetic nervous system, a number of hormones have been found to induce a rise in FFA and glucose in the blood, together with an increase in blood ketone levels. Among these, somatotropin (STH) or growth hormone (GH) plays an important role. Growth hormone is produced by the pituitary gland under the influence of a growth hormone releasing factor (GHRF) from the hypothalamus. It has been known for a long time that growth hormone is involved in the formation of cartilage and skeletal bone and therefore in the growth of the body. Too much GH leads to gigantism and acromegaly in adults. Apart from growth stimulation of skeletal bones, GH has many other anabolic properties such as stimulation of protein synthesis together with insulin in nearly all tissues. On the other hand, GH also has many catabolic effects such as stimulation of lipolysis and inhibition of insulin-induced glucose utilization, so that the nutrients become available for growth processes.

a

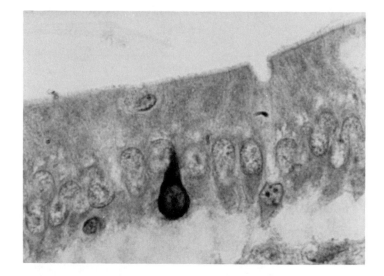

b

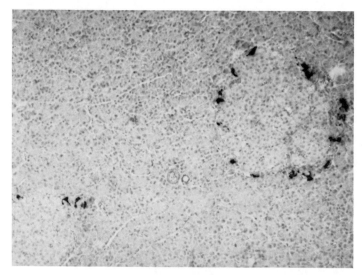

FIGURE 8.14
Histological section of intestinal wall and islet of Langerhans
a Section of intestinal wall indicating a somatostatin secreting cell between the other epithelial cells.
b Islet of Langerhans with somatostatin secreting cells (dark cells).
Staining with a specific antibody against somatostatin.

The blood insulin level increases together with the glucose level. When GH is injected over a longer period, degeneration of β-cells may set in because of the continuous stimulation of the enhanced glucose level, which may eventually lead to diabetes mellitus. This only occurs in adults and not in young individuals.

The stimulating influence on bone growth occurs indirectly via *somatomedins* which are released from the liver upon stimulation by growth hormone. There is some similarity in chemical structure between somatomedin and insulin so that it is argued that they have developed from a common prohormone during evolution.

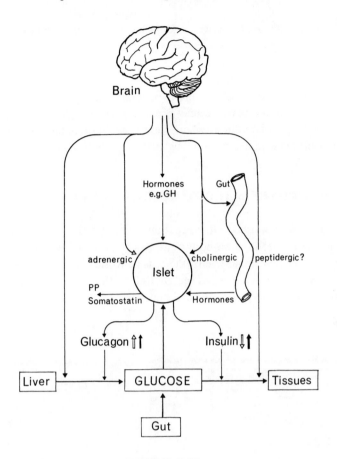

FIGURE 8.15

Diagram summarizing CNS and hormonal control of islet and glucose metabolism

a Influences of the central nervous system on the secretion of hormones from the islet of Langerhans. b Effects on blood glucose homeostasis. Symbol ⇨ indicates sympathetic actions and ➡ more parasympathetic and other influences.

This is why in the literature somatomedin-C is referred to as *insulin-like growth factor (IGF-1)*. It is produced in many tissues although the liver seems to be the main source of circulating IGF-1. There is some evidence that IGF-1 production is impaired in type 1 diabetes mellitus, whereas growth hormone levels are elevated. A feedback relationship between IGF-1 and GH is probably responsible for the increased GH levels. In summary, the effect of somatomedins is more directed at stimulation of growth, whereas insulin is more directed at glucose utilization. Both the presence of GH and proper feeding conditions are necessary for the release of somatomedins.

In 1973, when searching for a growth hormone releasing factor, Brazeau found a hypothalamic GH-inhibiting factor called *somatostatin*. Somatostatin is not only produced in the hypothalamus but also at other locations in the central nervous system where it may act as neurotransmitter. Other sites are the intestinal tract (see Figure. 8.14a) and the δ-cells of the islets of Langerhans which surround the β-cells, at least in the rat (Figure 8.14b). Somatostatin is a general inhibiting factor. It inhibits the release of almost all known gastro-intestinal peptides and pancreatic hormones, probably via paracrine routes. Further, it inhibits exocrine pancreatic activity, stomach emptying rate and duodenal motility. The result of these processes is that somatostatin inhibits the absorption of nutrients in the intestinal tract. Figure 8.15 summarizes the control by the CNS and the hormonal control of the islets and of glucose metabolism.

The processes described above and their interactions play a role in the adaptation to changing feeding conditions which may occur during the day-night cycle and in the long term during fasting. In conditions where food availability is high overfeeding may occur, leading to obesity. The next section discusses the metabolic adaptations to such conditions.

8   CONDITIONS RELATED TO NUTRIENT METABOLISM

The regulation of substrate flow, i.e. mainly that of glucose and FFA in the blood to cover the metabolic needs of all body cells, has to be extremely accurate. It needs to be adapted to several demands laid on by the outside world. The following side will show how an animal is capable of adapting its metabolism to the irregularity of food supply under varying conditions such as day-night rhythm, fasting and hyperphagia.

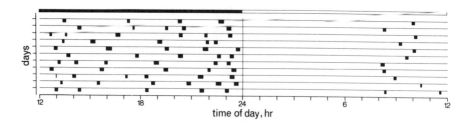

FIGURE 8.16
Pattern of food intake of one representative rat on successive days
(Source: Strubbe *et al.*, 1987)

8.1   DAY-NIGHT RHYTHM

Feeding does not go on continuously during the day-night cycle. A night-active animal, such as the rat, takes about 80 per cent of its daily food intake during the night in 8-10 meals. Only 1-2 meals are taken during daytime (Figure 8.16; Strubbe *et al.*, 1987). As was shown in Chapter 7, the intestinal tract adapts to the incoming nutrients by increasing its activity. This can be effected by an increased activity of the parasympathetic nervous system by acting on the muscarinic receptors of the intestines via acetylcholine. Evidence exists that the vagus nerve can also exert an effect on the release of intestinal hormones such as gastrin and, by doing so, increase the motility and secretory activity of the intestinal tract.

When rats take a test meal during nighttime, the increased vagal activity results in a steeper increase in glucose level, probably caused by an increased gastric emptying rate resulting in increased absorption (Figure 8.17; Strubbe *et al.*, 1987).

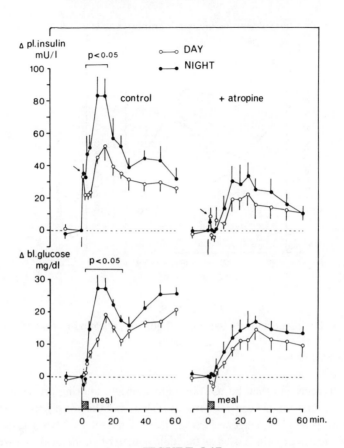

FIGURE 8.17
**Comparison of plasma insulin and blood glucose responses to food intake between the
light and dark phase with and without premedication with atropine**
(Source: Strubbe *et al.*,1987)

Apart from the vagally controlled pre-absorbtive insulin response (PIR), plasma insulin levels follow the glucose concentration simultaneously. After pharmacological blockade with atropine the night response is strongly weakened and shows more similarity to the daytime response. Therefore, this experiment provides evidence for an increased, vagally controlled absorption rate during the active period. These meals are probably eaten in anticipation to tide over the (day)light phase without incurring a large energy deficit (see Chapter 10).

8.2 METABOLISM DURING FOOD DEPRIVATION

Several circumstances exist where the outside world does not provide enough nutrients to cover metabolism and to replenish the reserve tissues. Therefore, animals dispose of many metabolic adaptations to survive short- and long-term fasting.

After one night of fasting the carbohydrate stores in the liver, delivering glucose from glycogen, are depleted. The blood glucose level drops while, at the same time, the free fatty acid level rises, probably due to a gradual decrease in insulin secretion (see Figure 8.18; Strubbe & Prins, 1987).

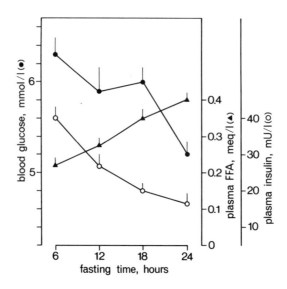

FIGURE 8.18
Effect of varying fasting periods on basal blood glucose, plasma insulin and plasma free fatty acids

(Source: Strubbe *et al.*, 1986)

The free fatty acids can then be used in many energy-requiring processes. However, glucose is the main fuel for the central nervous system and this must now be synthesized from the amino acids by gluconeogenesis in the liver. Since the circulating protein store is extremely small, muscle protein is used as the main source of amino acids. However, using muscle protein does not suffice for an animal to survive a long fasting period, so that a number of adaptations need to take place. To save muscle protein, all protein from damaged cells is used in gluconeogenesis. Moreover, the glycerol released during lipolysis is used for carbohydrate synthesis. This is not enough though, and an alternative source of fuel for the central nervous system is then tapped in the form of ketone bodies. As we have discussed above, excessive fatty acid utilization leads to excess acetyl-CoA. The liver transforms acetyl-CoA into acetoacetic acids and β-hydroxybuteric acid, the so-called *ketone bodies*. The central nervous system disposes of enzymes that can utilize these acids. If the concentration of ketone bodies is high enough, the central nervous system uses them as an energy source. In humans this already occurs in the first week of fasting.

The physiological mechanism controlling these metabolic changes is not yet fully re-solved, but some evidence exists that hormones such as GH and glucocorticoids from the adrenals are involved. A decreased responsiveness of insulin secretion caused by impaired cAMP synthesis in the β-cells, is probably one of the factors responsible for a drop in basal insulin (Figure 8.19; Strubbe & Prins, 1987). This enables the rat to shift from carbohydrate to fat metabolism. These adaptations enable mammals to survive fasting for a long time. For how long depends on the size of the reserves and the intensity of the metabolism. Large species can cope better with fasting than small species, since the latter have a higher metabolic rate in relation to their body weight.

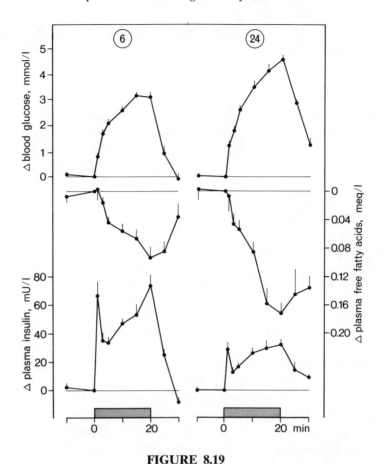

FIGURE 8.19

Effect of fasting periods of 6 and 24 hours on plasma insulin, FFA and blood glucose responses to intravenous glucose infusion during 20 min (10 mg/min)

(Source: Strubbe *et al.*, 1986)

Hibernating animals decrease their metabolism to an absolute minimum. Moreover, they fill their body reserves in anticipation of the winter to survive a long fasting period.

8.3   OBESITY

Obesity or adiposity is characterized by an increase in adipose tissue in relation to lean body mass, which is mainly caused by extra food intake which is not matched to the energy expenditure. Two phases can be distinguished. Firstly, the phase in which body weight increases continuously *(dynamic phase)* and secondly, the phase in which body weight remains stable at a high level *(static phase)*. Although there are many causes for the development of obesity, the primary cause may be a disturbed internal control of food intake and/or a deviating control of nutrient metabolism in the body. Both may partly be caused by a genetic background. Also, reduced energy expenditure may be involved. In the following chapters body weight regulation will be dealt with extensively.

Hyperphagia in humans and animals causes an increased flow of incoming nutrients. Not all nutrients can be burned directly and therefore parts of them are stored in the reserve tissues, of which adipose tissue is the main component. In adult animals the fat cells do not increase in number but in content. Insulin is the main key factor in this process as it stimulates the uptake and conversion of glucose and free fatty acids to triacylglycerols. However, when a fat cell increases in size, insulin can no longer incorporate glucose at the same rate.This so-called *insulin insensitivity* leads to extra production of insulin in the pancreatic islets and to an increased release of it, probably caused by the increased blood glucose levels. The islets of Langerhans increase in size by increased mitotic activity so that in obese subjects the islets are relatively big. The mean size of these islets is closely correlated to the degree of obesity. In aged humans with a predisposition for diabetes mellitus, the excessive filling of adipose tisue may lead to exhaustion of insulin in the islets. This is probably the reason why diabetes mellitus type 2 develops. This type is also called maturity onset diabetes and the best treatment for it is to reduce body weight so that the insulin action becomes more effective.

9   SUMMARY

The islets of Langerhans differ in composition depending on the part of the pancreas; in the 'head' of the pancreas there are relatively more D-cells secreting pancreatic polypeptide and somatostatin, whereas in the 'tail' of the pancreas relatively more A-cells secreting glucagon are found; in all parts the insulin secreting B-cells are dominant in number.

The secretion products of the different cell types are able to influence each other, either via the general circulation or via a paracrine way.

The pancreas is innervated by parasympathetic fibers from the vagal and sympathetic fibers from the splanchnic nerve; peptidergic innervation also exists.

Diabetics have either a relative or an absolute deficit of insulin production; after insulin treatment the most striking effect is lowering of blood glucose levels; it is important to prevent both a hypoglycemic and a hyperglycemic coma.

Food intake, more specifically the rising blood glucose level, is commonly regarded as the main stimulus for insulin secretion.

The insulin response to a glucose load is biphasic; the rapid insulin response at the first minute (PIR = preabsorptive insulin release) is mediated by the central nervous system; the vagus nerve plays an important role in the causation of this PIR; absence of the PIR results in an enhanced blood glucose and insulin response. Hormonal influences are also involved in the fine regulation of hormone secretion from the islets of Langerhans; for example cholecystokinin (CCK) and gastroinhibitory polypeptide (GIP) are able to stimulate and potentiate insulin secretion.

The substrate regulation in the blood to cover metabolic needs of all cells is extremely accurate; this metabolic regulation needs to be adapted to several conditions such as day-night rhythm, fasting and obesity.

9

Regulation of food intake

In man and animals body weight is often maintained within narrow limits on a variety of diets and under various external conditions. After temporary starvation or overfeeding the disturbed body weight returns to its normal level through compensatory adjustment of the food intake. This indicates that the energy content of the body is subject to homeostatic control by the regulation of food intake. Apart from this regulation the animal can adapt its internal physiology to the lack of incoming nutrients during starvation. The approach presented here is different from that of Chapter 8: now the focus is, on the regulation of food intake by positive and negative feedback on the central nervous system, i.e. a homeostatic approach.

It is generally held that the regulation of food intake operates through negative feedback control from the periphery. Feedback signals may arise from different parts of the periphery reporting to the central nervous system (CNS) on the current energy content of the body. Disturbances in this feedback control may lead to disorders in food intake such as hyperphagia (overeating) and, as a consequence, to the development of obesity. Already in 1840 Mohr described a case of severe obesity in a woman with a tumour in the hypothalamic area. Similar reports appeared later, dealing with changes in food intake and body weight which were often associated with disturbances in the hypothalamic area.

In 1940 Hetherington and Ranson described how, in their experiments, small lesions in the ventral hypothalamus (VMH) caused hyperphagia and obesity in animals. However, they suggested that the obesity was mainly due to hypo-activity of the animals and not primarily to hyperphagia. In 1943 Brobeck et al. emphasized hyperphagia as the main cause of obesity induced by hypothalamic injury. As soon as the animals woke up from anesthesia after a lesion in the VMH, they started to eat. Brobeck referred to the first weeks, which are characterized by hyperphagia and body weight gain, as the *dynamic phase*. After this period, the body weight stabilizes at a high level and food intake is reduced. This phase is known as the *static phase*. It is possible that in this phase signals related to adipose tissue inhibit food intake via other routes in the CNS. The observed hyperphagia after lesioning, and the reduction of food intake during electrical stimulation of an intact VMH led to the hypothesis that the VMH is the main target for feedback control of satiety signals. When Anand and Brobeck (1951) damaged a region which was located more lateral from the VMH, they obtained hypophagic animals showing a drop in body weight. When this same region was electrically stimulated, feeding behavior could be induced. On the basis of these data Stellar (1954) postulated the so-called *dual centre hypothesis* in which the lateral hypothalamus (LH) is the feeding center and the VMH the satiety center (Figure 9.1) Although many other brain areas were discovered to be involved in the regulation of food intake the original hypothesis has not been abandoned, but it has been adapted and refined in view of the many new findings. In the following sections the involvement of the VMH and LH in the regulation of body weight and food intake will be discussed in more detail.

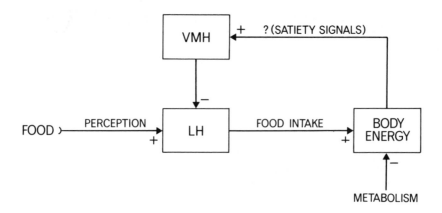

FIGURE 9.1
Simplified model of the control of feeding behavior
(From: De Ruiter *et al.*, 1985)

1   SET-POINT HYPOTHESIS

Food intake in animals depends on several internal and external variables. It is known that animals can adapt to changes in the caloric content of their diet so that over a wide range of energy contents the intake matches energy expenditure and the body weight remains essentially unchanged. On a diet with a constant caloric content, an increase in energy expenditure results in increased food intake. Examples of this phenomenon are the hyperphagia resulting from a lower ambient temperature when the animal has to compensate for its heat loss, exercise, and lactation. In most of these cases the level of energy content of the body remains constant since the animals eat the same amount of calories they spend. This is one of the arguments in favour of the theory that the energy content of the body and therefore the adipose tissue mass is regulated.

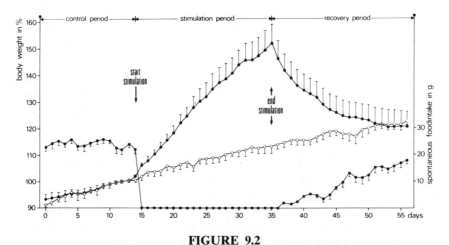

FIGURE 9.2
Influence of LH stimulation on feeding behavior
Mean and *SEM* of body weight (•) and spontaneous daily food intake (■) of 8 rats subjected to 30 min electrical LH stimulation from day 15 through 35. For comparison, the mean body weight of 4 non- stimulated controls (o) is added. (From: Steffens, 1975)

Another example is derived from the precise return to the original or control body weight after a forced deviation from its constant level. This is illustrated by an experiment in which rats were force-fed by stimulating the lateral hypothalamus (LH) electrically (Steffens, 1975; Steffens et al., 1972). The food the experimental animals ate was more than the daily intake of controls. They became obese and stopped all voluntarily feeding activities (Figure 9.2). After termination of stimulation, the body weight gradually returned to the level of that of the controls. As long as the rats were overweight they showed reduced food intake until their body weight had reached the control level. When an animal was starved, the opposite pattern of losing weight was seen. After offering food again *ad libitum*, the animal overate until the original body weight was reinstated. From these data the message is clear that the amount of body fat is regulated at a precisely prefixed point, a so-called 'set-point' (Keesey et al., 1976). In adult animals and man this set-point can hardly be influenced. Only during aging there is a tendency to increase the fat content.

There is some evidence from experiments that the ventral hypothalamus (VMH) and LH have a controlling influence on the set-point. Lesions in either VMH or LH lead to a new set-point for body weight which is defended as accurately as the original one. Partial food deprivation over a period of 14 days resulted in exactly the same decline in body weight in control rats and as in rats with an LH lesion. *Ad libitum* feeding after termination of the deprivation period caused an increase in food intake in both groups of animals such that the exact body weight of non-deprived control rats was reached again (Figure 9.3; Keesey *et al.*, 1976). Obese rats with VMH lesions maintained their new higher set-point of body weight in the same way.

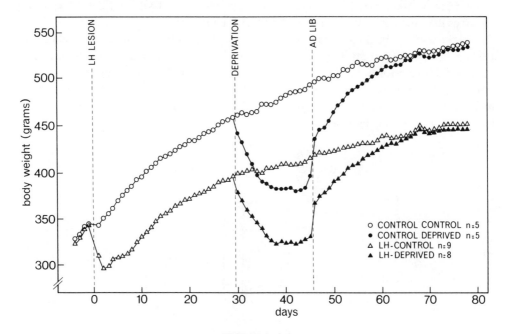

FIGURE 9.3

Recovery of body weight in LH-lesioned and control animals following a period of food restriction

Body weight of the LHA-deprived and control deprived groups was first reduced to 80 per cent of the value each rat normally maintained; they were then returned to an *ad libitum* feeding regime. (From: Keesey, 1976)

Summarizing, these data provide convincing evidence that the size of the body energy content, of which the adipose tissue mass forms the major component, is continuously monitored. The question remains, which factors are involved in this control.

Nervous as well as humoral factors may play a role in this monitoring function. At the moment the only humoral factor known to be strongly related to the amount of adipose tissue is the pancreatic hormone insulin, the mean concentration of which is positively correlated to the degree of obesity (Woods et al., 1986a;). Insulin, then, might provide the signal setting the motivational level in the brain for regulating food intake. If insulin plays a role in informing the brain on the amount of adipose tissue, there should be insulin-sensitive target areas somewhere in the central nervous system. Local application of insulin in these areas should then suppress feeding in a similar way as happens during overfeeding.

When very low doses of insulin were infused into the lateral cerebral ventricles of baboons and rats, suppression of food intake was observed (Woods *et al.*, 1986a). Therefore, the location of the sensors monitoring insulin may well be neural structures associated with the cerebral ventricles. Since the VMH contains many insulin receptors, the hormone may leak from the ventricles to the VMH. Another hypothesis is that insulin is taken up actively from the cerebrospinal fluid by cells in the ventricle wall that have connections with the VMH. When specific antibodies against insulin were locally infused into the VMH to bind insulin molecules at that site, feeding behavior was stimulated only in the dark phase but not in the light phase in which the general feeding motivation in the rat is relatively low (Strubbe & Mein, 1977). Thus, a change in insulin concentration in the VMH may modify feeding motivation. Injection of these antibodies into the LH was without effect which indicates that the main insulin target is the VMH (Figure 9.4).

In conclusion, insulin may well provide an important determinant in the regulation of feeding behavior by controlling the size of the adipose tissue mass. This lipostatic or ponderostatic control holds for the regulation of food intake mainly in the long term, setting the motivational background for feeding behavior. Superimposed on this long-term regulation are short-term signals that operate during individual meals and a few consecutive meals. This will be discussed in the next section.

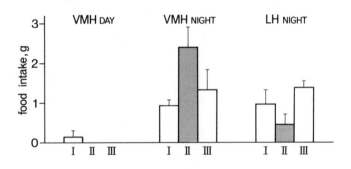

FIGURE 9.4

Effect of injection of antibody on insulin in the VMH or LH on food intake (g)

I, II, III: Successive one hour periods; the injection is made at the beginning of II.

(From: Strubbe & Mein, 1977)

2   SHORT-TERM REGULATION OF FOOD INTAKE

The feeding behavior of animals and humans is strongly related to the light-dark cycle. Rats, for instance, eat 80 per cent of their daily food intake during the dark phase (Figure 8.16). This means that within each day the energy content of the rat varies a great deal, as indicated also by the alteration in net lipogenesis during the dark phase and net lipolysis during the light phase. Body weight is only constant on a long-term scale.

This rhythmic control of food intake by oscillators in the CNS which are synchronized by the light-dark change, determines to a large extent the temporal organization of food intake and interacts with the short-term regulation of energy intake (see Chapter 10). These rhythmic factors are therefore modulating factors and not regulatory factors in the sense that they are involved in the regulation of body energy content. Rather, they constitute one of the constraints within which the energy regulation of food intake must operate. There are many constraints of this kind, ranging from learned habits in feeding behavior (e.g. adjustment of meal size to expectations of caloric content of the diet based on gustatory experience), to inhibition of feeding by demands of other urgent behaviors like sleeping behavior or social confrontations.

Since so many non-regulatory factors determine the start and finish of a meal, the question "Do short-term mechanisms for energy regulation exist?" is relevant.

A simple experiment carried out on rats demonstrates that there are indeed such short-term mechanisms. Sham feeding (in which ingested food leaves the esophagus or stomach via a fistula) results in overeating. This means that the animal lacks the feedback information from the ingested food by means of satiety signals.

There are a wide variety of *negative feedbacks*, or *satiety signals* acting in the short term as well as in the long term:

1 *Gastro-intestinal and oropharyngeal signals.* When food is ingested it passes through the mouth. Caloric content, measured by oropharyngeal receptors, may elicit satiety signals here (Epstein, 1967). The food is transported to the stomach and intestines where stretch receptors can estimate the volume ingested. There is also some evidence for chemoreceptors in the digestive tract, which may signal the caloric value of its contents. The afferent pathways to the central nervous system may be either nervous or hormonal (Deutsch, 1990; Kissileff & Van Itallie, 1982).

2 *Signals from absorbed nutrients.* The absorbed nutrients pass through the intestinal wall into the blood. It has been postulated that glucose, together with insulin (the so-called *glucostatic theory*, Mayer, 1955; Strubbe *et al.*, 1977) and also amino acids *(aminostatic theory)* (Chapter 2) give satiety signals to the central nervous system. Metabolism of these nutrients induces heat production. The subsequently increased body temperature could inhibit feeding activities. This concept has been formulated as the *thermostatic theory* (Strominger & Brobeck, 1953).

3 *Signals from the reserves.* The reserve tissues may also inform the central nervous system about their content. The signals may arise in association with the size of adipose tissue (*lipostatic theory*, Woods *et al.*, 1986b). Other signals may be involved in reporting the size of glycogen content and therefore contribute to short-term regulation (Mayer, 1955; Russek, 1970).

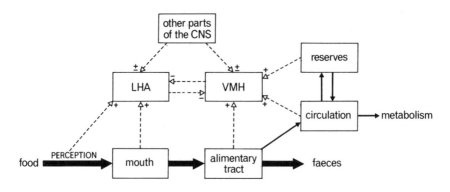

FIGURE 9.5
Simplified model of the control of feeding behavior

Apart from the negative satiety signals there are also *positive feedbacks* originating from the oropharyngeal region. These do not operate before feeding has started, and therefore cannot contribute to its initiation. Once eating has started, these signals keep the feeding motivation at a relatively high level (Wiepkema, 1971). They therefore stabilize feeding in the sense that they postpone feeding termination, but there is no reason to assume that termination of eating is due to a change in these signals. This is why we shall not discuss positive signals any further in this context.

Summarizing, the tendency to perform feeding behavior at any particular point in time, as observed in experiments on rats, will depend at least on:

(a) constraints such as day-night rhythm, and all kinds of learned habits and behavioral interactions;

(b) the current level of energy expenditure;

(c) possible shifts in the level of body energy content;

(d) the resultant of all above-mentioned satiety signals pooled in the central nervous system. (See Figure 9.5 for a summary.)

The main theories on feedback mechanisms will be discussed in more detail in the following sections.

3   GLUCOSTATIC THEORY

In contrast to many other tissues, the central nervous system is, to a large extent, dependent on a constant availability of glucose in the blood. As discussed in Chapter 8, a low blood glucose level will lead to a dangerous hypoglycemic coma. A high blood-glucose level will eventually lead to the equally dangerous complications associated with diabetes mellitus, e.g. glycolysation of cell membrane proteins. To a certain extent the nervous system can utilize ketone bodies in the case of a long fasting period in addition to the always required glucose to cover its metabolic needs. In order to avoid the problems associated with glucopenia (lack of glucose for the tissues), animals possess many defense mechanisms to keep blood glucose within very narrow limits despite a fluctuating food intake over the day-night cycle (see Figure 8.16). The liver plays an important role in this adaptation by supplying enough glucose to the brain during the inactive period of the day-night cycle. In the short-term this can be affected by glycogenolysis and in the long term by gluconeogenesis when the glycogen stores become depleted. However, it will be obvious that it is of advantage to the animal, in anticipation of sudden glucopenia, to keep the glycogen stores adequately filled. Only food intake can sufficiently replenish these stores. In view of the regulation of food intake by negative feedback signals, it seems a likely prediction that signals reporting the availability of glycogen in the liver could control food intake. These signals could then reach the central nervous system via afferent nervous pathways from the liver. They might, however, also derive from the rate of utilization of glucose in tissues, which is supposed to be a measure of the available glucose.

As early as 1955, Mayer formulated this concept in the *glucostatic theory*. According to this theory, food will be taken when the utilization of glucose by the various organs is insufficient. The rate of glucose utilization depends, of course, on the presence in the blood of glucose itself, but for many kinds of cells it also depends very much on adequate levels of hormones facilitating the uptake of glucose. Of these, insulin is the most important. The uptake of glucose is estimated from arterio-venous (AV) differences in glucose concentrations. Many groups of researchers have reported the AV differences to be inversely correlated to certain measures of feeding motivation, such as gastric hunger contractions and verbal reports of hunger feelings in human subjects (Mayer, 1955).

Experimental manipulation of glucose utilization also appeared to give results that are compatible with the glucostatic theory. For example, in the rat insulin injection caused hypoglycemia, hyperphagia and obesity.

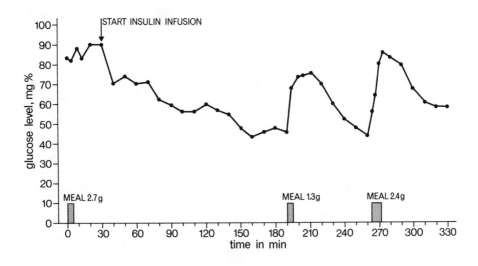

FIGURE 9.6

Effect of continuous insulin infusion on feeding pattern and blood-glucose concentration in a rat

Infusion rate 1.50 IU in 9 hours.

(From: Steffens, 1969)

Even more interesting is the fact that when insulin infusions were given at a low rate, the animal started to eat each time the glucose level was reduced to about 0.5 mg/ml (Figure 9.6; Steffens, 1969).

Furthermore, systemic administration of 2-deoxy-D-glucose (2-DOG) caused the rat to eat (Smith & Epstein, 1969). This analog of glucose antagonizes glucose utilization. This leads to a reduced net uptake of glucose by the cell. It has been suggested that in these studies the glucopenia triggers the eating. However, the induced glucopenia was very severe. The question arises therefore, whether glucopenia constitutes a factor in physiological circumstances in the free-feeding animal. In order to investigate the role of glucopenia, glucose and insulin levels were measured in the free-feeding rat (Strubbe *et al.*, 1977). During the entire meal cycle there was no evidence of hypoglycemia (Figure 9.7). The insulin level, however, was lower just before a meal than at any other time from the start of the previous meal, so that glucose utilization reached its lowest value in the normal range of variation just before each meal. This finding is compatible with the possibility that low glucose utilization might initiate feeding, but does not prove that this is so. Manipulations (by means of intravenous glucose infusions) of both glucose and insulin levels - so that relatively high levels were maintained during the inter-meal period - did not postpone the next meal as was predicted. Thus, the data so far do not support the view that decreasing glucose utilization contributes to the initiation of eating under normal *ad libitum* conditions. It remains possible, however, that interference with the time control of feeding disturbed the outcome of this experiment (see Chapter 10). Some reports suggest that a small drop in blood glucose level just before a meal could be a factor in feeding initiation (Campfield & Smith, 1990). However, the testing of this causality is not convincing. The small dip could be the result of increased muscular activity that advances feeding.

It is also suggested that increased glucose utilization provides a signal to stop feeding. Food-deprived dogs stopped eating as soon as large loads of glucose were rapidly injected into the portal vein (Russek, 1970). In guinea pigs, intraportal glucose infusion resulted in an increase in afferent discharges of the vagus nerve (Niijima, 1989).

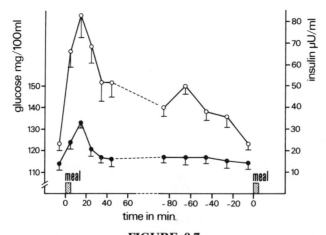

FIGURE 9.7

Average meal and interval cycle of rats

Plasma insulin (o—o), Blood glucose (•—•).

(From: Strubbe *et al.*, 1977)

However, when intraportal glucose infusions were given to food-deprived rats during the first meal after a fast, this affected neither the size nor the duration of the meal. Yet, the glucose concentration in the portal vein was raised much more rapidly and to far higher levels than were seen without infusion under otherwise similar conditions (Strubbe & Steffens, 1977).

Although there is no evidence available for the glucostatic control of feeding behavior in rats under normal *ad libitum* conditions, this does not imply that glucostatic mechanisms are of no value in nutritional homeostasis. There are two points where they may be important.

1 As mentioned above, there is convincing evidence that under conditions of dangerous glucopenia, glucostatic triggering of feeding behavior acts as an emergency mechanism. The glucopenia is sensed by so-called glucoreceptors. There is electrophysiological evidence that these are located in the LH and VMH. Increased metabolic activity of VMH neurons showed an increased glucose discharge rate (Oomura, 1983). The VMH also contains insulin receptors.

Another line of evidence is based on the hyperphagia seen after injury to the VMH caused by administration of goldthioglucose (GTG). It has been suggested that VMH glucoreceptors, unable to distinguish between GTG and glucose, take up GTG and are subsequently destroyed by its toxic effect. This view is supported by the fact that a normal insulin level is needed for the entrance of GTG into the VMH neurons, since no lesions were found in diabetic animals (Debons *et al.*, 1968).

2 The fact that the energy balance is maintained on a wide variety of diets containing varying caloric densities, proves beyond doubt that the responsible signals are calibrated in terms of calories. It is conceivable that glucose utilization signals provide the yardstick for this calibration.

4   THERMOSTATIC THEORY

Homeothermic endotherms keep their body temperature constant by a complex of physiological mechanisms. This constancy is necessary for the proper functioning of their physiology and behavior, and is accomplished by temperature regulation with feedbacks from all compartments of the body. The regulation involves a complicated interaction between heat-conserving and heat-dissipating mechanisms (vasodilatation and constriction, sweating, panting) and the heat-producing mechanisms. To the latter belong shivering (muscle activity) and non-shivering thermogenesis from brown adipose tissue (BAT activity). Brown adipose tissue is located at the dorsal side of the mammalian body especially around the heart and aorta. The fat cells contain much more mitochondria than white fat cells and the tissue is much more vascularized. It has a high metabolic rate which is very important for the generation of heat in the cold and, particularly in humans, during the early period after birth. In rodents and hibernators it is also important for heat generation during adult life. In rats, BAT is capable of generating 50 per cent of the non-shivering thermogenesis. Basically, all metabolic processes produce heat. Which fuel is used, depends on the food taken in.

From the above it will be clear that the regulation of the temperature is closely related to that of the energy balance. Brobeck used this relationship as the basis for his thermostatic theory on the regulation of food intake. He performed energy-balance studies and observed that high ambient temperatures were associated with decreased food intake while the opposite was true in the cold. For obvious reasons the animal must take extra food to keep warm in the cold, otherwise it would have to use too much of its reserves and thus risk a negative energy balance. Strominger & Brobeck (1953) suggested that in these cases the body temperature provides the satiety signal for the adjustment of energy expenditure to intake by determining the initiation and termination of eating.

The body temperature is monitored by thermoreceptors which are located in the preoptic area (POA) of the anterior hypothalamus. Some of these receptors control the stability of body temperature within very narrow limits (set-point control). Another part of the hypothalamus is probably involved in the control of feeding behavior, since cooling of this area in goats stimulated food intake whereas the opposite occurred after heating (Andersson & Larson, 1961). Apart from these central receptors, there are many thermo-receptors located in all peripheral organs, informing the CNS on the local temperature so that adequate control can be carried out. Temperature control of liver and intestines is very important in order to avoid irreversible damage to these organs. It has been reported that such damage may already occur at a core temperature above 40 °C (Falk, 1983). In anticipation of such dangerous core temperatures, rats stop all their heat-generating activities, including feeding behavior, at a core temperature of about 39.3 °C. It is seen in rats, but also in other rodents, that at 39.3 °C the thermoregulatory control on vasomotion causes an increased blood flow in the tail and ears to promote heat dissipation. In this way the combination of physiological and behavioral thermoregulation by stopping activities, allows optimal functioning of the animal and its organs.

Moderate changes in ambient temperature only affect meal *size* and not meal *frequency* (Davies, 1977). From this it can be deduced that the temperature-associated satiety signals only influence meal termination and not the start of a meal. The increase in temperature during feeding superimposes the general basal body temperature, which is the resultant of all metabolic and thermo-regulatory processes. Of these, the basal utilization of fuels like glucose is important in providing heat. Therefore the combined influence of glucose and insulin could play a role in producing heat for the basal temperature level while the meal-related temperature responses control the termination of feeding adequately. There are several examples of cases of obesity and overweight diabetics in whom insulin insensitivity is associated with a reduction of thermogenesis. In these subjects obesity is often associated with hyperphagia. Therefore, as discussed in the previous section, under certain conditions factors related to glucose utilization may contribute to the control of food intake, but probably in combination with behavioral thermoregulation.

5   GASTRO-INTESTINAL CONTROL OF SHORT-TERM FOOD INTAKE

Strong feedback signals appear to report the size and caloric content of food in the stomach and intestines (Kissileff & Van Itallie, 1982). This can be illustrated by the following experiments. When animals were fed bulk non-digestible material, this caused immediate satiation due to distention in various regions of the intestinal tract. This distention may act on gastric stretch receptors that activate the central hypothalamic satiety area. However, the animals were hungry again sooner than when the material was enriched with nutrients. This indicates that calories also play an important role. Indeed, recent views indicate that a strong feedback signal reports the caloric content of stomach and intestines (Deutsch, 1990). As already mentioned in Section 9.3, during sham feeding of liquid food leaving the body via a fistula, the animal adjusts immediately by overeating. On the other hand, when liquid food is infused into the stomach of a free-feeding rat, the next meal is postponed in proportion to the size of the infusion, but there is no effect on the size of the meal (Strubbe *et al.*, 1986). However, the adjustments in experiments using stomach infusions is not always complete (Kissileff & Van Itallie, 1982). This may be caused by the absence of oropharyngeal signals after infusion into the stomach (Epstein, 1967; Kissileff & Van Itallie, 1982). The speed of the effect is in contrast with the lack of effect when glucose is infused intravenously and indicates that the effects on food intake are solely due to feedback signalling from the gatrointestinal tract. But how is the information transferred from the GI tract to the CNS?

The afferent pathways to the CNS may be either nervous or hormonal or both (Deutsch, 1990). Among the latter, the intestinal hormones cholecystokinin (CCK) and bombesin, somatostatin and glucagon have recently attracted much attention as potential satiety factors. When infused intraperitoneally (in the abdominal cavity) or intravenously, the hormones invoke the normal behavioral sequence of satiety, i.e. drinking and sleeping (Gibbs *et al.*, 1973). The reducing effect on food intake is probably not due to a generalized effect of malaise. Since CCK is the most investigated hormone in these studies it will be described and discussed in some more detail.

During the last two decades increasing evidence has been obtained that CCK forms an important factor in the feeling of satiety after a meal. Moreover there is now increasing insight into the way in which CCK exerts its physiological effect in the induction of satiety.

CCK is a peptide which is released from the mucosa of the duodenum as a response to food coming from the stomach. It was isolated for the first time in 1967 from the intestinal tract of a pig. There are different types of CCK, ranging from a molecule with 39 amino acids to CCK-4 with only 4 amino acids. The most used molecular forms are CCK-33 and CCK-8. As discussed in Chapter 7, CCK influences bile secretion and the secretion by the exocrine pancreas. CCK was shown to induce insulin secretion when infused systemically. A new dimension was added when CCK was found in the CNS in relatively high concentrations. It was thought in the beginning that CCK could be a part of the gastrin molecule since its structure shows some similarity with that of the C-terminal part of that molecule. As the receptors are also found in the intestinal tract, in the vagus nerve and in the CNS, this suggests that CCK fulfils a coupling and an integrating function between CNS and intestines. As mentioned above, peripheral CCK causes a dose-dependent inhibition of food intake in many species including man. In the rat this appeared to occur only after intracardial infusion of CCK-8 but not after intraportal infusion (Figure 9.8; Steffens, 1969).

This may indicate that the liver removes most of the CCK. Since intraperitoneal administration was also effective, it is suggested that CCK-8 has a more paracrine action in the intestinal tract (Strubbe *et al.*, 1989). It is possible that CCK-33 is not removed by the liver and may thus still exert its hormonal function in suppressing food intake. This brings us to the question: Where are the targets for CCK? There is some evidence that injection in the ventricles and VMH suppresses feeding behavior, but it is difficult to understand that CCK should be able to pass the blood-brain barrier. There is evidence that CCK can act on the afferent pathways of the vagus nerve since subdiaphragmatic cutting of the nerve appeared to abolish the suppressive effects of CCK on food intake (Deutsch, 1990).

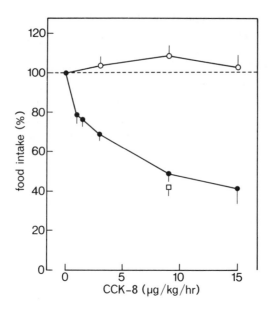

FIGURE 9.8
Satiety effect of graded loads of CCK-8 expressed as percentage of control intake when infused in the right atrium (•—•), into the portal vein (o—o) or intraperitoneally (◻—◻)
(From: Strubbe *et al.*, 1989)

When the area of entrance of the vagus nerve in the CNS (the nucleus tractus solitarius) was damaged, similar effects were obtained. In conclusion, since CCK-containing neurons and CCK receptors are found in the whole chain from intestine via afferent vagus to the nucleus tractus solitarius and from there via the parabrachial nucleus to the VMH, it is possible that CCK influences feeding motivation at all of these levels (Figure 9.9).

Other intestinal hormones like glucagon, bombesine and somatostatin also influence satiety. When antibodies against glucagon were infused into the portal vein, food intake was stimulated. This effect failed to occur after cutting the hepatic branches of the vagus nerve. As is the case for CCK, this is another example of feedback regulation by a combined action of hormones and the peripheral autonomic nervous sytem (for a review, see Deutsch, 1990).

All intestinal hormones mentioned here are released respectively ('in concert') along the intestinal tract after specific nutrients have been sensed. Apart from regulating adequate processing (transport, digestion and absorption) of the food, they can contribute to the whole effect of satiety by potentiation or acting synergistically.

6   CONCLUDING REMARKS

We have discussed many factors that play a role in the control of energy homeostasis in the body. Homeostatic control was shown to depend on internal and external conditions. Although this chapter discussed evaluations of long-term and short-term regulation of food intake, a precise definition of both types of regulation is not given.

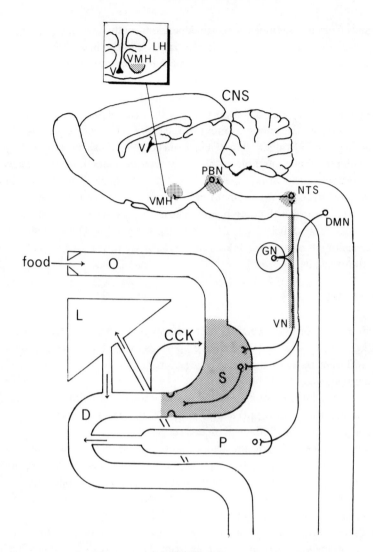

FIGURE 9.9
**Schematic presentation of anatomical structures where CCK is transmitted and where the
CCK receptors are located (dotted areas)**
S = stomach; D = duodenum; P = pancreas; L = liver; O = oropharyngeal cavity; CNS = central ner-
vous system; VN = vagus nerve; PBN = parabrachial nucleus; NTS = nucleus tractus solitarius; GN =
ganglion nodosum.

In fact, long-term regulation consists of a succession of short-term regulations and the
long-term signals discussed are also operating at all times. It is therefore impossible to
make a clearly defined distinction between these regulations.

The oropharyngeal and gastro-intestinal satiety signals probably play a role in the very
short-term regulation by tasting and measuring the volume. However, adequate adjustment in the long-term can only occur after a phase of learning about the texture and
caloric contents of the different diets. It is possible that signals from glucose utilization
play a role in calibration of central set-points so that a wide variety of diets can be taken
without incurring energy deficits or gaining body weight leading to obesity.

The signals from body temperature are dependent on the rate of thermogenesis, which is again dependent on the rate of incoming nutrients and the heat dissipation to the environment. The latter depends on the current ambient temperature. Although body temperature may play a role in short-term regulation by terminating eating, the change of the 'thermostat' may result in a disturbed energy balance in the long run. It should be pointed out, however, that such signals may report energy flow (input, output), but certainly do not unambiguously indicate the energy content of the body. The factors involved in this regulation are acting over a longer time scale. Conceivably, these messages do not act by directly switching feeding on and off, rather, they provide a motivational background level governing the effectiveness of the internal and external signals. The average insulin level may well prove an important determinant of this motivational background.

However, the control of feeding behavior is not necessarily achieved by the fact that signals reporting energy deficits trigger feeding responses. As will be evident from Chapter 10, the influence of circadian pacemakers can be quite dominant over the above-discussed control by satiety signals.

7   SUMMARY

The maintenance of body weight within narrow limits is controlled by the regulation of food intake; the regulation of food intake operates through positive and negative feedbacks from different parts of the periphery reporting the central nervous system (CNS) about the energy content of the body.

The size of body fat content is regulated at a so called 'set point', which can hardly be influenced; the ventromedial hypothalamic areas (VMH) and the lateral hypothalamic areas (LHA) exert a controlling influence on this set point.

Insulin may well be an important determinant in the regulation of feeding behavior by controlling the size of the adipose tissue mass; this control holds for the long-term regulation of food intake, setting the motivational background for feeding behavior.

Constraints such as day-night rhythms, habits and behaviors are modulating factors - not regulatory factors - in the regulation of food intake.

A wide variety of negative satiety feedbacks - gastro-intestinal and oropharyngeal signals, signals from absorbed nutrients, signals from the reserves - as well as positive feedbacks from the oropharyngeal region can be distinguished in both the short-term and long-term regulation of food intake.

The 'glucostatic theory' is one of several theories to explain the regulation of food intake: according to this theory food will be taken when the utilization of glucose by the organs of the body is insufficient as a result of non-adequate levels of glucose itself in the blood or of the hormone insulin; however, there is no evidence for glucostatic control of feeding behavior in rats under normal *ad lib* conditions.

Feedback signals from the gastro-intestinal tract to the central nervous system may be either nervous or hormonal, and the gut hormone cholecystokinin (CCK) is an important factor in the feeling of satiety after a meal.

10

Circadian rhythms
of food intake

Living organisms are influenced by many external rhythms ranging from tidal, daily and lunar to seasonal and annual rhythms. Although animals can directly adapt their physiology to periodically changing conditions, it can be of advantage to anticipate the events of the near future. The current view is, that during the geological era, species have optimized these anticipatory adaptive strategies by the process of natural selection so that animals can hibernate, migrate, escape from predators and perform optimal seasonal productivity. These adaptive strategies are controlled by endogenous innate programs of behavior and physiology which are determined by so-called external signals ('Zeitgebers'). However, as will become evident from this chapter, it is very difficult to ascertain that such endogenous innate programs actually exist.

One of the most striking innate programs is that governing the 24-hour rhythm of almost all physiological processes and behaviors. Since these rhythms have the same periodicity as the earth's rotation, it is rather difficult to ascertain whether they have an endogenous innate program that determines them: a so-called circadian rhythm (circa = approximately, dies = day). For instance, during the periodically changing environmental light conditions it is relatively easy for organisms to synchronize their physiology and behavior to this light-dark variation. Since day length changes only gradually in the course of the seasons the entrainment will seldom give any problem. There are at least two conditions where the presence of endogenous time control of sleeping, feeding and drinking becomes obvious, namely when we try to reverse our activity rhythm as happens by doing shift work or when travelling long distances by plane. In the latter case a transmeridian flight causes a so-called jet lag. The disturbed behavioral rhythms interact strongly with the internal physiological processes which are more strictly determined by endogenous programs entrained by the light-dark rhythm. This is why, apart from tiredness, gastrointestinal disturbances often arise when the normal rhythms are upset.

As mentioned above, there are many biological rhythms, each with its own characteristic functional adaptation (for a review see Rusak, 1981). However, since the circadian rhythm is the most important one as regards the control of food intake, the present survey will restrict itself to this rhythm. The characteristics of the endogenous timing for food intake will be discussed first. Next, the influence of circadian pacemakers and food-entrainable oscillators will be dealt with, followed by a discussion on the interaction between the circadian control and the external and internal conditions. (In the context of this book, the terms pacemaker, oscillator and master clock will have the same meaning.) Finally, the consequences of feeding at times other than predicted by the endogenous programs will be looked at.

1   TEMPORAL ORGANIZATION OF THE CONTROL OF FEEDING BEHAVIOR

The feeding behavior of most animals is strongly associated with their activity periods, which in turn are related to the light-dark cycle. In the following studies the emphasis will be on the rat, since this animal shows many characteristics of the general mammalian timing system in the circadian control of activity and food intake.

In general, rats are mostly active at night. When they are kept under artificial light-dark rhythms in the laboratory they are also nocturnal. Under *ad libitum* conditions rats keep to a 12-12 hour light-dark (LD) schedule and eat most of their food in the dark with peaks at the beginning (dusk) and towards the end (dawn) of the night. Each rat has its own meal pattern which is more or less repeated from day to day. As discussed above, the time-related pattern of feeding is probably under the influence of selective pressures acting within the ecological niche (Strubbe, 1982). Therefore, the function of this daily pattern is probably to keep the rat from venturing into the open to feed during the day, when the risk of predation is high.

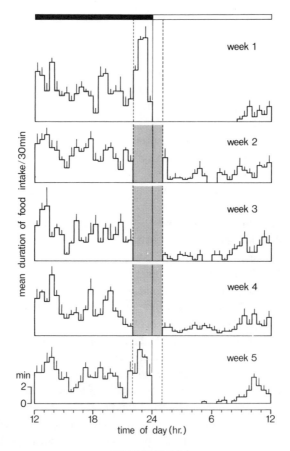

FIGURE 10.1

Mean duration of food intake per 30 minutes in rats during dark (black horizontal bar) and light periods (white bar)

The grey area indicates periodic food deprivation during weeks 2 to 4. (From: Kersten *et al.*, 1980)

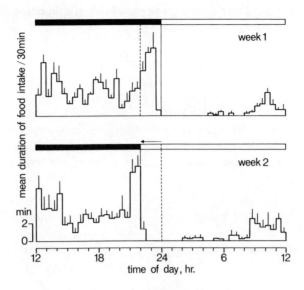

FIGURE 10.2
Mean duration of food intake per 30 minutes
At the beginning of week 2 light onset was advanced by 2 hours.(From: Kersten *et al.*, 1980)

In particular the dawn peak enables the rat to tide over the day without incurring an energy deficit. However, is the system that regulates energy intake and body weight also involved in causing this dawn peak, or is this initiated by the endogenous time control? The following experiments will shed some light on this.

In one experiment, with 12 hours light and 12 hours dark (12-12 LD), the dawn peak was prevented day after day by the removal of food 2 hours before lights on (Figure 10.1 and 10.3). The rats never advanced their dawn peaks to the final hours of food availability. Instead, they compensated by eating more in the daylight. Yet, in another experiment where food was available throughout the entire dark period, but the schedule was changed to 14-10 LD by switching the light on 2 hours earlier, the rats advanced their dawn peak by 2 hours after a single experience with the new regime (Figure 10.2). The conclusion is therefore that the dawn peak is governed, not by fluctuations in energy content of the body, but by the experience of the light-dark change itself (Kersten et al., 1980).

It seems likely then that an endogenous program tells the rat to feed in anticipation of the onset of light. This is even clearer from the experiment that started on a schedule of 14-10 LD (Figure 10.3). Shifting the LD regime to 12-12 with simultaneous food deprivation over the last 2 hours of the dark phase resulted in a gradual disappearance of the dawn peak (Spiteri et al., 1982). However, the motivation to feed at dawn persisted even when the access to food was restricted, and showed accompanying shifts with changes in the onset of light. This became clear from the quick reappearance of food intake after discontinuing food restriction (Figure 10.4). Again, the rats compensated for their caloric deficits by eating earlier in the daytime, when in nature the predation pressure would be high. The rapid shift in dawn feeding upon changes in light onset and food restriction, and its quick reappearance after discontinuing food restriction, argue against habit formation as being responsible for the occurrence and maintenance of dawn feeding? The question remains, which physiological mechanism is responsible for the initiation of dawn feeding. The following section will shed more light on this.

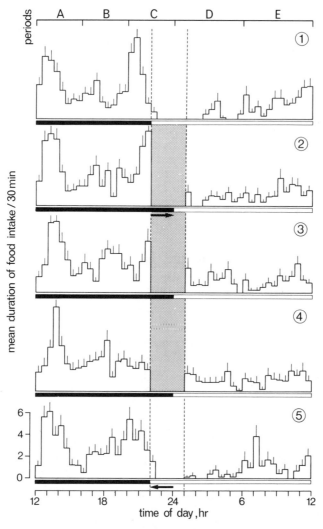

FIGURE 10.3

Mean duration of food intake per 30 minutes for weeks 1 (control), 2, 3 and 4 (delayed light onset with food restriction at dawn), and week 5 (post-experimental week with advanced light onset, from LD 12-12 to LD 14-10)

Periodic food restriction at dawn is indicated by the grey area.

(From: Spiteri *et al.*, 1982)

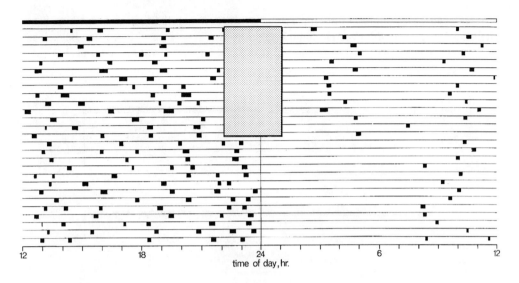

12   18   24   6   12
                                        time of day, hr.

FIGURE 10.4

Diagram of representative daily meal pattern of one rat during 14 days of delayed light onset (from LD 14-10 to LD 12-12) in combination with food deprivation at dawn (indicated by the gray area), and 13 days of *ad lib* food intake

Notice that on return to free feeding the rat ate immediately in the period in which it was previously deprived of food.

(From: Spiteri *et al.*, 1982)

2   THE CIRCADIAN OSCILLATOR OR LIGHT AVERSIVENESS?

Light can be very aversive to several nocturnal animals. Bright light can even damage parts of the retina in albino rats.

If rats are kept in large cages provided with a box, they prefer to stay in the box during the light phase. In this kind of experiment, food and water are only available outside, in the large cage, presented in a food hopper. During the dark phase the rats eat in front of the food hopper, whereas during the light phase they make rapid excursions to the food hopper, and, with every bite of food, they return to their burrow. This behavior is dependent on the intensity of the light and is therefore a good measure of light aversiveness. This leads to the question: Does light aversiveness play a role in the expression of the daily rhythm of feeding?

To find an answer to this question an experiment was carried in which a so-called skeleton photoperiod was introduced (SPP). SPP is a condition in which total darkness is interspersed with light pulses (15-30 min) at the original changes from light to dark and reverse. This method leaves the endogenous component of light-induced behavior intact. Starting from an LD situation, not many changes were observed after the introduction of SPP (Figure 10.5). The rats ate only one meal less during the light phase of the LD condition which indicates that light aversiveness does not play a major role in the expression of the daily rhythm. Apart from this, a few other points deserve attention. During the first 6 days on LD there was no stable rhythm but a shifting dusk period of feeding activity which was stabilized at 2-3 hours after lights off. The reason was that the LD rhythm of the experimental room was shifted 6 hours from the original rhythm of the animals. However, feeding activities started immediately as soon as the light went off. This feeding activity is probably caused by the small energy deficit incurred during the light phase, which is replenished as soon as the aversive light is taken away. Therefore, this feeding peak is no longer seen during the SPP condition. From these results in the SPP condition it is clear that the light pulses (Zeitgebers) can keep the rhythm at a fixed position in the daily cycle.

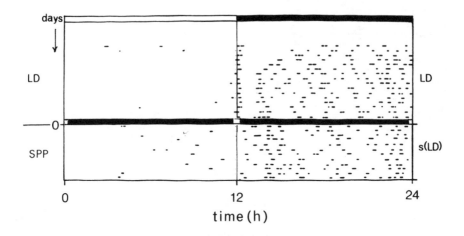

FIGURE 10.5

Daily meal pattern of one representative rat under SPP conditions

The lighting conditions were changed from LD to SPP on day 0.

(From: Strubbe, 1990)

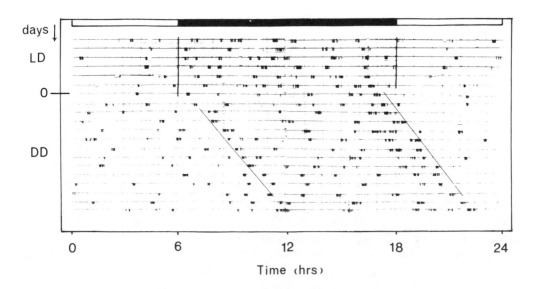

FIGURE 10.6

Daily meal pattern of one rat under constant lighting conditions

The lighting conditions were changed from LD to DD on day 0. From that day onwards a clear free-running rhythm appears.

(From: Strubbe, 1990)

When the external light conditions were kept constant, for instance continuous light (LL) or continuous dark (DD), the animals were unable to synchronize their oscillators to the environment. Therefore the behavioral expression will be the result of the control by the endogenous oscillators. A so-called circadian-free running rhythm then develops with a daily period which is close to that of one single earth rotation (Figure 10.6). The periods may be shorter or longer than 24 hours. The length of the period depends on the species and on the individuals. There is some evidence that on average day-active animals have shorter periods than night-active animals.

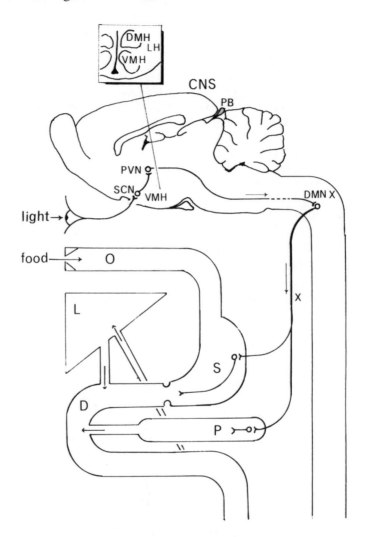

FIGURE 10.7
Schematic presentation of the different brain areas involved in transferring circadian rhythmicity to other parts of the body

O = oropharyngeal cavity; L = liver; S = stomach; P = pancreas; D = duodenum; X = vagus nerve; DMNX = dorsal motor nucleus of the vagus nerve; PVN = paraventricular nucleus; SCN = suprachiasmatic nucleus; VMH = ventromedial hypothalamus; LH = lateral hypothalamus; DMH = dorsomedial hypothalamus; PB = pineal body.

These endogenous rhythms must be generated somewhere in the body with the central nervous system as a major candidate. It is not surprising then that during the last decades many investigators have been looking for such a circadian master clock. They succeeded by discovering that in reptiles and birds this master clock is located in the pineal gland (see Figure 10.7). When this gland was removed the animals became arrhythmic. However, extirpation of the pineal gland in mammals did not result in arrhythmicity. Subsequently, all pathways through which light information enters the brain were checked by lesioning them. Finally, the suprachiasmatic nucleus (SCN), a small area above the chiasma opticum, was identified as the major master clock generating the circadian rhythm in mammals (Figure 10.7; Rusak, 1981; Turek, 1985). Lesioning of this area caused immediate disruption of the circadian rhythm of food intake so that the animals became arrhythmic (Figure 10.8; Strubbe et al., 1987). The same lesion also disrupted the circadian rhythm of many other behaviors and physiological processes. Although the master circadian pacemakers appear to be located in this area, there is evidence that many other areas contain suboscillators that are also involved in generating rhythmic processes which are superimposed on the circadian rhythm (Oatly, 1971). Among them, the ultradian rhythms (period shorter than 24 hours) can be mentioned. The role of the pineal gland in the circadian organization of mammals is still under investigation. There is increasing evidence that pinealectomized rats entrain their rhythms faster than intact control rats (Turek, 1985). Therefore, the current hypothesis is that the pineal gland and probably its hormone melatonin fulfils a more stabilizing function in the circadian rhythm governed by the SCN. In addition, melatonin has been found to play a role in reproduction by modulating the release of gonadotrophins from the pituitary.

Even when a rodent has a perfect lesion in the SCN with a total disruption of rhythmicity, the animal still has some memory for time. When SCN-lesioned rats are fed one single meal per day during daytime, they learn to increase their activity thus advancing the time of feeding. It is suggested that these meal-associated rhythms are governed by a food-entrainable oscillator (Boulos et al., 1980; Rosenwasser et al., 1984; Stephan, 1984).

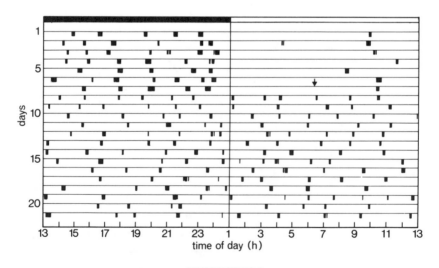

FIGURE 10.8

Disruption of the daily pattern of feeding in lesioned rats

On day 7, the arrow marks the time of electrolytic lesion of the suprachiasmatic nucleus (SCN). (From: Strubbe et al., 1987)

3   INFLUENCE OF THE SUPRACHIASMATIC NUCLEUS

The circadian pacemaker in the SCN influences many behavioral and physiological processes. Therefore, the pathways from the SCN to other brain areas and reverse have been under major investigation during the last decade. The presence of 20 different neurotransmitters and neuropeptides in perikarya and axons within the SCN is evidence for a huge exchange of information between the SCN and other brain areas (Turek, 1985). Electrophysiological recording of unit activity in the SCN shows the presence of neurons which are active either during the dark phase or during the light phase. Most of them, however, are active during the light phase. This is in accordance with the metabolic activity measured by means of 2-^{14}C-deoxyglucose autoradiography. This increased activity during the light phase occurs irrespective of whether the animal is day-active or night-active (Schwartz et al., 1983). This rhythm in activity remains, even when all other nervous connections are disrupted, but it is free-running now. This self-generating rhythm is rather unique for the central nervous system. It implies that the SCN cells need information from the environment to synchronize their activity. The experiments described above showed that this information is provided by light. It reaches the SCN cells via the retina and the retinohypothalamic tract. A second pathway from the retina is that via the lateral geniculate nucleus. Avian pancreatic polypeptide and acetylcholine, and probably also hormones like the pineal hormone melatonin, may fulfil an entraining function in the SCN (Turek, 1985).

From the circadian pacemaker in the SCN the rhythm needs to be transferred to other regions in the brain which possess endogenous programs for sleeping, feeding, drinking and autonomic functions. The SCN has many projections to the dorsomedial hypothalamus (DMH) and the *paraventricular nucleus* (PVN), from which areas several autonomic functions are controlled (Figure 10.7). As will be shown in Chapter 12, the PVN is probably involved in the circadian control of food choice. The classical feeding areas, the *ventromedial hypothalamus* (VMH) and *lateral hypothalamus* (LH), are probably also links in the chain of projections from the SCN.

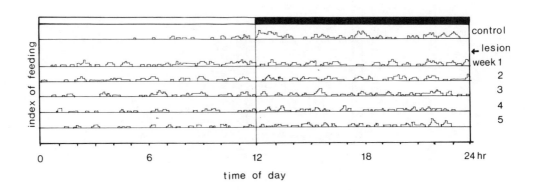

FIGURE 10.9
Effect of a lesion in the ventromedial hypothalamus
on the activity of feeding behavior in a rat

For instance, a large lesion in the VMH which did not damage any part of the SCN resulted in a loss of circadian rhythmicity in food intake (see Figure 10.9). However, very small lesions in the VMH did not result in loss of rhythmicity. It is suggested therefore that in the case of a large lesion the pathways from the SCN to the dorsomedial hypothalamus are destroyed, and that the VMH itself is probably not a link in the chain of neuronal connections from the SCN to the feeding areas. A lesion in the LH caused extremely aphagic animals which had to be fed artificially during the first week after lesioning to survive. When the animals recovered from the lesion they started to eat again of their own accord. However, the recordings of the feeding activity showed that feeding was exclusively during the dark phase. An explanation is hard to give but the animals probably have an increased aversion to eating during the light phase.

4   INTERACTION BETWEEN FEEDING AND OTHER BEHAVIORS

Not only feeding activity but also drinking activity shows a bi- or trimodal distribution over the dark phase with peaks at dawn and at dusk (Figure 10.11, days 1-10). Under normal conditions a very close temporal association exists between feeding and drinking behavior, so that 70-90 per cent of that daily food intake occurs during a meal (Strubbe et al., 1986). Water intake is very important for the processing of food intake of rats. When there is no water available, the intake of food is strongly reduced. Although these two behaviors occur in very close temporal association, the circadian characteristics of the one behavior do not change significantly in the absence of the other. This was shown in experiments where food access was prevented during the dark phase so that the animals were forced to eat during the light phase (Figure 10.10).

In such a situation 17.5 per cent of the daily water intake was food-associated compared to 71 per cent occurring ad libitum. The circadian characteristics of the temporal distribution of drinking activity and nest occupation were retained as well (Spiteri, 1982).

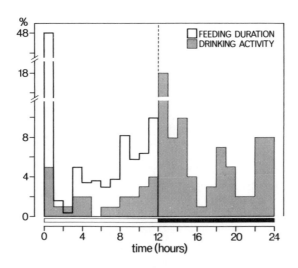

FIGURE 10.10
Percentage distribution of feeding behavior during the LD cycle with food access during the light phase

(From: Strubbe *et al.*, 1986)

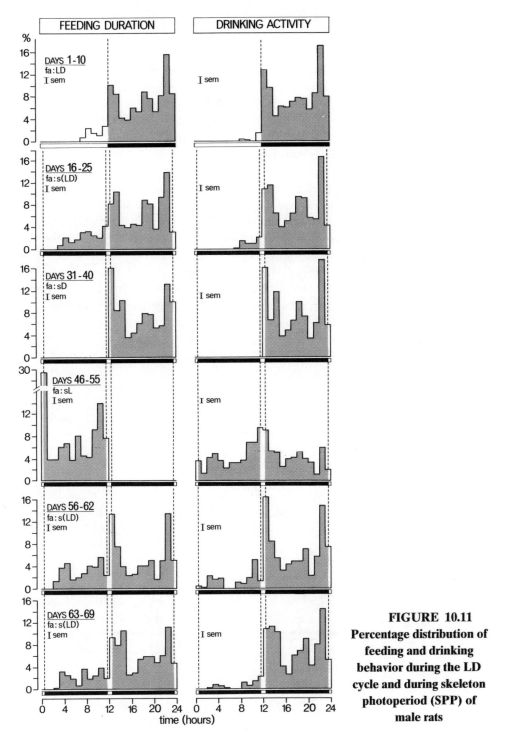

FIGURE 10.11
Percentage distribution of feeding and drinking behavior during the LD cycle and during skeleton photoperiod (SPP) of male rats

Days 1-25 and days 56-69: normal access to food and water; days 31-40: with food access during the 'subjective' dark phase, days 46-55: with food access during the 'subjective' light phase

fa = food availability; s = subjective; *SEM* = averaged standard error of the mean in all observed periods of 30 min. (From: Strubbe *et al.*, 1986)

These experiments show indeed that although feeding and drinking may be causally related, they need not occur in close temporal association. However, it may be possible that light intensity has some aversive properties that 'force' drinking and most of the other activities to take place in the dark phase despite feeding during the light phase.

Therefore, the following experiment was carried to examine to what extent light is involved in the dissociation of the expression of drinking and feeding behavior. The experiments were performed under an LD regime employing skeleton photoperiod (SPP) and with food deprivation during the subjective night. Figure 10.11 (days 1-10) shows that feeding and drinking were closely associated during the normal LD cycle, but under SPP conditions (days 16-25) a small increase in feeding activity during the subjective day was not accompanied by an equivalent increase in water intake. This indicates a stronger coupling of drinking to the subjective night. Restriction of food availability to the subjective day caused partial but not permanent desynchronization between feeding and drinking behavior (days 46-55). Synchrony was re-established within one day once the food was available *ad libitum* from day 56 onwards. A complete return to the original feeding and drinking pattern took 3 days (Figure 10.12). Since the ambient conditions during the subjective day and night are exactly similar, this rapid return to the original rhythm indicates that endogenous oscillators control the daily rhythm. It is suggested therefore that separate slave oscillators controlling feeding and drinking are governed by a hypothetical circadian 'master' oscillator which remains definitely entrained to the original rhythm by the light pulses of the SPP condition (Strubbe et al., 1986a).

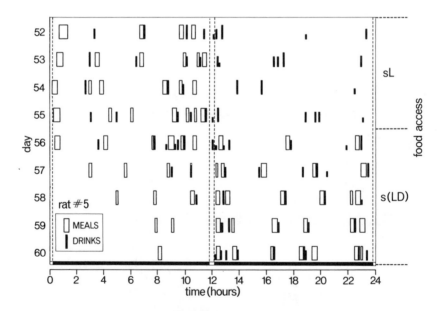

FIGURE 10.12
Re-estabishment of feeding and drinking pattern after SPP
See Figure 10.11 for abbreviations.
A larger black bar means a larger drinking bout.
(From: Strubbe *et al.*, 1986)

Since the master oscillator did not shift during the daily cycle, one might expect that sleeping goes on during the subjective light phase, thereby interacting strongly with the feeding and drinking behavior during this phase. It is possible therefore that besides the master oscillator, sleeping also forced the animals to feed during the subjective dark phase as soon as food was available again.

Generally speaking, rats are social animals and live in colonies consisting of dominant and submissive rats. Monitoring the individual feeding activity showed that the dominant rats ate according to the patterns described above, whereas the submissive rats did so during the light phase. So far, it is not clear whether the submissive animals were forced by the dominant animals or that they were able to shift their endogenous pacemakers so that the rhythmic expression of feeding activity is the reverse of that of the dominant rats.

5   INTERACTION BETWEEN CALORIC AND CIRCADIAN CONTROL OF FEEDING BEHAVIOR

We have shown that the temporal control of feeding can be phase-shifted and entrained by light-dark changes but not by food deprivation schedules. These results provide evidence that a circadian pacemaker, which is entrained by the light-dark changes, controls the time pattern of the rat's motivation to feed at certain times and in particular at the end of the dark phase. It is possible therefore that during these times the pacemaker activity interacts with satiety signals. In Section 9.5 the strong influence of feedback signals reporting the caloric content of the stomach or of the intestines was discussed. We can now put the question of how strong this interaction is at different times in the light-dark cycle. In other words, what is the relative contribution of the influence of satiety signals from the intestine and the influence of the circadian pacemakers to the control of feeding? This is investigated in the following experiment in which the amount of food taken voluntarily after infusion of liquid food directly into the stomach, is measured during several parts of the light-dark cycle. This method of food administration avoids the possible influence of oral sensations (Strubbe et al., 1986).

The study demonstrates that intragastric infusion of food is more effective in suppressing food intake during the light phase and first part of the dark phase than during the second half (Figure 10.13). This effect is mainly due to a delay of the first meal after the infusion, by which the inter-meal interval during the infusion period is increased. No significant effects on meal size were observed in the various periods. Therefore, these results suggest that mainly meal initiation is affected by the gradual infusion of calories into the stomach and not termination of the meal. The adjustment of food intake was not complete and ranged from 65 to 90 per cent during the light phase and the first part of the dark phase (Figure 10.13). This incomplete suppression might be the result of the absence of oropharyngeal signals during and after the infusion. In addition, the different nature of the liquid food compared to that of ordinary laboratory food may induce a less precise adjustment.

The experiment also shows that during the second part of the dark phase the suppression is far less complete. During the last 3-hour period the first meal after infusion was not even postponed. From this it can be deduced that signals reporting the amount of calories infused at dawn are neglected or overruled by other signals forcing the rat to advance eating prior to the light phase, a period of aphagia and sleep.

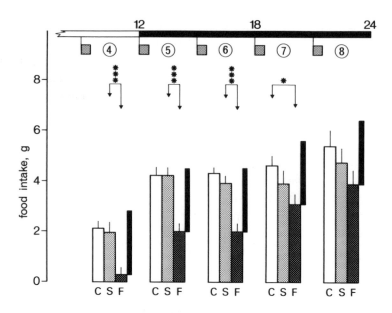

FIGURE 10.13
Food of intragestive food infusion on food intake

Food intake (g) per 3-hour period in controls, i.e. no infusion (C), solvent infusion (S) and food infusion (F). The black bar indicates the amount of liquid diet infused as grams of normal food yielding the same amount of energy. Black bar on top indicates the dark phase. The time of infusion is expressed as the gray bars on top. *** $p < 0.005$.
(From: Strubbe *et al.*, 1986)

This possible anticipatory feeding during the end of the night is supported by others who found that the stomachs of rats were hard-packed with food at the end of the night but not during the beginning of the dark phase (Armstrong, 1980).

Section 10.3 elucidated the mechanism behind such anticipatory feeding. Based on the results of experiments described, it was suggested that the occurrence and maintenance of the feeding peak at the end of the night is controlled by a circadian pacemaker which is entrained to light onset. The present results confirm this view and suggest that a circadian pacemaker dominates feeding motivation at the end of the dark phase thereby overruling signals reporting the amount of calories in the intestines. The latter signal may contribute more to satiety during other parts of the light-dark cycle where feeding is dependent on the rat's immediate energy requirement. Increased energy expenditure will require increased food intake and may therefore interact with the circadian control of food intake. This will be discussed in the next section.

6   INTERACTION BETWEEN ENERGY REQUIREMENTS AND CIRCADIAN CONTROL OF FEEDING

As discussed before (Chapter 9) animals can adapt to changes in the caloric content of the diet so that over a wide range of energy contents intake matches energy expenditure, leaving the body weight essentially unchanged. On a diet with a constant energy density an increase in energy expenditure in animals results in increased food intake. Examples of this phenomenon are the hyperphagia resulting from a decrease in ambient temperature, the diabetic state and lactation. The increases in daily food intake may be effected by an increase in meal frequency, individual meal size or both.

Extra heat dissipation due to a drop in ambient temperature causes an increase in meal size but not in meal frequency. This indicates that in this situation it is meal termination that is regulated rather than meal initiation (see also Chapter 9). The same strategy is followed in the diabetic state which is associated with energy loss via the kidneys. Diabetic rats compensate for their loss of glucose via the urine by ingesting an equal amount of food. On average, this is twice as much as under healthy control conditions. The rats' metabolism is totally disturbed by the increased glucose levels and changes in fatty-acid metabolism. The concomitant polyuria caused by osmotic diuresis forces the rats to drink a volume of 20 times more than normal. However, in spite of these disturbances they retain the circadian characteristics of their meal pattern. A clear dusk and dawn peak of feeding activity is apparent (Figure 10.14). Again, the increased energy expenditure is compensated for by an increased meal size and a small increase in feeding during the light phase, whereas all other factors determining the meal pattern are unaffected.

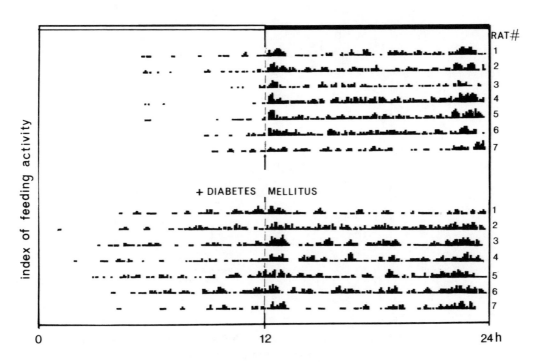

FIGURE 10.14
Feeding activity of rats made diabetic with alloxan

Alloxan destroys the insulin secreting β-cells (lower part).

(From: Strubbe, 1993)

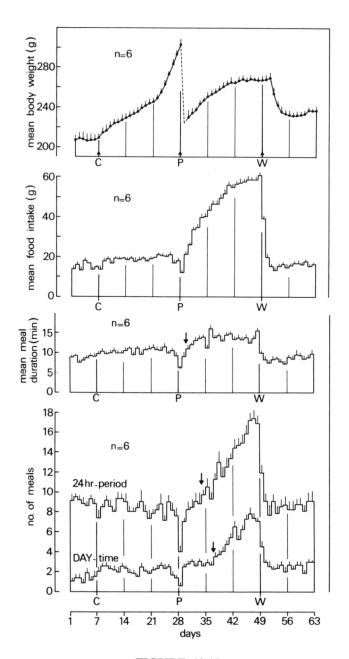

FIGURE 10.15

Time course of body weight, food intake, meal duration and frequency in female rats

The values are daily mean values of 6 female rats during a control week, 3 postconception weeks (C = conception), 3 weeks postpartum (P = partus) and 2 weeks after weaning (W). Arrows mark changes in food intake strategy.

(From: Strubbe & Gorissen, 1980)

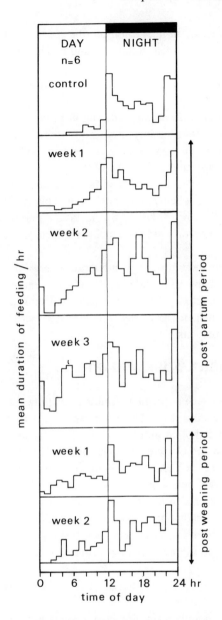

FIGURE 10.16
Pattern of food intake postpartum
Food intake is expressed as the mean duration of feeding per hour, respectively, a control week, three weeks of lactation and two weeks after weaning. (From: Strubbe & Gorissen, 1980)

These studies show that with moderate changes in energy expenditure, food intake is regulated by the regulation of meal termination. In terms of food intake regulation it is of interest to see how feeding behavior changes with higher energy requirements. This was investigated by measuring the meal patterns of a lactating rat feeding a litter of 10 pups.

During the estrus cycle food intake varied on average between 17 g in diestrus and 13 g in estrus. There is some evidence that increased levels of estrogens are responsible for the suppression of feeding activity during estrus. On the second day after conception food intake was established at diestrus levels and remained high for weeks. Body weight gradually increased during this period. From giving birth onward, food intake increased rapidly reaching on average 58 g during the third week postpartum. During the first postpartum week food intake rose from 20 to 40 g. This was accomplished through an increase in meal *size*, not in meal *frequency*. In contrast, the further rise in intake during the second and third postpartum weeks occurred through an increase in meal frequency and not in meal size. In particular the increase in number of daytime meals was prominent (Figure 10.15). The typical bimodal pattern persisted during the three postconception and postpartum weeks (Figure 10.16; Strubbe & Gorissen, 1980).

As regards the physiological causation of the extra food intake during lactation, several possibilities deserve consideration. It is obvious that the increased food intake is the result of the high energy requirement associated with the process of milk synthesis. When this is increased by means of increasing the litter size, the food intake of the dam also increases. Although the suckling stimulus may play a role, the results so far are equivocal. Factors signalling the rate of energy expenditure are probably more important. This study shows again that, although the feeding strategy changes during the lactation period, thereby interfering with the sleeping behavior, no alteration of the circadian characteristics of the meal pattern occurred.

Summarizing, the experiments described here show that, with moderate energy demands, food intake in the rat is regulated through changes in meal *size*. With higher energy requirements, however, meal *frequency* is also affected. The physiological mechanisms underlying this causation are not known, but one can safely say that internal factors signalling the high rate of energy expenditure, are responsible.

7   PHYSIOLOGICAL DISTURBANCES: SHIFT WORK AND JET LAG

As mentioned in Section 10.5 several behaviors such as feeding, drinking, sleeping and social behavior may strongly interfere with the circadian clock. There are two possibilities for interaction. Firstly, the clock is entrained at a fixed point in time and the behavior may be forced in a direction different from that which the clock determines. The behavioral adaptation of such experiments were discussed in Sections 10.1-10.4. In human society shift work is an example of such an interaction. Secondly, behavior is habituated to a rhythm and the clock shifts due to alteration of light conditions. In nature, the day to day change in light conditions during the seasons is only gradual so that an animal can easily adjust the circadian pacemakers to this shift. However, problems arise after a long transmeridian flight. The long period of physiological adaptation to the new day-night rhythm is known as *jet lag*.

Shift work

In the Netherlands 20 per cent of the working population is involved in some kind of shift work, i.e. work at other times of the daily cycle than most other people do. They often work during (part of) the dark phase and sleep during (part of) the light phase. Apart from problems related to social behavior within family life, many of them develop several disturbances in their normal physiology. Some of these disturbances may even lead to a pathological situation.

Disturbances may occur in sleeping behavior, but also in the processing of food which may lead to stomach bleeding and other disorders of the intestinal tract. Depression and the occurrence of cancer are more often seen in shift workers than in non-shift workers. The reason is that the circadian pacemakers that determine the activity level of many physiological processes do not shift with the activity rhythm. The fact that the pacemaker does not shift with behavior was clearly demonstrated by the experiments on rats discussed in Section 10.5.

During shift work the activity of gastro-intestinal processes related to digestion, absorption and storage is probably less optimal. Examples from rat physiology discussed here are insulin secretion and bile flow. The responsiveness of insulin secretion during the active phase is dependent on food intake and on the influence of vagal nerve activity stimulating the effective transport and absorption of food. The activity of the vagal nerve is controlled by the SCN and determines optimal processing of food (Strubbe et al., 1986). Since the influence of the SCN is not shifted, the vagal nerve activity will not be optimal during shift work which may lead to the above-mentioned disturbances. A similar effect can be observed in bile flow. Although, bile flow is closely linked with food intake (Figure 10.17) the rhythm still persists during food deprivation lasting 2 days (Vonk et al., 1978). Therefore, it seems that a food-dependent as well as a food-nondependent factor stimulates bile flow. Again, the food-nondependent factor is probably coupled to the peripheral output of the SCN determining the rhythm. This may cause insufficient digestion of the food taken during the inactive phase which may lead to gastro-intestinal disturbances. Feeding an adapted diet at optimal times and light conditions can probably prevent some of these disorders. Optimal light conditions may eventually shift the pacemaker influence, however, repeated changing from shiftwork to the original rhythm may cause even more adverse effects.

Jet lag

The effects of jet lag are similar to those caused by shift work but they disappear sooner and are often less severe. A transmeridian flight causes a rapid shift in time and location. This causes a strong external and internal desynchronization of the subject. The external synchronization is caused by the change in external time schemes with regard to the endogenous programming of biological rhythms.

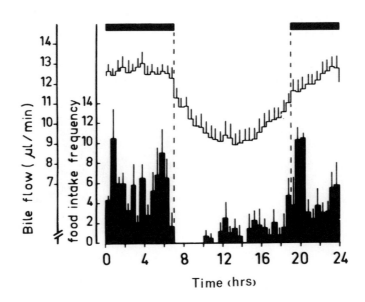

FIGURE 10.17

Food intake patterns of rats (vertical solid bars) and bile flow (top graph)

(From: Vonk *et al.*, 1978)

The internal desynchronization is caused by the different resynchronization times of the various biological rhythms. The rhythm of the physiological variable with the longest synchronization time determines the total duration of resynchronization of an individual. Repeated desynchronization, which takes place in the crew members of planes travelling on transmeridian flights, causes many disturbances in sleep and gastro-intestinal physiology. Future research into this problem can possibly reveal how light conditions and eating specific diets at optimal times might help to circumvent these problems.

8   SUMMARY

The circadian rhythm is the most important biological rhythm in the control of food intake; rats eat most of their food in the dark with peaks at the beginning (dusk) and toward the end (dawn) of the night.

The dawn peak is governed, not by fluctuations in energy content of the body, or by habit formation, but by experience of the light-dark change itself light aversiveness does not play a major role in the temporal distribution of feeding behavior.

The suprachiasmatic nucleus (SCN) is the master clock generating the circadian rhythm; the pineal and probably the hormonal output of melatonin fulfils a more stabilizing function on the circadian rhythm governed by the SCN. The SCN has many projections to the dorsomedial hypothalamus (DMH) and to the paraventricular nucleus (PVN); the PVN is probably involved in the circadian control of food choice.

Drinking activity shows a bimodal distribution over the dark phase with peaks at dawn and at dusk; under normal conditions a close temporal association exists between feeding and drinking. Light intensity seems to have some aversive properties 'forcing' drinking to occur in the dark phase despite feeding during the light phase; it is suggested that separate slave oscillators controlling feeding and drinking are governed by a hypothesized circadian 'master' oscillator.

Several studies suggest that a circadian pacemaker dominates feeding motivation at the end of the dark phase thereby overruling signals reporting the amount of calories in the intestines.

With moderate energy requirements food intake regulation in the rat occurs through changes in meal size; with higher energy requirements, however, meal frequency is also affected.

11
Neuro-endocrine factors

Chapter 9 discussed a few aspects of neurotransmission and the influence of neural circuits on food intake, notably the peptidergic pathway of cholecystokinin (CCK). This chapter focuses on the interplay between the different neurotransmitters and their receptors involved in the control of food intake. In the last decade several detailed reviews have appeared on this subject. This chapter will restrict itself to summarizing the effects of neurotransmitters, (neuro)hormones, receptor agonists and antagonists in their function as inhibitors and stimulators of food intake, as they appear from experiments on test animals (usually rats). Some of the neurotransmitters also play a role in the regulation of food preferences and the same can be said for the circadian rhythm. Understanding the many influences of neurotransmission in the control of food intake is important for those dealing with the treatment of food intake disorders such as anorexia nervosa, bulimia nervosa and obesity.

The interrelationship between food intake, food selection, hormonal effects and the central nervous system is schematically visualized in Figure 11.1.

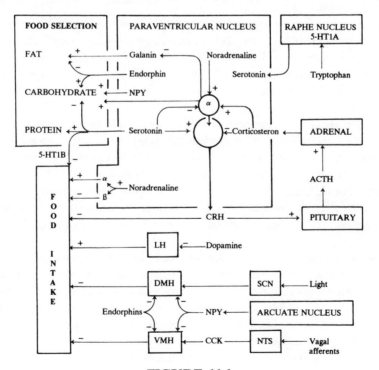

FIGURE 11.1

Schematic presentation of several neuro-endocrine processes in the control
of feeding behavior

1   NEUROPEPTIDES AS INHIBITORS OF FOOD INTAKE

A number of peptide hormones are known to inhibit food intake. The most important of these, cholecystokinin (CCK) has already been extensively discussed in Chapter 9. However, other peptides such as somatostatin, bombesin, pancreatic polypeptide, thyrotropin-releasing factor (TRH), corticotropin-releasing factor (CRF), glucagon and calcitonin can also suppress feeding. With a few exceptions, the effects of the hormonal peptides from the intestine are mostly suppressive.

The main questions that arise concerning the effect of hormones are: 1) Does the hormone have a physiological effect or is the effect a pharmacological one caused by an unphysiologically high dosage? 2) What/where are the targets? As regards CCK the targets are probably to be found all along the tract from stomach to hypothalamus but its action starts with a major effect on the stomach (Chapter 9). Apart from changing the activities of the feeding areas via these routes, it is also possible that CCK modifies the stomach-emptying rate and/or intestinal motility via efferent routes from the CNS.

To answer the first question, it is important to develop assays for measuring the hormonal blood levels. Once such assays are available, the hormone levels under different feeding conditions can be compared with those obtained after injection or infusion of the hormone at certain sites. If the concentrations in the latter case are equal to or lower than those without hormone treatment, it may be that the levels are within the physiological range. If they are higher they may be unphysiological and therefore the dose is probably pharmacological. When animals show disturbances in their behavior or physiology it is possible that higher hormone concentrations are circulating than under control conditions. The hyperinsulinemia in obesity (see Chapter 8) is an example of such a deviation.

The problem can also be approached by treatment with the hormone's antagonists. For example, many studies have been carried out on proglumide, an antagonist of CCK, that causes an increase in meal size (Shillabeer & Davison, 1987). *Glucagon* is another example of a peptide that suppresses feeding activity. Intraperitoneal infusion of antibodies against glucagon was found to reverse the satiety effects of glucagon. Moreover, cutting the vagus nerve to the liver lifted the effect of glucagon, indicating that glucagon may act as a short-term satiety signal to the liver (Woods, 1986). Therefore, glucagon may be an important factor in the regulation of food intake.

On this subject the influence of corticotropin-releasing factor (CRF), which is released during *stress* situations, should also be mentioned. CRF stimulates the secretion of adrenocorticotropic hormone (ACTH). Apart from this, it regulates gastro-intestinal processes, the immune system, and feeding. Local injections of CRF in the paraventricular nucleus (PVN) proved to cause suppression of feeding and this cannot be attributed to the release of ACTH or to corticosteroids from the adrenals. The inhibitory influence was reversed by administering CRF antagonists (Woods, 1986).

In conclusion, treatment with antagonists leads to more insight into the effects of physiological hormone release and the effects of hormones on food intake. However, the treatment with antagonists is, to a certain extent, unphysiological and may have unknown (toxic) side effects in the complicated physiological system (Shillabeer & Davison, 1987). Therefore, the question whether a hormone has food-intake regulating properties must be investigated from various angles. Firstly, the causality of the hormonal effect must be studied and this can be done by infusing it peripherally. Then the plasma levels during the infusions can be measured and compared with those of normal controls. Finally, pharmacological antagonists are applied and their effects measured.

One of the side-effects of unphysiological high doses of the hormones in question or the effects of pharmaceuticals may be the feeling of some nausea. Under untreated conditions this may also occur during overfeeding. It might be caused by the release of intestinal hormones, since sometimes injected intestinal hormones also cause nausea and sickness, and obviously, food intake will then be suppressed. Although the genuine satiety feelings are not known so far, in these tests one should be aware of inhibition of feeding caused by nausea.

2   NEUROPEPTIDES AS STIMULATORS OF FEEDING

Although some kinds of stress *suppress* feeding, for instance by the potent influence of CRF, there are also types of stress in which an increase in feeding activity can be observed. This stress-induced feeding can be the cause of certain kinds of obesity. For instance, when the tail of a rat is pinched, this results in increased food intake, and continuous tail pinching in obesity. Since subsequent administration of the opiate antagonist naloxone lifts the hyperphagia, the endogenous opiate system seems to be a major candidate as causative agent of stress-induced feeding. Many studies in this field suggest that endogenous opioids play a role in determining meal *size* rather than meal *initiation*. Endogenous opioids like β-endorphin, dynorphin and several enkephalins were tested by intracerebral administration to rats. When infused at a low dose into the ventromedial hypothalamus (VMH) and paraventricular nucleus (PVN) food intake was increased. The dorsomedial hypothalamus (DMH) appeared to be involved also, since lesions of this area resulted in a lack of effect of the morphine antagonist naloxone on food intake. Moreover, many other areas in the hypothalamus appeared to be sensitive to the regulating influence of opioids and their antagonists. Although central administration of β-endorphin induced feeding, the peripheral administration suppressed it. It is unclear whether the intraperitoneal route used is the disturbing factor here, or whether the effect is specific. In any case, the dosage of the opioids is very important; high dosages cause suppression. This also appeared from studies in which morphine was administrated over a longer period (Morley, 1987; Woods, 1986).

In humans similar effects were observed, but, apart from the modulating role of opioids and their antagonists on total food intake, there also appeared to be an influence on the selection of the different food components. The opioid antagonists induce a reduction in the preference for sugars and an increased preference for fats. Although stress-related modulation of feeding is probably mediated by endogenous opioids, these may be not the only factors.

During the last decade two peptides have been discovered that both stimulate feeding behavior: neuropeptide Y (NPY) and galanin (Morley, 1987; Woods, 1986). *Neuropeptide Y* is quite a common peptide in the central nervous system and shows many effects upon intracerebral administration. Many NPY terminals are located in the PVN, VMH and dorsomedial nucleus (DMN); they originate mainly from the arcuate nucleus which is located close to the VMH. When injected into the PVN and VMH, NPY stimulates feeding and drinking behavior. The effect appears not to be mediated by noradrenergic stimulation since an α-adrenergic blockade with the drug phentolamine could not inhibit the effect (Morley, 1987). When test animals were injected with NPY in the PVN, they showed an increased preference for carbohydrate. The stimulating effect of NPY could not be inhibited by peripheral treatment with CCK. This may be an indication of the central signals being overruled by the peripheral signals. Although NPY is the most powerful natural orexigenic (appetite stimulating) peptide, so far its precise physiological functioning in normal feeding is unknown. It could possibly play a role in the causation of hyperphagia in obesity and bulimia. If this is so, the development of antagonists may be a fruitful approach in the treatment of these disorders.

Galanin is a peptide which is commonly found in the intestinal tract and in the islets of Langerhans where it is co-localized with noradrenalin. There are also galanin receptors in the central nervous system. Galanin injected into the PVN stimulated feeding activity and, in contrast to NPY, it induced an increased preference for a fat diet. An α-adrenergic blockade attenuated the galanin-induced feeding which may indicate that part of the effect is mediated by α-noradrenergic transmission (Woods, 1986). This noradrenergic modulation of feeding behavior will be discussed in the next section.

3   ADRENERGIC NEUROTRANSMISSION

In 1962 Grossman measured an increase in food intake after administration of nora-drenalin to the ventromedial hypothalamus. Since lesioning of this area causes hyperphagia (see Chapter 9) it is concluded that noradrenalin acts as an inhibiting neurotransmitter in the VMH. α-Adrenergic agonists were found to mimic this effect and antagonists reversed it. Although the effect was very significant, current views are that the receptors are located in the PVN (Leibowitz & Shor-Posner, 1986; Woods, 1986). Electrolytic lesioning of the PVN also caused an increase in feeding activity and body weight. Certain other brain areas inhibit feeding activity upon noradrenergic stimulation. This is probably due to a β_2-adrenergic influence. Although the noradrenergic innervation of the PVN modulates feeding, it is still unclear what the precise function is in the whole system of food intake regulation (Morley, 1987; Woods, 1986).

The number and sensitivity of adrenergic receptors are dependent on other hormonal factors originating from the CNS itself and from the periphery. Adrenalectomy performed by Leibowitz resulted in a reduction of the noradrenergic stimulating effect in the PVN (Morley, 1987). Adrenalectomy not only caused the loss of peripheral adrenalin pro-duction but also that of adrenal steroids. Administration of glucocorticoids reversed the response. The increased food intake at the beginning of the dark phase may be causally related to the increase in corticosterone level at that time by increasing the number of α-receptors, a phenomenon known as 'upregulation'. The density of the α-receptors was indeed found to be high at the beginning of the dark phase and low during the light phase. Hence, treatment of the PVN at the onset of the dark phase is more effective in inducing food intake than at other times. Perhaps the corticoids need to be present for optimal functioning of the adrenergic receptors. From the upregulation of α_2-receptors and the inhibitory influences of corticosterone on CRF neurons one may expect an inhibitory function of the α_2-receptors on CRF release. However, other studies showed stimulation of CRF release by the α- and β_2-receptors. Evidence exists that the release depends on the concentration of noradrenalin, i.e. low doses have an inhibiting and high doses a stimulating effect. Therefore, the influence of adrenergic receptors on the control of CRF release may depend on the experimental conditions and the state of the test subject.

There are more transmitters in the central nervous system that need to be present for optimal functioning of normal food intake. One of them is *dopamine* which is widespread in the central nervous system and has either a stimulating or an inhibiting effect depending on the location where it is administered (Leibowitz & Shor-Posner, 1986; Morley, 1987; Woods, 1986). Another very important neurotransmitter is serotonin. Figure 11.1 presents a schematic overview of neuro-endocrine processes and the control of feeding behavior. The regulating function of serotonin will be discussed in the following section.

4   SEROTONERGIC CONTROL

Changes in serotonin (5-hydroxytryptamine, 5-HT) metabolism are found in many disorders of feeding behavior such as anorexia nervosa, bulimia, obesity and maturity onset diabetes mellitus. (Several of these disorders are associated with depression; see Chapter 12.) This suggests an important role for 5-HT in the control of food intake. A key factor in this respect is the amino acid tryptophan which is a precursor of 5-HT. From the Raphe-nuclei in the lower brain stem serotonergic projections reach the hypothalamus and other parts of the central nervous system. Since many 5-HT receptors are located in the PVN and VMH, the hypothalamus seems to be an important target for 5-HT.

Administration of serotonin to the PVN suppresses food intake by reducing meal size and subsequently body weight. On the other hand, selective lesioning of 5-HT neurons with p-chlorophenyl alanine results in hyperphagia and obesity. In addition, 5-HT antagonizes α_2-noradrenergically stimulated feeding when administered in the PVN (Leibowitz & Shor-Posner, 1986). Therefore, in this condition a stimulating influence of α_2-receptors on CRF release may be expected (see also Section 11.3). In general, these results point at inhibition of food intake by increased serotonergic neurotransmission. Therefore, drugs are being developed which cause increased concentrations of 5-HT at the nerve terminals. One such drugs is *fenfluramine* which is an anorectic agent and releases 5-HT from presynaptic terminals and blocks the re-uptake of 5-HT, thereby activating 5-HT_{1B} receptors (see Figure 11.1) However, the precise targets are not known, since more 5-HT receptors are found in the gastro-intestinal tract than in the central nervous system (Woods, 1986). Therefore 5-HT might influence feeding behavior not only via central mechanisms, but also via the periphery.

5   FOOD PREFERENCES AND NEUROTRANSMISSION

In previous chapters most of the studies on food intake regulation have been performed by using standard laboratory food. In Chapter 9 it was stated that the intake of energy is regulated independently of whether the animals had ingested carbohydrates, protein or fats. However, in the last decade evidence has been accumulating to support the notion that the intake of fats, proteins and carbohydrates is also regulated individually. In view of the treatment of several feeding disorders, research on this may proof to be very important.

As already mentioned, serotonin administered to the PVN and VMH suppresses food intake. This is achieved mainly by the suppression of carbohydrate intake, whereas the intake of protein remains unaltered. Since tryptophan is the precursor of serotonin, the synthesis of serotonin depends on the uptake of tryptophan by the neurons. The regulation of the availability and the uptake of tryptophan, leading to the synthesis of serotonin, is and has been the subject of many studies and controversies (Blundell, 1986; Fernstrom, 1988; Wurtman & Wurtman, 1986). In the circulation, tryptophan occurs either bound to albumin or in the unbound form. The latter form is taken up from the blood with the aid of a transport carrier in the blood-brain barrier. This carrier can also transport other large neutral amino acids (LNAA) such as leucine, isoleucine, valine, tyrosine, and phenylalanine. This leads to competition among the amino acids for the carrier. Therefore, the uptake of tryptophan in the neurons is dependent on the levels of the other LNAA's.

The uptake of tryptophan can be enhanced by raising the tryptophan level via food intake or by lowering the levels of the other LNAAs. The latter may be achieved by increasing the blood insulin level (Wurtman & Wurtman, 1986). Tryptophan itself is relatively insensitive to the effect of insulin and so its availability to the neurons will increase. One of the assumptions in these experiments is that carbohydrate diets would stimulate insulin secretion more than protein-rich diets. As a consequence, a carbohydrate diet would induce more tryptophan to be incorporated than a protein diet. The subsequent increase in serotonin would then enhance the preference for protein and suppress that for carbohydrate. However, experiments on rats have shown no differences between the insulin responses after carbohydrate-rich and carbohydrate-free food ingestion (Figure 11.2) (De Jong *et al.*, 1977; Strubbe & Steffens, 1975). Therefore, it is expected that insulin lowers all LNAAs to the same level. According to the above-mentioned theory this will not lead to differential 5-HT synthesis. It is possible that differences do occur after a longer period of fasting when the insulin levels become lower (Strubbe & Steffens, 1975). So far there is no strong evidence for an important role for insulin in this process.

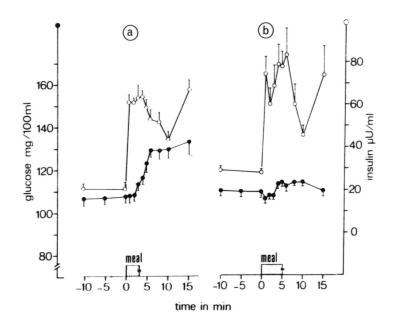

FIGURE 11.2

Patterns of blood glucose (•) and plasma insulin (o) during feeding of (a) a carbohydrate-rich (50 per cent W/W) meal and (b) a carbohydrate-free meal

Notice the higher glucose response in (a) whereas insulin responses did not differ.
(From: Strubbe & Steffens, 1975)

Addition of tryptophan led to a decrease in food consumption consisting mainly of a decreased carbohydrate intake whereas the protein intake was unaffected (Silverstone & Goodall, 1986). There are also several pharmacological studies in which the transmission of serotonin is modulated. For example, treatment with fenfluramine, the 5-HT releaser and uptake inhibitor, not only suppressed carbohydrate intake but also that of fat (Fernstrom, 1988). In other experiments it proved possible to differentiate food selection within a meal showing that treatment with fenfluramine increased the time spent on eating protein compared to that on carbohydrate intake (Blundell, 1986). Although the evidence available so far shows that 5-HT can regulate the preference for carbohydrate and protein, the factors controlling the rate of serotonin transmission are still unsolved.

The link between food intake and serotonin can be extended to the regulation of activity of other monoamine systems, for instance of the catecholamines and their precursor tyrosine. A carbohydrate-rich meal suppresses the release of noradrenalin in the PVN. There is also evidence for a stimulating influence of protein intake on tyrosine levels (Anderson, 1979; Leibowitz & Shor-Posner, 1986). Comparable relationships exist between food intake and the supply of choline precursor for the synthesis of acetylcholine. Moreover, the presence of several co-factors like vitamins (B_6) is needed for an optimal transmitter function in the central nervous system.

6   CIRCADIAN CONTROL OF FOOD SELECTION

From the previous it has become apparent that in rats the peptides neuropeptide Y and galanin can induce preferences for carbohydrate and fat respectively after injection into the PVN, and also, that serotonin can alter the preferences for diets: when injected into the PVN, protein is selected and, at the same time, carbohydrate preference is suppressed. Thus, a rat can alter its food preferences for the major energy-delivering nutrients by selectively activating the transmission of these three neurotransmitters in the PVN (Tempel & Leibowitz, 1990). In addition, the noradrenergic mechanisms may play a role in the selection of food. These preferences are dependent on the external and internal conditions of the animal. One of these conditions is the daily rhythm of food intake. It has been shown in rats that marked differences exist in food preferences over the day-night cycle when offered three different macronutrients (Leibowitz & Shor-Posner, 1986; Tempel & Leibowitz, 1990). During daytime and at the beginning of the dark phase rats prefer to eat carbohydrate, whereas later in the dark phase protein is also eaten (Leibowitz & Shor-Posner, 1986; Tempel & Leibowitz, 1990).

These experiments were repeated by offering rats three more complete diets which were enriched with the three macronutrients (Figure 11.3). It then appeared that carbohydrate is indeed preferred during daytime and at the beginning of the dark phase. In this experiment the preference for protein was less clear than in the pure macronutrient-selection experiment. During the dark phase there was a gradually increasing preference for fat, reaching its highest level during the dawn peak. There is evidence in the literature that the serotonin levels are relatively low at the beginning of the dark phase and that, together with the increased α-adrenergic activity, carbohydrate preference is induced. Subsequently, the serotonin level is increased via the mechanism described in Section 11.5, and the preference for protein is increased. Since galanin in the PVN increased the preference for fat, it may well be an important candidate for the stimulation of the increased fat consumption towards the end of the dark phase (Tempel & Leibowitz, 1990).

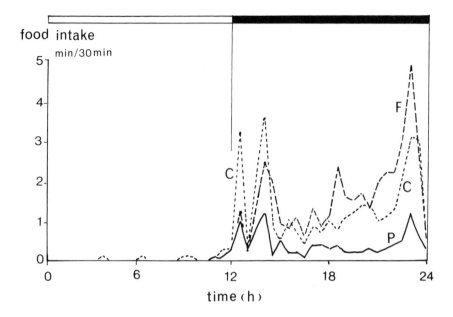

FIGURE 11.3

Daily rhythm of food preferences in the rat

C= carbohydrate, P= protein and F= fat.

As regards the functionality of these changing preferences the following can be hypothesized. Carbohydrate-eating at the beginning of the dark phase can possibly replenish the depleted glycogen stores in the liver more rapidly. In contrast, the consumption of fat at dawn may be in anticipation of incurring energy deficits during the light phase (see also chapter 10). The animals may opt for fat since fat contains twice the amount of energy per gram as compared to carbohydrate and protein. Therefore, the temporal control of varying food preferences induced by changes in neurotransmission in the PVN, is probably adaptive for energy and body weight regulation.

7   SUMMARY

A number of peptide hormones, of which CCK is the most important, are able to inhibit food intake; other peptides (for example opioids) stimulate food intake. Injection of the neuropeptide Y into the PVN induces an increased preference for carbohydrate, while injection of galanine induces an increased preference for fat. Several neurotransmitters in the CNS are needed to be present for optimal functioning of normal food intake; the most important ones are noradrenaline, dopamine and serotonin.

Changes in serotonin metabolism are found in many disorders of feeding behavior such as anorexia nervosa, bulimia and obesity; increased serotonergic neurotransmission inhibits food intake.

There is increasing evidence that the intake of fat, protein and carbohydrate is regulated separately; these food preferences are dependent on the daily rhythm of food intake. It has been shown that during day-time and the beginning of the dark phase rats prefer to eat carbohydrate, whereas later in the dark phase protein and fat are also eaten.

12
Obesity

This chapter reviews the various aspects of obesity in humans and, since they are commonly used as models for human obesity, also in animals. The animal models in question are often based on genetic aberrations, and are especially important in studying the causes of obesity. Apart from the genetic factors that, together, determine the predisposition, environmental causes such as stress-inducing factors play a role in the occurrence of obesity.

Obesity develops when the intake of energy surpasses its expenditure. The energy surplus is then stored in the reserve tissues of which the adipose tissues form the main component.

Increased food intake may be the result of disrupted feedback control, as discussed in Chapters 9 and 10. Moreover, the endogenously determined food preferences may result in selecting more fattening diets (Chapter 5). All these behaviors are based on an aberrant physiology that may partly be caused by primarily genetically determined defects. A decrease in energy expenditure will also result in an increase in energy content of the reserves. Again, this may be the result of genetically determined defects in physiology, but also in behavior. Thus, both food intake and expenditure determine the final energy content of the reserve tissues.

Regulation of the energy balance can take place at different levels (see also Chapter 9). The nature of regulation may differ among people as well as among animals and depends on meal frequency (Fábry, 1967). There is some evidence that at high meal frequency, nutrients are utilized immediately, whereas at low meal frequency the consumed nutrients are stored first (lipogenesis) and released later (lipolysis). Summarizing, taking fewer, larger meals (gorging) increases the body fat content in both animals and humans in comparison to consuming the same energetic amount in many small meals (nibbling) (Fábry, 1967; see also Chapter 21).

The environmental factors causing stress in the individual play a special role. They may result either in stress-induced feeding (see Chapter 11) or in anorexia nervosa and bulimia nervosa (see Chapter 13). It will be obvious that the causes for obesity are multifactorial.

Although obesity can be regarded as a risk factor for many serious diseases in humans, such as coronary heart disease and maturity onset (type 2) diabetes mellitus, the occurence of gall stones, arthritic problems and varicoses, it is not a disease by itself. Yet, many physiological abnormalities develop in obese subjects, such as hypertension and hyperinsulinemia due to insulin resistance and excessive insulin secretion. A rare exception is Cushing's syndrome, in which obesity is the secondary result of overproduction of cortisol. Thus, even though the majority of obese people may be healthy, they run a higher risk of incurring disorders by secondary changes in their physiology. On the other hand, the possibility to store energy for periods of food shortage, implies that obese bodies survive periods of famine more easily. This may be one of the selection pressures in genetic determination. In hibernating animals the increased fat content just prior to wintertime allows the animals to survive the long non-feeding period. Obesity can even be an advantage in providing the energy necessary during illness. Long-lasting obesity, however, may result in a higher risk of disorders such as atherosclerosis and ensuing coronary problems, and diabetes mellitus. Since in the present society the disadvantages of being obese are in the majority, treatment of obesity is important. This may vary from dieting to taking drugs and, only in very severely obese people, gastro-intestinal surgery or lipectomy.

1   CAUSES OF OBESITY IN HUMANS AND IN ANIMAL MODELS

1.1   GENETIC PREDISPOSITION

Obesity cannot be regarded as a disease except for severe cases. Well-filled adipose tissue can even be of survival value in overcoming periods of food shortage or famine. It may well be, that in some parts of the world this adaptation has developed more than in others. This is, for example, the case for obesity in Pima indians. Similar to the situation in hibernation (for instance marmots growing very fat in advance of wintertime), in ancient times humans seem to have filled their adipose tissue in times of plenty to use their stores in times of famine (Wendorf & Goldfine, 1991). Selection pressure may thus have lead to an adapted 'thrifty' genotype which is capable of surviving unfavourable external conditions like fluctuations in the supply of food in periods of drought or low ambient temperatures. Mammals possess an extraordinary capacity for storing triglycerides in adipose tissue. Although the range over which the body fat stores are filled under normal *ad lib* conditions may vary greatly between individuals, each individual or group of individuals has its own genetically determined set-point of adipose-tissue size which is defended, depending on age and external conditions (Van Itallie & Kissileff, 1990; see also Chapter 9). There is some evidence showing that body mass indexes (BMI), i.e. body weight in kilograms divided by the square of the height in meters (kg/m^2), of children correlate with those of their biological parents and not with those of their adoptive parents. Moreover, when both parents are obese, the children have a higher chance of becoming obese as well (Stunkard, 1990). These phenomena are an indication that human obesity is partly genetically based. However, the behavioral component, which is often very similar for parents and children, should also be taken into account, since behavior and genetic background may act in the same direction.

The mechanism of the set-point defense, i.e. the regulation of food intake, has already been discussed in Chapter 9. Genetically determined disturbances in this regulation may lead to an imbalance between intake and subsequent storage and expenditure. Obesity research aims at finding out about the nature of the genetic background of obesity and the site of action on behavior and physiology. Animal models of genetic obesity may help to find the origin of such defects although it is very difficult to distinguish between the primary and secondary effects. In one such model, the feeding behavior of the obese (*fa*/*fa*) Zucker rat has been compared with that of a non-obese (*Fa*/–) litter mate (Prins *et al.*, 1986). Since one of the first effects was increased food intake, the following study was undertaken to investigate the difference in feeding strategy of obese (body weight 570 g) and non-obese lean (370 g) rats. The feeding patterns on normal laboratory food were recorded as described in Chapter 11.

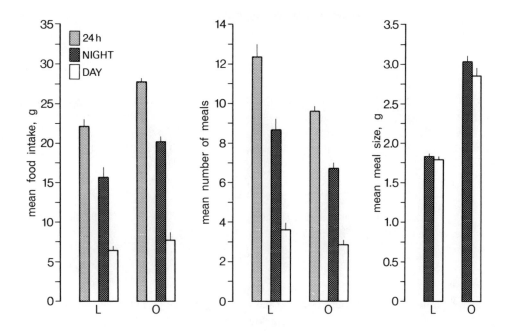

FIGURE 12.1

Mean food intake (g), number of meals and meal size (g) per 24 hr and during the light and dark period

Mean ± *SEM* in lean (•) and obese (o) Zucker rats.

(From: Prins *et al.*, 1986)

Figure 12.1 shows an increase in food intake by the obese rats as compared to that in the lean rats. This difference is not caused by an increased number of meals but by an increase in meal size during both light and dark phase. Since the circadian characteristics of the feeding pattern are only slightly different for lean and obese rats, the circadian rhythm probably does not play an important role as causal factor (Figure 12.2).

Therefore, the deviating factor in food-intake regulation may not be found in the *initiation* of meals but in their *termination*. In fact the strategy is quite similar to that in lactating and diabetic rats or in rats kept on low ambient temperatures, although there may be different mechanisms underlying the control (see Chapter 10). There are indications of a decreased feedback control in the short term (CCK, body temperature) or in the long term (insulin) in obese rats. All these deviating aspects for the regulation of food intake could play a role in the development of obesity. Among them, the disturbances in feedback regulation on the hypothalamus might prove to be very important. These will be discussed in the next section.

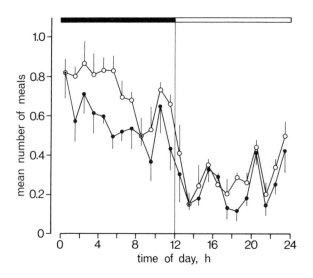

FIGURE 12.2

**Mean number of meals per hour over the whole light-dark cycle for lean and obese
Zucker rats**

Mean ± *SEM*) • lean o obese.
(From: Prins *et al.*, 1986)

1.2   HYPOTHALAMIC DEFECTS

As reported in Section 9.1, hypothalamic defects in the form of tumours may lead to
hyperphagia and obesity. Smaller defects, i.e. a decrease or increase in the number of cells
or receptors and in neurotransmission, are difficult to distinguish in humans. In animals,
research on neurotransmission is easier to perform. Therefore, during the last decade such
research has been carried out on the obese Zucker rat and on other obese animal models.
Manipulation by changing neurotransmission is one of the approaches in this field (see
also Section 11.3). Such manipulations lead to temporary changes in body fat content by
overeating. A more permanent induction of obesity is seen after lesioning of the VMH
electrolytically or chemically by goldthioglucose (GTG). In Chapters 9 and 10 the role of
the hypothalamus has been clearly demonstrated using this method. In this context the
role of the paraventricular nucleus (PVN) deserves still more attention. As pointed out in
Chapter 11, the PVN plays a more central role in the control of feeding behavior. Similar
to lesioning of the VMH, lesioning of the PVN led to excessive food consumption, eventually resulting in obesity (Woods *et al.*, 1986). Cutting the nerves around the PVN caused a
similar effect, suggesting that the nervous signals are relayed in the PVN. The precise
mechanism has not been elucidated so far, but it has been reported that noradrenalin
could play a causal role, since obese Zucker rats have lower noradrenalin levels in their
PVN. Since more or less effects on food intake can be observed after manipulations in the
PVN and VMH, it is possible that they are linked in the same chain of neuronal circuits.
Future research is necessary to unravel the precise wiring and function of all the neuronal
elements involved in the regulation of food intake and body weight. The genetic and hypothalamic factors discussed here are components of the internal substrate on which the

regulation of body weight rests. Diet can be seen as an external factor in the causation of obesity and is the subject of the following section.

1.3   DIET-INDUCED OBESITY

The environmental factor 'diet' plays an important role in the increase of adipose-tissue mass. In particular the amount of fat consumption seems to be an important determinant. For instance, the Japanese life style of eating low-fat diets leads to a lower body mass index than when the Japanese follow the American eating habits. Epidemiological studies in the USA, in which the feeding habits are followed over the last century, show a general trend of a decreasing carbohydrate intake (from 60 per cent of total energy intake in 1920 to 45 per cent in 1980) which is compensated for by an increased fat consumption (from 30 per cent to 45 per cent of total energy intake). The protein intake remained relatively constant. There was also a decrease in crude fiber intake from 7 to 4 gram per day per individual. During the 20th century the body mass index in the USA is found to be positively correlated with the increased fat consumption but not with the total energy intake (Van Itallie & Kissileff, 1990). However, the causality of this relationship has not been proven yet.

One of the questions that arises, is why the intake of fat is more fattening than that of an equicaloric amount of carbohydrate. One reason could be that excessive fat intake causes a more efficient storage in adipose tissue than excessive carbohydrate intake. This aspect will be further explored in Chapters 20 and 21. The difference may, for instance, lie in the increased heat production of the conversion and subsequent storage of carbohydrate. The taste of food may also be an important determinant in the development of obesity. Obese people have stronger preferences for fat and sweet foods than lean people (Drewnosky, 1990). It has been reported that humans who had been given a high-fat diet for several months, decreased their energy intake in response to a low-fat diet (Van Itallie & Kissileff, 1990). Fat diets may be more tasty than non-fat diets. Also, the satiety signals arising from the ingestion of fat may be different from those induced by carbohydrate. The latter has the highest satiety value. Thus there is an indication that extra fat does not contribute as much to short-term satiety as proteins and carbohydrates (Blundell & Burley, 1990). In this respect the endogenous circadian control of food selection of fats and carbohydrates (Chapter 11), in combination with taste control and insulin secretion (Chapter 8), may play an important role in storing the right food at the right time. Despite the negative results in the Zucker rats on normal standard lab food the disturbances in the timing of macronutrient intake and storage may still be determinants in the causation of obesity.

Among the other external determinants related to diet, food advertizing, the increasing number of supermarkets, fast-food restaurants and vending machines should be mentioned. These determinants may contribute to the wide variety of diet choice and the increase in fat consumption which are certainly the two major environmental factors in the causation of obesity.

1.4   ROLE OF ENERGY EXPENDITURE AND THERMOGENESIS

One of the explanations for obesity is the expenditure of energy. Obesity may develop as a consequence of low energy expenditure as compared to energy intake. Therefore, in obesity research much attention is paid to the former.

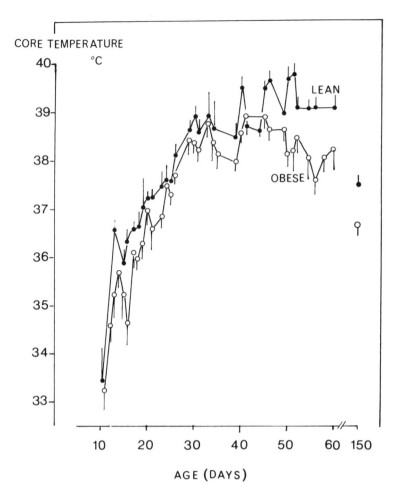

FIGURE 12.3
Core temperature of young and adult lean and obese Zucker rats
(From: Strubbe, 1988)

Expenditure is made up of three major components (see Chapter 16). Firstly, the *basal metabolic rate* (BMR) that requires about 60-70 per cent; secondly, the *thermic effect of food* (TEF) which accounts for 10 per cent; and thirdly, *physical activity* that accounts for the remaining 20-30 per cent in humans. The latter may vary strongly depending on the rate of voluntary activity (Van Itallie & Kissileff, 1990). In obese subjects the physical activity could be lower, due to aversiveness to moving the heavier body. On the other hand, obese people need more energy for their physical activity, so that the total energy spent on physical activity is mostly equal to that of non-obese people. Interpretation and comparison of this kind of study is therefore extremely difficult. Humans who have previously been obese, seem to have lower caloric needs once they have reached their normal weight than humans who have never been overweight at all (Leibel & Hirsch, 1984). This may be due to a relatively low BMR in the former group. Also the thermogenic effect of food in obese humans is often reduced. It has been reported that in animal models for obesity, such as the obese *ob/ob* mouse, the rate of thermogenesis is decreased (Trayhurn & James,

1983). Also in the obese Zucker rat, a decreased thermogenesis has been measured.

Figure 12.3 shows that in adult Zucker rats, but also in pups, the core temperature in animals developing obesity is lower than that in the non-obese litter mates. This lower body temperature already became manifest on day 34, when the first differences in body weight could be measured. The first visual signs of obesity became apparent from day 40 onward. According to the thermostatic theory discussed in Chapter 9, obese animals may eat more since they reach their thermostatic set-point later. This could be one of the explanations for the development of hyperphagia and obesity in rodents. As discussed in Chapter 7, the thermogenesis by brown adipose tissue (BAT) plays an important role in mammals, and in particular in rodents. A decrease in sympathetic activity will lead to decreased thermogenesis by BAT. In some obese animals there is indeed some evidence for decreased sympathetic activity (Trayhurn & James, 1983). However, the role of BAT in thermogenesis in adult humans is probably relatively small. As mentioned above, also in some human families showing obesity a decreased metabolic rate is an indication for a decrease in thermogenesis. The decreased metabolism is probably the cause of the weight increase (Leibel & Hirsch, 1984), although it is probably not the only explanation for the occurrence of obesity (Van Itallie & Kissileff, 1990). So far, it is unclear how the 'thrifty' processes cause a decreased metabolic rate. Presumably, less futile metabolic cycles such as lipolysis and subsequent reesterification, and a decreased thermogenesis are involved (Leibel & Hirsch, 1984). Part of the thermogenic response is related to insulin insensitivity, which is a phenomenon seen in obese subjects as well as in diabetic and overweight diabetic animals and humans. The consequences of obesity for maturity onset diabetes mellitus and the risk for other diseases will be reviewed in the next section.

2   OBESITY AND ITS COMPLICATIONS

2.1   OVERWEIGHT TYPE 2 DIABETES MELLITUS

One of the risks the obese who have the predisposition may incur is developing diabetes mellitus. This is the case for type 2 diabetes mellitus or non-insulin dependent diabetes mellitus (NIDDM) which constitutes 90 per cent of all diabetics. Ten per cent of diabetics suffer from juvenile insulin-dependent type 1 diabetes mellitus. Eighty per cent of the type 2 diabetics are overweight and the remaining 20 per cent are defined as a non-obese separate group, type 2a.

Type 2 diabetes mellitus is characterized by:

1   Higher blood glucose values during fasting and glucose intolerance after a standard oral glucose load.

2   Insulin levels may be normal, lower or enhanced, depending on the stage of the disease. The disease develops from the age of 35 years onward and is therefore also called maturity onset diabetes.

3   Insulin secretion is abnormal.

4   Insulin resistance accompanied by obesity in 80 per cent of the cases.

5   Increased serum LDL-cholesterol and triglycerides levels and a decrease in HDL-cholesterol.

In a later stage the disease is accompanied by all known secondary complications that are also seen in type 1 diabetes such as retinopathy, neuropathy etc. (See Chapter 8; also Porte, 1991).

The prevalence of diabetes mellitus differs between the various human races. There is an extremely high incidence of NIDDM in the Paleo-indians in North America. One sub-group of these, the Pima indians of Arizona, have an incidence of more than 50 per cent. It is suggested that in this group thrifty genes have developed (Saad *et al.*, 1990; Wendorf & Goldfine, 1991). Some 11,000 years ago the ancestors of these indians migrated from North Asia to North America and during this period they were exposed to a greatly fluctuating supply of food. A thrifty genotype is supposed to have developed then by natural selection. This genotype prevented the development of hypoglycaemia in times of fasting, whereas in times of food abundance, excess food intake was stored in adipose tissue and liver.

Today, the times of plenty are continuously with us and so are the accompanying characteristics that are disadvantageous to the thrifty genotype.This may lead to the development of obesity, insulin resistance, hyperglycaemia and dysfunctioning of the β-cells.

In the early stages of diabetes an increased insulin level is found, followed by a lowered level accompanied by increasing hyperglycaemia. Since there does not appear to be a decrease in the number of β-cells, the decreased insulin level must be the result of decreased insulin secretion (Porte, 1991). Possible explanations are decreased glucose transport across the membrane, or a defect in a signal-transducing enzyme (phospholipase *c*). In addition, a mitochondrial defect and a toxic effect of glucose have been mentioned. All these possibilities boast some evidence. One other explanation is the development of so-called amyline plaques, which are found in 70-90 per cent of the IDDM patients. These plaques are probably formed by high concentrations of the hormone amyline in insulin-dependent diabetes mellitus (IDDM) which is co-localized with insulin in the same secretory granules. The plaques may delay the transport of stimulating factors to the β-cells and of insulin. That may be true for glucose, since this caused a delayed secretory response, but arginine stimulation showed no difference with controls (Porte, 1991). There are still other possible explanations for the decreased insulin response. It is, for example, suggested that amyline plays a role in peripheral insulin resistance. However, amyline can only do so in very high concentrations. At the moment of writing there is no evidence for a physiological role.

From the various possible explanations it will be clear that the expression of the disease is multifactorial and that the contribution of the thrifty genes is difficult to establish. One thing is clear and that is that reducing body weight by food restriction is one of the best therapies against further development of the disease which reduces insulin resistance. However, the question remains as to what causes the mild hyperphagia in overweight diabetics. Is it possible that the decreased thermogenesis, caused by insulin insensitivity, plays a role?

Treatment of the disease is by oral antidiabetic drugs and, at a later more severe stage, with insulin.

From this section it will be clear that overweight diabetes, with its glucose intolerance and abnormal lipid metabolism, forms a certain health risk. Also, hypertension by increased peripheral resistance often develops. This is one of the main risk factors of cardiovascular disease and strokes and will be discussed in the following section.

2.2   HYPERTENSION AND CARDIOVASCULAR COMPLICATIONS IN RELATION TO BODY-FAT DISTRIBUTION

Obesity is often closely associated with the occurrence of hypertension, at least in the majority of European obese humans. It has been suggested that insulin resistance forms a causal factor in this respect. However, in Pima indians such a relationship was not found to exist: systolic and diastolic blood pressure appeared to be independent of insulin levels in the blood.

The *body mass index (BMI)* is generally adopted as a good operational measure for the degree of obesity. Three different obese groups are recognized: normal weight *non-obese* humans (*BMI* 20-25); *mildly obese* people with an overweight of 20-40 per cent and a prevalence of 90 per cent (*BMI* 25-27); *moderately obese* with an overweight of 41-100 per cent and a prevalence of 9 per cent (*BMI* 27-35); and *severe obese* with an overweight of more than 100 per cent and a prevalence of 0.5 per cent (above *BMI* > 35). There are other classifications, for instance that in which overweight is defined as 25 < *BMI* < 30 and where severe obesity starts from *BMI* = 30 onwards. The prevalences may depend on the gender. There are some limitations to this classification, because it applies to the average population. Body composition should also be taken into account, as well as fat distribution. For example, in athletes the lean body mass is extremely high and instead of fat, the muscle mass contributes relatively more to the total body weight.

Of course *BMI* does not include the distribution of fat. In the last decade the latter has become more and more important. The following example may illustrate this. The *BMI* is found to be equal for two groups of people: people with hypertension and people with normal blood pressure. Careful examination of the body fat distribution showed more abdominal fat in the individuals of the hypertensive group than in those of the normotensive group in whom the fat was distributed subcutaneously. Thus, abdominal fat appears to be more correlated with hypertension than subcutaneous fat, although the precise causal relationship is not known so far.

Differences between sexes show more fat deposition in thorax and abdomen in males, whereas in women relatively more fat is located around the hips. In order to characterize the various fat distributions, another measure, the *Waist to hip girth ratio (WHR)* has been developed. So, a complete image emerges from data on *BMI*, body composition, i.e. percentage of fat, and *WHR*. In males the *WHR* is positively correlated with hyperinsulinemia, atheroscleroses and cardiovascular complications and the occurrence of strokes. In female-type obesity however, these disorders are less pronounced. Apart from these risks and diseases, the following abnormalities can be mentioned. In women, an increase in adrenal androgens is found to lead to an increased plasma testosterone level. This, in turn, may cause disturbances in the female reproductive physiology resulting in amenorrhea (delay of menstruation) and hair growth on body, arms and legs (hirsutism). Although in males the testosterone levels remain normal, impotence and loss of libido are often associated with obesity. Moreover, the fat content around viscera and waist may cause irregularities in breathing during sleep so that sleep disturbances occur.

The various disorders and risks of diseases form the motive for treatment of obesity in Western society and these are discussed below.

3   TREATMENT OF OBESITY

Moderately but more so severely obese persons deserve treatment. There are several possibilities ranging from dieting to drug treatment. In severe cases gastro-intestinal surgery and lipectomy may be performed.

3.1   DIETARY TREATMENT

Energy restriction is one of the main approaches in the treatment of obesity although it is also one of the most complicated ones (Van Itallie & Kissileff, 1990). Another option is to replace specific nutrients in the diet, such as fats. These days, the trend is to compose low-energy diets in which the fat content is reduced and (partly) replaced by carbohydrates. The reason is the earlier mentioned higher caloric cost of carbohydrate conversion and the low satiating effect of fats. The use of more complex carbohydrates with a lower absorption rate has also become very popular. Also, the covert replacement of fat by artificial, non-absorbable fat, or by protein, as it is used in commercially available 'light' products, reduces the energy intake.

The amount of prescribed energy is largely dependent on the age and general activity of the patient (Van Itallie & Kissileff, 1990). The main problem with the therapy of energy restriction is the continuation of dieting after the patient has lost weight. Patients often increase their intake once the therapy is discontinued, to return to their original body weight (weight cycling). For both clinician and patient this kind of therapy is often very disappointing and stressful.

Changing the life style often has more effect. Such *behavioral therapy* includes a set of behaviorial adaptations such as self-monitoring of eating habits, reducing the fat content of the diet, attending a specified weight-reduction class including nutritional education, following an exercise program, goal setting and reducing stress. The success will also depend on social support and whether the program is followed in closed groups. This approach is sometimes accompanied by drug treatment.

3.2   DRUG THERAPY

When dietary treatment is not successful, drug therapy is another possibility for the treatment of obesity. Most drugs interfere with aminergic and serotonergic neurotransmission. There are two groups of possible drugs, thermogenic and serotonergic drugs. Examples of the former group are β_3-agonists, cholecystokinin agonists, lipase inhibitors and antisteroids, and amphetamines. Not all these drugs are used for the treatment of human obesity. Some of them are still under investigation and others are not allowed to be used by law in several countries. *Serotonergic* drugs are the most effective, with fenfluramine as a well-known example. Fenfluramine is regarded as a serotonin releaser and re-uptake inhibitor. The result is an increase in serotonin neurotransmission and this interferes with meal size. Moreover, it acts centrally on the feeding areas and also on peripheral regions such as serotonergic transmission in the gastro-intestinal tract. This may be responsible for the differing rates of gastric emptying and intestinal transport. Side effects of the interference with serotonergic neurotransmission by fenfluramine are intermeal carbohydrate craving and depression. Together with behavioral therapy and dieting, it may be a helpful drug as support (Guy-Grand, 1990).

The *thermogenic* drugs belonging to the family of β_3-agonists such as ephedrine-caffeine are supposed to act centrally and on brown adipose tissue (Astrup *et al.*, 1992). They restrict food intake by their direct anorectic or thermogenic effect.

The drug therapies mentioned all have some disadvantages since they often have side effects. Another disadvantage is that after stopping the drug treatment, the body weight will soon return to its original value if the therapy is not accompanied by behavioral therapy.

3.3   SURGICAL TREATMENT

For severely obese patients gastro-intestinal surgery is sometimes used as treatment (Sugerman, 1990). The different possibilities vary from suturing the cheeks together to gastric bypasses, gastric banding and inserting gastric balloons. All treatments are based on the physical inhibition of the intake or absorption of food. Although from a technical point of view the surgery poses no problems and the required weight loss is reached, there are some complications and side-effects in bypass surgery and reduction of stomach size. This approach requires a change in the behavior and life style of the patients, because 'normal' food intake has become impossible. After the surgical treatment, diet adjustment is still necessary.

4   SUMMARY

Obesity occurs when energy intake surpasses energy expenditure so that excess of energy is stored in the reserve tissues, of which the adipose tissue is the main component.

Sometimes obesity can be of survival value to overcome periods of famine; under the force of selection pressures an adapted 'thrifty' genotype may be the one which is able to survive unfavourable external conditions like fluctuations in the supply of food.

There is evidence for a genetic background to human obesity, e.g., when both parents are obese, the children have also a higher chance to become obese; however, the behavioral component should also be taken into account. Hypothalamic defects may lead to hyperphagia and obesity; lesioning of the PVN or VMH leads to excessive food consumption resulting in obesity.

Food intake, in particular the amount of fat consumption, seems to be an important determinant in the development of obesity; it may be that excess fat intake is stored more efficiently than excess carbohydrated in adipose tissue.

The taste of food may important; obese people have stronger preferences for fat and sweet food than lean people have.

Normal-weight people that have been obese before dieting have lower caloric demands than people that have never been overweight. One of the reasons for this may be a decreased thermogenesis; a decreased sympathetic activity will lead to a decreased thermogenesis by brown adipose tissue (BAT); however, the role of BAT in thermogenesis in adult humans is probably small.

Mild obesity cannot be regarded as a disease but increases the risk getting diseases such as non-insulin dependent diabetes mellitus and hypertension.

Body mass index (BMI) is a good operational measure for the degree of obesity; one classification for obese people is: BMI 20-25, normal-weight; BMI 25-30, overweight; BMI 30-40, obese; BMI >40, morbidly obese. Body composition - measuring the amount of body fat - and fat distribution - measuring the waist to hip ratio - provide important additional information.

There are several possible treatments for obesity, ranging from dieting to drug treatment and in severe cases gastro-intestinal surgery and lipectomy. Caloric restriction is one of the main approaches in the treatment of obesity though it is one of the most difficult ones; changing the whole life style often has much more effect.

A few drugs are used in the treatment of obesity, most of them interfere with aminergic and serotonergic neurotransmission.

13

Anorexia and bulimia nervosa

As early as 1689 Richard Morton (1637-1698) reported on a young woman showing decreased feeding activity, accompanied by decreased thermogenesis and faintness. There were no fever or other appearances of this so far unknown disease. In his interpretation Morton suggested that the disease could be of nervous origin.

The concept of anorexia as a clinical phenomenon was adopted in the second half of the 19th century by William Gull in England and Lasèque in France. The latter author used the name 'l'anorexie hysterique' in his 1873 publication. Eight of his female patients between 18 and 32 years had a chronically decreased appetite and believed that feeding could harm them. Lasèque, and also Gull, thought that the main causes of the disease were of a psychological nature, and that the family of the patient played an important role during the course of the disease (Beaumont *et al.*, 1987). Since 1950 anorexia has been recognized as a psychopathological disorder with specific clinical aspects (Beaumont *et al.*, 1987; Yager, 1988). The most conspicuous of these is abnormal feeding behavior leading to chronical aphagia which in the untreated patient may lead to an extremely low body weight and finally to death.

Since the number of patients is on the increase, interest in the causes and treatment of the disorder is growing. Depending on country and lifestyle the incidences are estimated between 1 and 10 per cent in 16-18 year old women (Beaumont *et al.*, 1987). Other percentages may be mentioned, depending on the definition used.

Another psychopathological disorder, also caused by abnormal feeding behavior, is *bulimia nervosa*. This is generally seen in close relationship to anorexia nervosa, although the various expressions of the two make them seem to be separate disorders, each with its own symptoms and causes. Bulimia is characterized by periods of overeating in large binges, followed by vomiting. The general attention is more focused on anorexia than on bulimia since the latter is less visible in the short-term. Bulimic patients may have quite a normal body weight. The incidence of bulimia is much higher than that of anorexia. The estimates range from 1 to 19 per cent in young adults. There are reports of 6 out of 10 female American students suffering from the disorder.

During the last few decades much psychological and physiological research has been performed on both conditions. However, although our knowledge of them is growing, many questions still remain to be solved.

The following sections will discuss some psychological and physiological aspects of the disorders in greater detail. These include the general aspects and criteria of anorexia and of bulimia, their possible causes and possible therapy.

As mentioned above, major psychological issues are involved in anorexia and bulimia which are more often mentioned in the literature than their physiological aspects. This chapter emphasizes the physiological aspects, for psychological and social aspects see Chapter 5.

1   CRITERIA FOR ANOREXIA NERVOSA

The term anorexia nervosa means 'lack of appetite due to nervous factors'. The disorder is characterized by a decrease in food intake and, as a consequence, decrease in body weight. It occurs mainly between 14 and 17 years of age, predominantly in females, only one out of ten patients is male (Beaumont *et al.*, 1987; Crisp, 1984). Patients are continuously concerned about their body weight and consider their body weight subjectively as a weight problem. The causes of the disorder are multifactorial and unclear so far. It is suggested that genetic factors or hypothalamic dysfunctioning are involved. Moreover, it is generally accepted that psychological factors play a major role in the onset of the illness. Since individuals may vary a great deal regarding the different symptoms of anorexia, international criteria have been postulated in which the main symptoms of the disorder are described. These are the so-called 'Diagnostic and Statistical Manual of Mental Disorders' (DSM-III-R) criteria (Yager, 1988). In brief, the DSM-III-R criteria for anorexia nervosa are:

1 Refusal to keep body weight above the strict minimum appropriate for age and height.
2 Fear of gaining weight, even in the underweight patient.
3 Disturbance in the way the patient feels and estimates her/his own body weight and size: the patient experiences her/his weight as overweight whereas there is a clear underweight.
4 In women, amenorrhea (lack of menstruation) occurs during at least three consecutive periods.

The term 'anorexia' is not exactly correct since it remains a question whether there is, in fact, a real lack of appetite. The patient may experience a feeling of hunger but seems to suppress feeding (Beaumont *et al.*, 1987; Yager, 1988). The supposed lack of appetite is mainly perceived by the people around the patient. Patients sometimes lose weight to dramatically low levels of 60 per cent of their normal weight. They continuously feel cold and sometimes faint. There is amenorrhea and the patients think they behave normally.

2   CRITERIA FOR BULIMIA NERVOSA

Bulimia nervosa literally means 'hunger like an ox by nervous factors' (bous = ox, limos= hunger) The disorder is characterized by periods of binge eating (at least twice a week on average). Afterwards the patients forcefully rid themselves of the food by vomiting or by taking laxatives. Similar to anorexics, bulimics show obsessive behavior regarding their control of food intake and body weight. In about 50 per cent of the cases bulimia is preceeded by anorexia. Although there is some connection between the two disorders, they may be two different phenomena caused by similar pathologies, each with its own characteristics (Yager, 1988).

The DSM-III-R criteria for bulimia can be summarized as follows:

1 Repeated binge eating, often with episodes of restrained eating.
2 Feeling of loss of control over feeding behavior during a binge.
3 Self-induced vomiting, use of laxatives and diuretics, periods of fasting and specific

diets, performing extra physical exercise in order to reduce body weight.

4 At least two binges per week during three months.

5 Continuously worried about body weight and appearance.

Sometimes binges occur even five times per day. The 'meals' consist of highly palatable food with a high energy content. There are patients who eat 20-30 times the normal daily amount of calories in one single binge. Although there are overweight and underweight bulimics, the majority do not have an abnormal body weight. Therefore, the disorder is not as noticeable as anorexia. In contrast to anorexic patients, bulimics know their behavior is abnormal and often try to get professional help. Whereas anorexia is almost exclusively a female's disease, bulimia has relatively more male patients although here too, the majority are female. The male/female ratio is 1 to 4. The average age of bulimics is higher than that of anorexics. Bulimia occurs mostly between the ages of 20 and 30, although the onset can also be later.

The abuse of laxatives by bulimics (sometimes up to 40 pills per day) to rid the intestinal tract of the extra calories leads to a disturbed electrolyte balance. The patient loses potassium, sodium chloride and hydrogen ions. The low potassium levels cause disturbed heart functioning and epilepsy (Yager, 1988). Vomiting also interferes with the electrolyte balance. In addition, it causes erosion of the dental glaze by the passing gastric acid. Moreover, due to the continuous attempts to vomit tooth prints may cause smalls lesions on the fingers.

3   CAUSES OF ANOREXIA AND BULIMIA

The causes of anorexia and bulimia are assumed to be multifactorial. Since bulimia is often preceded by anorexia, this section discusses the factors causing the onset of both disorders together, bearing the previous chapters in mind. Not only psychological factors, but also physiological factors are involved. Moreover, the time factor in the development of the disease is of importance. For instance, the disorder may be initiated by a set of factors. These may then cause a change in behavior, so that the patient enters a new physiological state in which other factors may come into play. All these factors may act upon a substrate of environmental and genetic factors which together may constitute a certain predisposition to the condition.

3.1   GENETIC AND ENVIRONMENTAL FACTORS

There are a few reports indicating that there actually is a genetic predisposition to the development of anorexia nervosa, for example, reports of identical twins showing the same anorexic behavior. Moreover, it seems that in some families the prevalence is higher than in others (Crisp, 1984; Fishman, 1976). If genetic factors do play a role they may also be involved in changes in neurotransmitter activity, thereby eventually lowering the threshold for the initiation of the disorder (see below). However, it is also possible that in those cases environmental factors determine the predisposition to the disorder. In general, the predisposing factors for the development of a feeding disorder are: young and female, Caucasian, socio-economical status middle or higher class, perfectionist, intelligent, living in an environment with strong interactions with other people (parents, males), and moving in circles where being slim is thought to be important. Patients are also found in professions such as ballet dancer and fashion show model, where being slim is necessary. However, in the latter cases the motive for being slim is different from that of sufferers from the main type of anorexia nervosa and therefore the onset is also different. The environmental factors may to a large extent determine the psychological factors which in the literature are often regarded as the main causal factors for the feeding disorders. It is suggested that many of the other deviations, including disturbances in physiology, are secondary (Beaumont *et al.*, 1987; Crisp, 1984; Hsu, 1990). We will now look at the latter factors more closely.

3.2   ENVIRONMENTAL AND PSYCHOLOGICAL FACTORS

Social and cultural factors

In Western society there are two general trends as regards body weight. Firstly, there is an increase in mean body weight caused by an increased food intake and a decrease in infective diseases such as the common infant diseases. Secondly, the ideal body weight for females as, for example, dictated by women's periodicals, is decreasing. In the non-Western societies a positive correlation exists between income and body weight , whereas in Western society this correlation is generally negative (Hsu, 1990).

Throughout history, in different cultures a trend can be observed from underfeeding, to an ample supply of good quality food, to overfeeding. The problem of unintented underweight caused by low food supply is replaced by the luxury problems related to the always available foods of good quality, and to the diminished energy expenditure. This may lead to problems of body weight and food intake.

Behavior during puberty

Many causes can be mentioned for the fact that feeding disturbances mostly start during adolescence. During puberty sexuality develops and anorexia is often seen as a tool to escape from this development. The reasons for this may vary. There may be a fear of sexuality including a fear of becoming mature, and of gaining an identity and independence. To avoid this, the patient tries to keep her pre-puberal body weight down, i.e. between 35 and 40 kg. By doing so, there is a continuous reduction in body fat, eventually resulting in amenorrhea. In this way the patient continues her physiological status and postpones the onset of puberty. This is known as a 'phobic avoidance process'. Two main topics keep the mind occupied. Firstly, the fear of becoming fat and secondly, the current status induced by restricted feeding (Crisp, 1984). The anorexic patient is restless, but eating must be avoided at all times. She may show some displacement behavior, such as preparing the meal for the family, as is often seen. This restless behavior is also expressed in physical exercise. Although in the obese often ineffective, in the anorexic this is effective and her body weight is sometimes even reduced to below 30 kg.

Lack of self-confidence may be another cause of anorexia, especially in the female teenager. This may be caused by many factors, but in most cases the parents play a major role. When a slight body overweight shows and is commented on at home or at school, this can be a reason to start slimming. The patient loses weight (mainly body fat), on which she is complimented by her environment. This gives her more confidence and more reason to diet. Finally, the dieting becomes obsessive, resulting in anorexia nervosa.

It appears that parents play an important role in the onset of the disease. It is not surprising, therefore, that parents of patients are found to have many characteristics in common. Firstly, they can be described as successful socially and economically, and worried about the impression they make on the outside world. Secondly, they often adhere to strict rules. They often have an authoritarian life style that prevents open discussion in the family and this may lead to conflicts. Thirdly, the parents tend to exaggerate the protection of their child or pay much attention to one of the other children in the family, which may be handicapped, a genius, or a child that has died at an early age. Any of these circumstances may make it difficult for a child to gain its own identity. It may develop a fear of adulthood and, as a result of such a family situation, anorexia and amenorrhea may set in. Although many of these behavioral interactions with parents may take place during puberty, they may already have their onset soon after birth. Evidence for this has emerged from reports on children who developed anorexia during puberty and who had been relatively quiet and obedient much earlier in childhood (Crisp, 1984; Hsu, 1990). Although many of the factors mentioned so far can be regarded as motives, those causing and triggering the onset of the disorder are difficult to distinguish. All the conditions discussed here struggling with identity and dogmas imposed by parents and society, starving, dieting or languishing have one thing in common: they can be considered a stress situation for the patient. The following section highlights the physiological factors, in particular those related to stress.

3.3   PHYSIOLOGICAL FACTORS

The behavioral changes in anorexics lead to many secondary abnormalities in their internal physiology. In fact, the whole physiology changes due to the decrease of incoming nutrients. Here we will only discuss the main events leading to possible causes and maintenance of the disorder.

As regards feeding behavior, anorexics try to avoid eating carbohydrates and fats which are easily stored in adipose tissue (see Chapter 8). They often pay much attention to proteins, from which they take what is necessary. Hence, the situation of anorexics is not similar to that of people in developing countries suffering from protein deficiency. The change in food choice may be the result of a change in hypothalamic transmission (see Chapter 12) or it may have been learned in order to avoid gaining weight.

The non-stop feeding by bulimics during a binge indicates a disturbed feedback regulation. Stretch in the alimentary tract, which contributes to satiety in the short term, may be ignored, but a decreased CCK release may also be involved (see also Chapter 9). A decrease in CCK release on food intake in bulimics has indeed been reported (Geracioti & Liddle, 1988). However, this may also have been caused by the strange feeding pattern of bulimics. On the other hand, lower CCK levels may facilitate binge eating.

Endocrine factors

As mentioned earlier, one of the criteria for an anorexic patient is amenorrhea which is caused by dysfunctioning of the gonadotrophin-releasing mechanism in the pituitary gland. This release is under the influence of releasing factors from the hypothalamus. Amenorrhea can already occur before the body weight drops, so it may be the result of the stressful condition at the onset of the disorder (Russel & Beumont, 1987). However, there is evidence that continuation of amenorrhea is related to the diminished body fat content. When anorexics are treated and body weight increases again, menstruation resumes at a certain level of body fat content. This means that the body fat content is sensed somewhere in the body, with the hypothalamus as a major candidate. One of the factors reporting on body fat content is insulin (see Chapter 10). Another factor which is related to the female gender is estrogen. In anorexics the estrogen blood level is lowered, probably due to the reduced gonadotrophin release from the pituitary gland. Under normal weight conditions the liver and also the adipose tissues are involved in the metabolism of estrogens. The main estrogen, estradiol, is converted to estrone and finally to estriol which leaves the body via the urine. In normal weight and obese females this results in large quantities of estriol in the urine, but in anorexics none is found. There is evidence that in malnourished females estrone is converted to another component, namely 2-hydroxyestrone, which is known as catecholestrogen (see Figure 13.1). The latter process takes place in the brain, mainly in the hypothalamus, and in the liver. It is possible that in anorexics this shift in estrogen metabolism is caused by the absence of inactive adipose tissue. Therefore the production of catecholestrogens in females may form the link between the adipose tissue and the hypothalamus (Fishman *et al.*, 1975; Fishman, 1976). The abnormal estrogen metabolism may also be one of the causes of lanugo (down) hair growth in anorexics. The possible consequences of the hormonal deviations for the regulation of food intake and the neurochemical transmission in the hypothalamus are discussed in the following.

Hypothalamic hormonal factors

Patients suffering from a feeding disorder have been found to have abnormalities in their hypothalamic functioning. Some of them, having anorexia, showed hypothalamic tumors in post-mortem investigation (Beaumont, 1987). However, tumoral causation of anorexia only occurs in a small subgroup, and probably has nothing to do with the general phenomenon of anorexia.

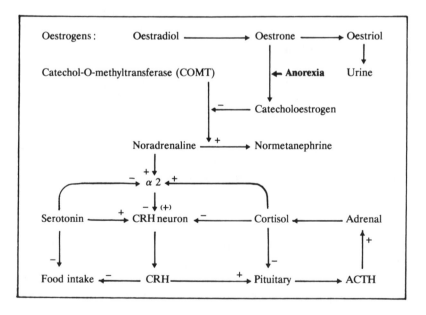

FIGURE 13.1

Schematic presentation of the mode of action of the main neurotransmitters and hormones involved in the initiation and maintenance of anorexia nervosa

As mentioned earlier, the gonadotrophin (LH and FSH) release which is induced by the hypothalamic gonadotrophin releasing factors (GRF), is almost abolished and insensitive to GRF action. This is probably secondary to the decrease in body weight. Also, the thyroid function is diminished which is expressed in lower levels of triiodothyronine (T3) and thyroxine (T4). However, levels of thyroid stimulating hormone (TSH) secreted by the pituitary gland were not different from normal. These results indicate that the primary disturbances are at the level of feedback regulation on the hypothalamus. In any case, the hypothyroidy can explain several features seen in anorexia such as bradycardia, hypothermia, dry skin, constipation, and decreased metabolic rate. Growth hormone levels are increased in 50 per cent of the patients.

The above-mentioned defects may be the result of decreased food intake and body-weight reduction. They also occur in bulimics after a fasting, i.e. weight-reduction period. Therefore, they may be related to a decreased blood-glucose level which is known to stimulate the release of growth hormone.

One very important hypothalamic factor which may play a role in the initiation of the first phase of the disorders is corticotrophin releasing hormone (CRH). Although there is a clear circadian rhythm in the basal CRH release, stress can cause strong stimulation of the releasing hormone. This stimulates the release of adrenocorticotropin releasing hormone (ACTH) by the pituitary gland. In its turn, ACTH stimulates the release of corticosteroids from the adrenal and these have a feedback effect on the hypothalamus. This feedback cycle is known as the hypothalamic-pituitary-adrenal axis (HPA) (see Figure 11.1). In anorexics the cortisol (glucocorticoid) levels are enhanced during the entire day-night cycle. When dexamethasone, a drug which acts on glucocorticoid receptors, is injected, normal control subjects show a reduction in their cortisol level. This so-called dexamethasone-suppression test (DST) proved to be ineffective in anorexic patients, i.e. dexamethasone did not reduce their cortisol levels. This indicates the existence of a defect in feedback regulation in anorexics. The dexamethasone-suppression test is also ineffective in patients suffering from depression and bulimia (Van Itallie & Kissileff, 1990). Although disturbances of the HPA-axis in anorexics may partly be caused by some degree of undernutrition, their combination with depression cannot be ruled out as a causal factor (Trayhurn & James, 1983). There are reports showing a positive correlation between the body weight of the bulimic and the response to the dexamethasone-suppression test. This may be an indication of the energy content of the body involved in determining the sensitivity of the HPA-axis (Sugerman, 1990).

Hypothalamic neurotransmission

Three major transmitters have been reported to be involved in anorexia and bulimia: noradrenalin, dopamine, and serotonin (5-HT). The general effects of these transmitter systems on the control of food intake have already been discussed in Chapter 12. It then emerged that deviations in serotonergic neurotransmission may be responsible for the changes in food intake and food selection as seen in anorexics. There is evidence that tryptophan (precursor of serotonin) and the large neutral amino acid (LNAA) levels are lowered in the brain of anorexics. Together with decreased 5-HT and 5-HIAA (metabolite of 5-HT) levels, this indicates a decrease in serotonergic neurotransmission leading to depression as well as binge eating and long periods of non-feeding. However, the causation of the decreased amino acid levels and the subsequent decrease in serotonin level is hard to interpret, since anorexics eat less and virtually no carbohydrates. As discussed in Chapter 12, after decreased carbohydrate intake the tryptophan/LNAA ratio decreases so that less serotonin precursor becomes available.

The noradrenergic activity also appears to be affected since lower noradrenalin-metabolite levels are measured. The decreased uptake of tyrosine due to decreased food intake may also be an explanation of the reduced noradrenergic transmission leading to decreased food intake (see Chapter 11). So far, there is no evidence of a dysfunctioning of the dopamine system.

Also in patients suffering from bulimia a decreased serotonergic transmission was observed. In particular the presynaptic release seems to be decreased. This decrease in serotonergic transmission may explain the occurrence of binge eating, depression and impulsive behavior often seen in bulimics. Moreover, endogenous opioids seem to maintain a binge since the opioid antagonist naloxone was capable of attenuating binge-eating behavior (Mitchell *et al.*, 1986).

Before summarizing these physiological phenomena, the treatment of the disorders will be discussed since successful treatment provides additional information on the nature and location of the defects.

4   TREATMENT

A very important treatment is psychotherapy, which focuses on changing both mind and behavior. If such treatment is not given, the patient will soon return to the original state of suffering from the disorder. In fact, psychotherapy will reduce some of the events, culminating in the final high level of stress. In addition to psychotherapy, it is possible to apply pharmacotherapy. The first goal for anorexics is, however, to regain some body fat, since the low body weight also leads to depression. Below a certain dangerously low body weight parenteral feeding has to be supplied. This should immediately be followed by psychotherapy, in order to re-establish the patient's self-perception. However, psychotherapy will only be effective if the proper body weight has been regained. Simultaneously, good feeding habits should be taught, although the initiative to learn has to lie with the patient. Psychotherapy and the altered attitude and behavior concerning food intake and body weight have to work together, and completely synchronized, for a long time.

Pharmacotherapy can be a help to regain weight and to reduce the fears for body-weight gain. In bulimics it can be used to reduce the pattern of binge eating. In general, treatment with pharmaceuticals is aimed at reducting depression. About 50 per cent of patients score high on a depression test.

Treatment with tricyclic antidepressants like imipramine, chlomipramine and amytryptiline results in a stimulation of serotonergic neurotransmission (Jimerson *et al.*, 1990; Judd *et al.*, 1987; Mitchell *et al.*, 1990). Such treatment not only results in an improvement of mood but also in a short-term increase in body weight. In the long term the effects are not so clear. Treatment with lithium carbonate also appears to be effective, but this substance has dangerous toxic side effects.

In summary, the following steps are taken in the treatment of anorexia:

1   Increasing body weight to a minimum level.
2   Teaching good eating habits.
3   Changing the attitude of body versus food.
4   Treatment of the physical consequences of the disorder.
5   Prevention of a recurrence of the disorder.

In bulimics similar goals can be formulated with greater emphasis on inducing a regular eating pattern. In the pharmacotherapy of bulimics the antidepressants score highly. Mood and binge eating are inversely correlated. Therefore, in bulimics also the body-weight reducing 5-HT releaser and reuptake inhibitor fenfluramine are effective.

5   CONCLUDING REMARKS

Although difficult to distinguish, there are primary and secondary factors involved in the initiation and secondary consequences of the disorders respectively. These are presented schematically in Figure 13.2. The secondary effects are mainly due to the abnormal feeding habits and subsequent deviations from the normal body weight. As shown, the background factors have genetic, environmental and in particular family components in a society where to be slim is favoured. Together, all these environmental factors accumulate to form a hypothetical stress 'ladder' in which each step is capable of surpassing the final threshold, thereby triggering the onset of the disorder. Psychotherapy is mainly focused on removing some of the steps in the ladder. One of the main factors related to stress is the release of CRH which reduces food intake and modifies, via the release of corticoids, the receptor events in the central nervous system.

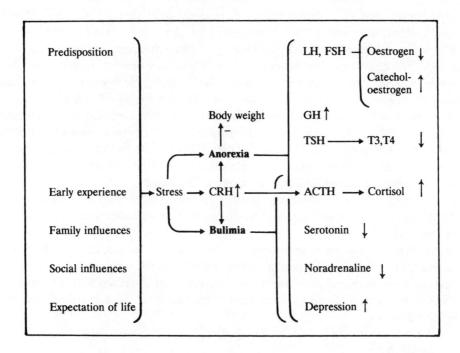

FIGURE 13.2
Schematic summary of the different events leading to anorexia and bulimia nervosa
The left-hand side of the figure represents the primary influences and towards the right-hand side more secondary effects and influences come into play.

Once anorexia sets in, the patient loses weight and menstruation stops. The latter is probably a more general adaptation to prevent conception in a time when the body energy reserves are reduced. Together with abnormal thyroid functioning, these symptoms form the secondary consequences of the disorder which cause many perceivable and externally visible effects. There are only a few factors which could possibly sustain the obsessive behavior in relation to the maintenance of a low body weight. One of them is the increased CRH level which results in increased corticoid levels. CRH release is strongly stimulated by many factors since it is impossible to suppress it in the dexamethasone-suppression test (DST). There are other factors inversely related to body-fat mass in females and these are the catecholestrogens. They competitively inhibit O-methylation of catecholamines by O-methyl transferase enzyme thereby extending the half-life of the catecholamines. In young women with anorexia nervosa estrone appeared to be converted to 2-hydroxyestrone at a much higher rate than in obese patients or normal-weight controls (Fishman *et al.*, 1975). Therefore, this relationship forms a direct link between brain catecholamines, estrogens and body fat. The precise mode of action in the control of food intake is not known, but it could be that estrone sustains the CRH release in addition to stress factors and catecholaminergic action (Figure 13.1). As soon as the body weight has returned to a certain level, the catecholestrogen levels drop, since estrone is then converted to oestriol, which leaves the body via faeces and urine. As a consequence of the increased body weight which probably induces neurochemical alterations of the just mentioned catecholamine and CRH neurotransmission, the patient is susceptible to psychotherapy and pharmacotherapy. Very important is the subsequent psychotherapy aimed at reducing the various steps on the stress ladder. It is also important to reduce some of the components of the stress ladder responsible for inducing depression which reduces serotonin transmission. Moreover, serotonin transmission is largely dependent on effective food intake in order to transfer tryptophan to 5-HT. In anorexics, and also in bulimics, most of the effective pharmacotherapeutic drugs act on 5-HT neurotransmission which is aimed at the relief of depression. The stimulating effect of 5-HT on CRH release, however, is not in accordance with the hypothesis that CRH is the main causal factor in the initiation and maintenance of the syndrome. The observed decreased noradrenalin output fits in with the concept in which CRH release is stimulated when α-adrenergic functioning is reduced. However, in contrast, the catecholestrogen hypothesis predicts an enhanced catecholamine transmission.

Little that we know about the causation of anorexia and bulimia, all pathways lead to serotonin, CRH, the catecholestrogens and the catecholamines acting as causal factors in the initiation and maintenance of the disease. Future research should focus on these factors and their interactions with the regulation of food intake in the causation of disorders in eating behavior.

6   SUMMARY

Anorexia nervosa means 'lack of appetite due to nervous factors'; it is characterized by a decrease in food intake and as a consequence a decrease in body weight; it occurs mainly in female ages 14-17 years.

Bulimia nervosa means 'hunger like an ox by nervous factors'; it is characterized by periods of binge eating; the patients get rid of the food by vomiting or by using laxatives; most patients have normal body weight; bulimia has relatively more male patients than anorexia although the majority are female; the average age is also higher than in anorexia (30-40 years).

The causes of anorexia and bulimia are probably multifactorial: genetic, environmental, psychological and physiological. Some studies show a certain genetic predisposition to the development of anorexia nervosa; social and cultural factors also seem to be involved; the onset of anorexia, mostly during adolescence, is often seen as a tool to escape from this development and the role of the parents is always mentioned as important in the onset of the disease.

The change in behavior of anorexics leads to many secondary deviations in the internal physiology; endocrine factors (estrogens, insulin), hypothalamic hormonal factors (hypothalamic pituitary adrenal axis) and hypothalamic neurotransmission (a decreased serotonergic and noradrenergic neurotransmission) are involved. In the treatment of eating disorders psychotherapy is important; however, in anorexia nervosa the first aim is to increase body weight to a minimum level.

14

Central nervous mechanisms during exercise

In order to survive, all cells of the body need a continuous supply of metabolic substrates in the form of glucose and free fatty acids (FFA). Since they need to be deaminated before they can be utilized, amino acids are not immediately involved. Deamination takes place in the liver only, after an excess of amino acids is ingested. Utilization of the deaminated amino acids parallels that of glucose and FFA.

Under resting conditions and on a normal diet (i.e. containing 55 carbohydrate energy%, 35 fat energy% and 10 protein energy%) the respiratory quotient (RQ), i.e. the ratio between CO_2 production and O_2 utilization is 0.85. This implies that both glucose and FFA contribute 50 per cent to the energy supply, as the RQ would have a value of 0.7 and 1.0 respectively if only FFA or glucose were utilized. For the resting rat it has been found that the blood-glucose and plasma-FFA concentrations amount to 100 mg and 5 mg per 100 ml respectively. This means that the individual amounts of glucose and FFA in the blood circulation of the animal are sufficient to cover its metabolic needs for about 2 min and 20 sec respectively. Therefore, the regulation of blood-glucose and plasma-FFA levels must be very accurate, especially if the increased needs during periods of activity are taken into account. This control is carried out by the central nervous system via the autonomic nervous system (Scheurink & Steffens, 1990; Steffens *et al.*, 1990).

1 REGULATION OF BLOOD-GLUCOSE AND PLASMA-FFA LEVELS IN THE RESTING RAT

Insulin is the major anabolic hormone. Its major effects are: (a) activation of glucose transporters in the cell membranes of many tissues, e.g. fat cells, resting muscle cells; (b) activation of glycogen synthetase in liver cells and muscle cells with simultaneous suppression of phosphorylase activity so that glucose is converted to glycogen; and (c) induction of fat synthesis in the liver and fat cells.

Neurons and active muscle cells do not need insulin for glucose transport across the cell membranes as their glucose transporters are insulin-independent. Also, the epithelial cells in the alimentary tract and in the nephron are insulin-independent. In these cells glucose is co-transported with sodium ions.

The major stimulus for the release of insulin from the β-cells in the islets of Langerhans is an increase in blood glucose above the basal value of 1 mg/ml. This increase is the consequence of the absorption of glucose in the alimentary tract after food intake. Besides glucose, amino acids like arginine, and intestinal hormones, for example cholecystokinine, also trigger the release of insulin. In addition, the autonomic nervous system exercises an important control on the release of insulin from the β-cells (see Figure 14.1). Stimulation of the parasympathetic nerves innervating the pancreas elicits the release of insulin via a muscarinic mechanism, whereas sympathetic activity suppresses insulin release.

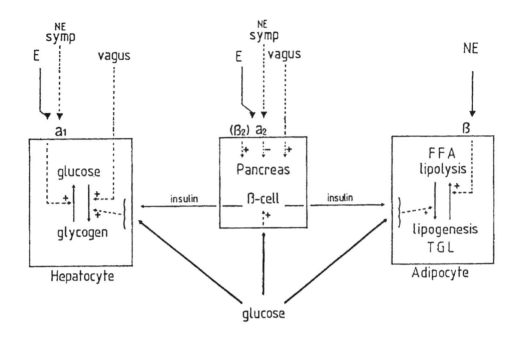

FIGURE 14.1

**Model showing the influences of the autonomic nervous system and circulating cate-
cholamines on the hepatocyte, adipocyte and β-cells of the islet of Langerhans**

The effects of glucose on β-cells and the combined effects of glucose and insulin on glycogenesis and
lipogenesis are also indicated.

(From: Steffens *et al.*, 1990)

This suppression occurs independently of the blood glucose level via an α_2-
adrenoreceptor-mediated mechanism. A β_2-adrenoreceptor-mediated mechanism, however, causes a slight stimulation of insulin release. Since only adrenalin released from the
adrenal medulla has an affinity to β_2-adrenoreceptors, only adrenalin is capable of stimulating insulin release. This reaction is important during strong activation of the sympathetic system. If there were only an α_2-adrenoreceptor effect, the insulin release might be
completely suppressed and this could interfere with the energy supply. To guarantee an
undisturbed supply of energy, a basal glucose transport across the cell membranes is necessary. This transport can only be realized if a basal amount of insulin is present.

In the resting condition, the balance between the sympathetic and parasympathetic
system modulates the amount of insulin released upon a glucose stimulus (Steffens *et al.*,
1990). Elimination of the α_2-adrenoreceptors on the β-cells by an α_2-adrenoreceptor antagonist leads to an increased insulin release after a certain glucose load. Elimination of
the muscarinic receptors, however, results in a diminished insulin release after the glucose
load (Kaneto *et al.*, 1974). A shift in the sympathetic/parasympathetic balance towards the
sympathetic system occurs during exercise, when increased metabolic needs have to be
covered. Firstly, the production of glucose and FFA must be increased. Secondly, the released substrates must be utilized by the cells that need them, i.e. active muscle cells and
neuronal tissue (Galbo, 1983).

Obviously the activity of the autonomic nervous system, also in its regulatory role in metabolism, is controlled by the central nervous system.

2   ACTIVATION OF THE SYMPATHETIC NERVOUS SYSTEM DURING EXERCISE

To find out about the role and activation of the sympathetic nervous system during exercise in relation to the expected increase in blood glucose and FFA utilization, experiments have been carried out on rats at a relatively high level of physical activity. The rats were subjected to swimming in a pool (length 3 m, width 40 cm, depth 90 cm) filled with tepid water (33 °C) against a current of 13 m/min (Figure 14.2).

The procedure was as follows. The rats were transferred from their home cage to the starting platform in the pool, where they were kept for 20 min. Then the platform was lowered to the bottom of the pool forcing the animals to swim against the current. After 15 min of swimming a removable resting platform was offered to the rat at the upstream end of the pool. The rats immediately climbed on it once it was presented (Scheurink *et al.*, 1988). Before the start of the actual experiment the rats were made well-accustomed to the experimental procedure. Plasma-FFA and blood-glucose levels were found to rise slightly and plasma insulin to decrease slightly while the rats were waiting on the platform (Figure 14.3). During swimming both FFA and glucose levels increased considerably while the plasma-insulin level decreased. After the swimming had stopped, the FFA and glucose concentrations declined and the insulin level returned to its basal value. The increase in plasma-FFA and blood-glucose levels during swimming are an indication of the production of FFA and glucose exceeding their utilization.

Throughout the experiment the activity of the sympathetic nervous system was determined by measuring the plasma noradrenalin (NA) and adrenalin (A) concentrations. The basal levels in the home cage were low: 40 pg/ml and 180 pg/ml plasma for A and NA respectively (Figure 14.4).

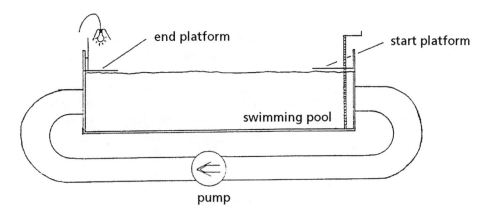

FIGURE 14.2
Model of the swimming pool with counter-current

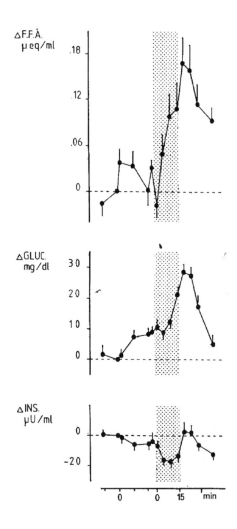

FIGURE 14.3

Plasma-FFA, blood-glucose and plasma-insulin concentrations before, during, and after exercise (●—●)

The values for FFA, glucose, and insulin are expressed as mean changes ±*SE* from the basal values. The basal value was measured at $t = -1$ min in the home cage, immediately before the rat was transferred to the starting platform in the swimming pool. The shaded area represents the swimming period.

(From: Scheurink *et al.*, 1989b)

The transfer of the rats to the waiting platform caused a brief increase in plasma-A and -NA levels, which returned to near-basal levels during waiting. Lowering the platform resulted in a steep increase in plasma-A concentration at the moment of immersion of the paws, whereas the plasma-NA concentrations did not change at all. The plasma NA levels started to rise at the moment of swimming and continued to do so during the entire swimming period. The plasma-A levels declined during swimming (Scheurink *et al.*, 1989a).

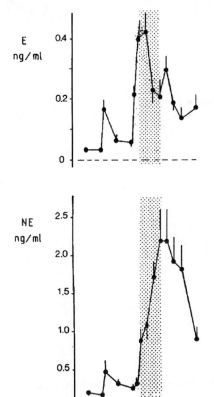

E
ng/ml

NE
ng/ml

FIGURE 14.4

Plasma adrenalin (A) and noradrenalin (NA)
concentrations before, during, and after
exercise (•—•)

The values for A and NA are expressed as means ±*SE*.
The dotted area represents the swimming period.
(From: Scheurink *et al.*, 1989a)

The A and NA profiles before, during and after swimming can explain the glucose, FFA and insulin profiles (cf. Figure 14.1). Adrenalin and NA suppress the release of insulin from the β-cells by an α_2-adrenoreceptor-mediated mechanism. Glycogenolysis in the liver is promoted by an α_1-adrenoreceptor-mediated mechanism (Shimazu, 1986). Lipolysis in white adipose cells is promoted by a β_3-adrenoreceptor-mediated mechanism (Hollenga & Zaagsma, 1989). Lipolysis is induced only by NA, because under physiological conditions A does not reach a sufficiently high plasma concentration to initiate lipolysis (Scheurink *et al.*, 1989b). Adrenalin suppresses insulin release and activates glycogenolysis in the liver. This is probably not brought about by an immediate effect of A, since A and NA have equal effects on α_1- and α_2-adrenoreceptors. However, A promotes the release of a neuropeptide, galanin, from the sympathetic nerve endings in the islets of Langerhans. In its turn, galanin suppresses the release of insulin (Scheurink *et al.*, 1992). In the liver glycogenolysis is promoted by prostaglandins, released by the action of A on cells lining the liver sinuses (Gardeman *et al.*). In addition, A induces the release of glucagon from the α-cells of the islets of Langerhans by activation of α_2-adrenoreceptors. Glucagon also causes glycogenolysis in the liver. Finally, A induces glycogenolysis in muscle cells via the activation of β_2-adrenoreceptors (Richter *et al.*, 1982). Glucose as such is not formed in the muscle cell due to the absence of the enzyme phosphatase which converts glucose-6-phosphate to glucose. Glucose-6-phosphate is immediately utilized in the glucolytic pathways.

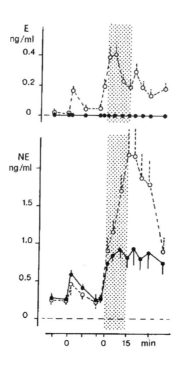

FIGURE 14.5

Effect of adrenodemedullation on plasma adrenalin (E) and noradrenalin (NE) concentrations, before, during, and after exercise (•—•). o-‑o = data of control experiment; dotted area = swimming period

The values for A and NA are expressed as means $\pm SE$.
(From: Scheurink *et al.*, 1989a)

However, in this condition muscle cells take up less glucose from the blood so that glycogenolysis in these cells saves blood glucose and this might contribute to the rise in blood glucose.

The asynchronization of the release of A and NA is quite remarkable. The level of A only increases during the immersion of the paws while during swimming it declines. At the same time, the level of NA continues to increase. Since the adrenal medulla is the only source of A, rats with the adrenal medulla removed were then subjected to swimming (Scheurink *et al.*, 1989a). During the exercise no adrenalin could be detected in these animals, whereas the NA levels attained a level of only 40 per cent of that of controls (Figure 14.5). The simplest explanation for this is that the adrenal medulla releases NA in addition to A. However, if this were true, it is difficult to explain why, during immersion of the paws, only plasma-A levels increased. Another explanation is based on the influence of A on β_2-adrenoreceptors present in the presynaptic membranes of the postganglionic sympathetic nerve endings (Langer, 1981). The release of NA from the activated nerve ending inhibits any further NA release by its negative feedback on the α-adrenoreceptors, which are present besides the β_2-adrenoreceptors. Contrary to NA, adrenalin also has an affinity to the β_2-adrenoreceptor. Stimulation of the β_2-adrenoreceptor induces the release of NA (Figure 14.6).

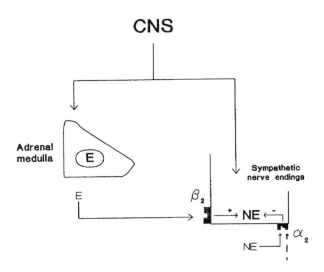

FIGURE 14.6
Adrenergic regulatory mechanisms on the presynaptic nerve endings of the sympathetic nervous system
(Source: Scheurink & Steffens, 1990)

Administration of the specific β_2-adrenoreceptor agonist ICI 118.551 to control animals just before the start of swimming, results in a plasma-NA profile which is exactly the same as that of a swimming adrenodemedullated rat (Figure 14.7).

Administration of the specific β_2-adrenoreceptor agonist fenoterol in an adrenomedullated rat just before swimming, resulted in a close to normal NA profile. Also, infusion of 20 ng A/min, starting at the moment of immersion of the paws and continued during the swimming period, resulted in completely normal NA profiles in adrenodemedullated rats (Figure 14.8).

The A profile in control rats during swimming was exactly comparable to the A profile in swimming adrenodemedullated rats during the administration of 20 ng A/min (Scheurink *et al.*, 1989a). The important role of α_2- and β_2-adrenoreceptors in the release of NA from postganglionic sympathetic nerve endings is confirmed by the extreme increase in plasma-NA levels when simultaneously the α_2-adrenoreceptor antagonist yohimbine and the β_2-adrenoreceptor agonist fenoterol are administered before the start of swimming in adrenomedullated rats (Figure 14.8; Scheurink *et al.*, 1989b).

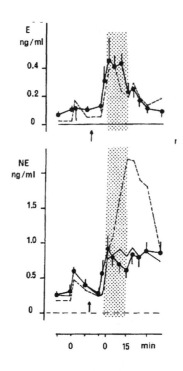

FIGURE 14.7

Effect of intravenous administration of β_2-selective adrenoreceptor antagonist ICI 188.551 on plasma A and NA concentrations before, during, and after exercise (•—•)

The data are expressed as mean values ± *SE*. The moment of injection is indicated by an arrow. Shaded area = swimming period.

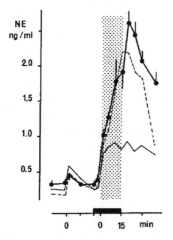

FIGURE 14.8

Effect of intravenous administration of 20 ng A/min on plasma-NE concentrations in exercising adrenodemedullated rats (•—•)

—- = values of control experiment with intact rats; ... = values of experiments with adrenodemedullated rats.

The period in which the infusion was given is indicated by a horizontal bar at the bottom of the graph. The data are expressed as in Figure 14.7.

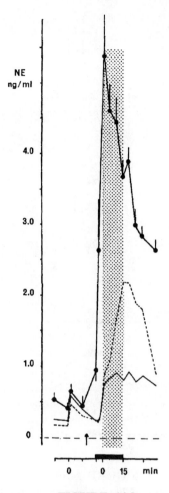

FIGURE 14.9

**Effect of intravenous administration of the α_2-selective adrenoreceptor antagonist yohim-
bine in combination with intravenous infusion of β_2-selective adrenoreceptor agonist
fenoterol on plasma-NE concentrations in exercising adrenodemedullated rats (•—•)**

The moment of injection of yohimbine is indicated by an arrow. The data are expressed as in Figure
14.8.

(From: Scheurink *et al.*, 1989a)

The strong interaction between A and NA, which is only released from the sympathetic nerve endings, also has consequences for the blood-glucose and plasma-FFA levels.
The former do not rise at all and the latter only slightly in adrenodemedullated rats while
swimming. Infusion of 20 ng A/min during swimming re-established the control values of
blood glucose and plasma FFA in adrenodemedullated rats (Figure 14.10; Scheuring *et al.*,
1989a).

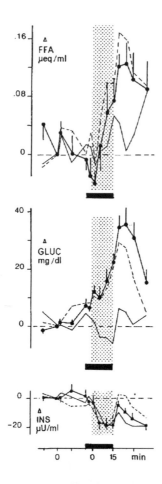

FIGURE 14.10

Effect of an intravenous infusion of 20 ng A/min on plasma-FFA, blood-glucose, and plasma-insulin concentrations in exercising adrenodemedullated rats (•—•)

—- = values of control experiments with intact rats; .... values of control experiments with adrenodemedullated rats. The data are expressed as in Figure 14. 7.
(From: Scheurink *et al.*, 1989b)

The swimming exercise was performed with rats that were well-accustomed to the procedure. When rats were confronted with swimming for the first time, however, the plasma-A level increased more and the plasma NA less than in well-accustomed rats (Figure 14.11; Scheurink *et al.*, 1989b). These results indicate that emotional stress, as occurs during the first confrontation with swimming, activates the adrenal medulla with a concomitant diminished activation of the neural branch of the sympathetic nervous system. The latter results in a lower spill-over of NA from the sympathetic nerve endings into the blood.

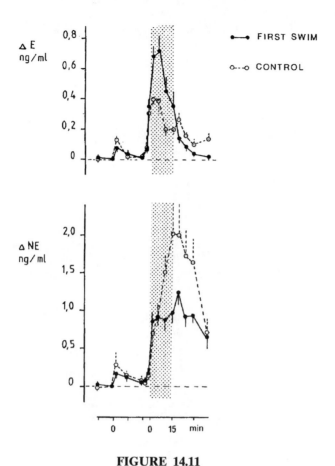

FIGURE 14.11

Plasma-E and -NE concentrations before, during, and after exercise in rats exposed for the first time to experimental exercise conditions (●—●) and rats well accustomed to the entire experimental procedure (o---o)

The values of A and NA are expressed as average changes ± *SE* from basal values.
The swimming period is indicated by the shaded area.
(From: Scheurink *et al.*, 1989c)

3   CENTRAL NERVOUS EFFECTS ON BLOOD GLUCOSE AND PLASMA FFA LEVELS DURING EXERCISE

Neuroanatomical studies have revealed that the hypothalamus can be considered as a main integration centre of the autonomic nervous system (Figures 14.6 and 14.12; Luiten *et al*, 1987). The hypothalamic paraventricular nucleus (PVN) projects directly and indirectly on the intermediolateral column of the spinal cord (IML), the motor neuron pool of the sympathetic nervous system. In addition, neurons in the PVN synthetize corticotropin releasing factor (CRF) which is released in the capillaries of the median eminence of the hypothalamus. CRF stimulates the release of ACTH from the pituitary which in its turn induces the release of corticosterone from the adrenal cortex. The latter hormone plays a role in the stress response so that the effects caused by stress remain controllable.

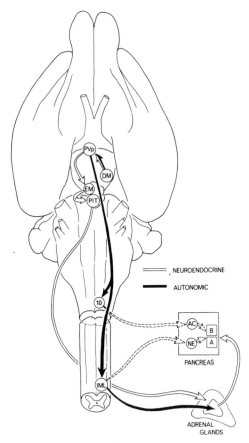

FIGURE 14.12

Diagram summarizing humoral and neural pathways from the paraventricular nucleus (PVN) of the hypothalamus to the pancreas and adrenal glands

The outflow pathways of the dorsomedial nucleus (DM) to the PVN are also indicated. Abbreviations: EM = eminantia mediana; PIT = pituitary gland; IML = intermediolateral column of the spinal cord; AC = postganglionic parasympathetic neurons of the pancreas indicated by their neu-rotransmitter acetylcholine; NA = postganglionic sympathetic neurons of the pancreas indicated by their neurotransmitter noradrenalin; A = glucagon producing α-cells of the islet of Langerhans; B = insulin producing β-cells of the islet of Langerhans.

    The PVN receives neural projections from other parts of the hypothalamus, mainly from the ventromedial and lateral areas (VMH and LHA respectively) via the dorsomedial hypothalamic area. In addition, the hypothalamus receives projections from several nora-drenergic, dopaminergic and serotonergic nuclei in the brainstem. Among others, many noradrenergic and serotonergic receptors are located in the PVN and in other parts of the hypothalamus (Scheurink *et al.*, 1989c). Since adrenergic and serotonergic receptors in the PVN are involved in the regulation of food intake and food selection (Leibowitz *et al.*, 1988; Tempel *et al.*, 1989) it is plausible that these receptors also fulfil a function in the regulation of glucose and FFA metabolism. To test this hypothesis various experiments were carried out (the results can be found in Figure 14.13; 14.14; Scheurink *et al.*, 1988). The α-adrenoreceptor antagonist phentolamine was administered into the VMH just before the start of swimming. This led to the elimination of the increase in blood glucose during swimming, while the increase in plasma FFA was exaggerated (Scheurink *et al.*, 1989b). When the β-adrenoreceptor antagonist timolol was administered in the VMH during swimming this caused nearly complete elimination of the increase in plasma FFA, whereas the blood-glucose level was temporarily suppressed. Administration of phentolamine into the LHA just before swimming also completely inhibited the blood-glucose level to increase during swimming. However, plasma-FFA levels did not differ from those in control animals. Timolol administration in the same situation resulted in a decreased basal FFA level whereas a nearly normal increase during swimming could be observed. The increase in blood glucose was briefly postponed during swimming (Scheurink *et al.*, 1988).

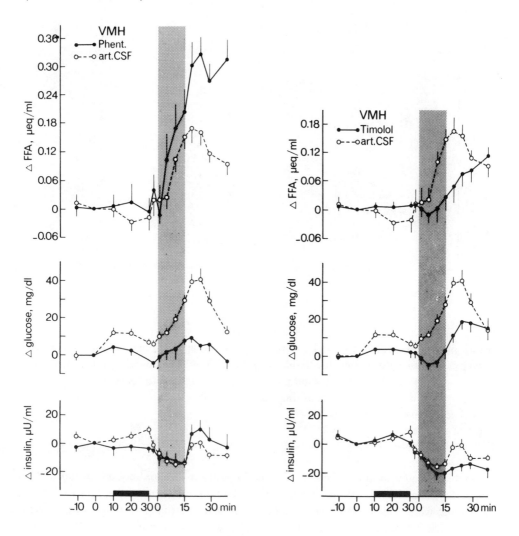

FIGURE 14.13

Effect of hypothalamic infusion of α-adrenoreceptor antagonist phentolamine (left panel) and β-adrenoreceptor antagonist timolol (right panel) into the ventromedial hypothalamus (VMH) on plasma-FFA, blood-glucose, and plasma-insulin concentrations during exercise

o—o = control experiment; shaded area = swimming period; horizontal bar at the bottom of each graph = period in which drug was administered. (From: Scheurink *et al.*, 1988)

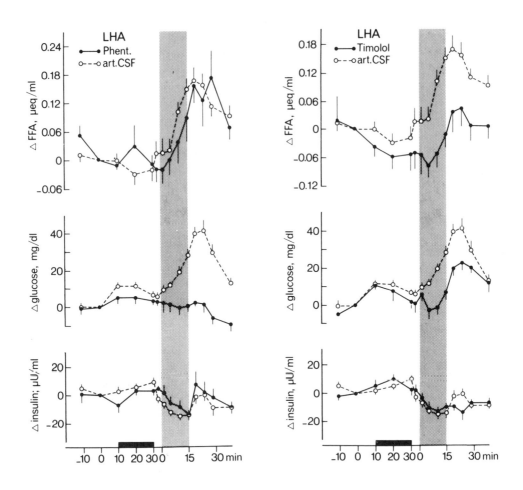

FIGURE 14.14

Effect of hypothalamic infusion of α-adrenoreceptor antagonist phentolamine (left panel) and β-adrenoreceptor antagonist timolol (right panel) into the lateral hypothalamic area (LHA) on plasma-FFA, blood-glucose, and plasma-insulin concentrations during exercise (o—o)

o—o data from control experiment.
The data are expressed as in Figure 14.13.
(From: Scheurink *et al.*, 1988)

Administration of phentolamine and timolol into the PVN did not affect plasma-FFA levels (Scheurink *et al.*, 1990b). Phentolamine inhibited the increase in glucose during swimming, while timolol transiently suppressed glucose during swimming. These results indicate that an extended area in the hypothalamus (encompassing the VMH, LHA and PVN) controls blood-glucose levels during exercise, apparently via α-adrenoreceptor mechanisms. The observed temporarily postponed increase in blood glucose during swimming after β-adrenoreceptor blockade in these hypothalamic areas might be due to increased glucose utilization during swimming. Control of plasma FFA concentrations is apparently exerted by β-adrenoreceptor mechanisms in the VMH.

Plasma-insulin concentrations appeared not to be affected after administration of α- and β-blockers into the VMH, LHA and PVN and did not deviate from the control values.

Now the question arises as to whether the changes in blood-glucose and plasma-FFA concentrations can be attributed to changes in plasma-NA and -A levels. The answer is simple: in most cases no relationship can be observed between changes in plasma-FFA and blood-glucose levels on the one hand and alterations in plasma-NA and -A on the other hand after α- and β-adrenoreceptor blockade in several hypothalamic areas (Scheurink *et al.*, 1989b; Scheurink *et al.*, 1990a,b). The reason why there is no such relationship may be as follows. Blood glucose and plasma FFA can only be replenished by glycogenolysis in the liver and lipolysis in the fat cells. Liver glycogenolysis is stimulated by sympathetic nerves in the liver (Shimazu, 1986). Lipolysis is stimulated by noradrenalin released from the sympathetic nerve endings innervating the smooth muscle cells in blood vessel walls in fat tissue. White fat cells themselves do not receive innervation. NA diffuses from these muscle cells to the fat cells where it causes lipolysis. The spillage of NA from the liver and the blood vessel walls into adipose tissue is only small compared to that of NA in the walls of the heart, i.e. the great veins and arteries, so that the contribution of the spillage from liver and adipose tissue is negligible. The large changes in plasma-NA and -A levels during exercise might be caused by the requirements of the cardiovascular system in which the hypothalamus also participates. If changes in plasma-NA and -A levels occur during swimming after the administration of α- and β-adrenoreceptor antagonists, these changes may be induced by cardiovascular demands and not be related to metabolic requirements.

Serotonergic mechanisms in the PVN are also involved in the regulation of plasma-FFA and blood-glucose levels. Administration of serotonergic agonists in the PVN proved to cause an immediate increase in plasma-A and blood-glucose levels in resting rats, whereas plasma-NA and -FFA levels did not change at all. The latter increased much less during swimming after administration of serotonergic agonists into the PVN, whereas plasma-A and blood-glucose levels reached higher concentrations. These results indicate that serotonergic mechanisms in the PVN are responsible for the selective activation of the adrenal medullary branch of the sympathetic nervous system with a simultaneous suppression of the activity of the neuronal branch. Activation of the neuronal branch of the sympathetic system which releases NA, is affected by noradrenergic mechanisms in the PVN as described in the previous sections.

4   CONCLUDING REMARKS

The increase in blood-glucose and plasma-FFA concentrations during exercise is caused by activation of the sympathetic nervous system. Adrenalin released from the adrenal medulla induces glycogenolysis in liver and muscle. Only glycogenolysis in the liver contributes to the replenishment of the blood-glucose concentration, because glucose-6-phosphate formed in the muscle cells by glycogenolysis is utilized only by these cells themselves. Glycogenolysis in the liver is stimulated by circulating adrenalin, by glucagon (also released by sympathetic activity) and by increased sympathetic activity in the liver. Simultaneously, insulin release is suppressed by sympathetic activity so that the glucose produced is only available to the active muscle cells and the nervous system. Noradrenalin, only released from the sympathetic nerve endings during exercise, causes lipolysis which is simultaneously stimulated by the suppressed insulin release.

Sympathetic activity is regulated at several levels: (a) at the sympathetic nerve endings themselves; (b) in the medulla oblongata and the brainstem; and (c) in the higher brain areas.

Adrenalin regulates the release of noradrenalin from the sympathetic nerve endings. Probably, many other humoral factors such as hormones and neuropeptides are involved in this regulating activity because many receptors for these factors are present at the sympathetic nerve endings.

Noradrenergic and serotonergic agonists and antagonists administered in several areas of the hypothalamus can individually alter the activity of the adrenal medullary branch of the sympathetic system and the activity of the neuronal branch. It may well be possible that parts of the neuronal branch of the sympathetic system can also be individually activated. This, and also the contribution of other areas of the medulla oblongata and the brainstem need further investigation.

The involvement of higher brain areas is evident if one takes into account that situations such as being aware of a stressful environment or the start of activity which is initiated by the motor cortex of the brain, lead to an immediate increase in sympathetic activity. This response is so fast that feedback from the activated tissues is impossible, so that it has to be considered as an anticipatory response.

5   SUMMARY

The regulation of blood glucose and plasma free fatty acids (FFA) levels must be very accurate both in rest and during periods of augmented needs; this control is exerted by the CNS via the autonomic nervous system.

The increase in blood glucose and plasma FFA concentrations during exercise is caused by activation of the sympathetic nervous system; adrenaline (A) released from the adrenal medulla elicits glycogenolysis in liver and muscle; insulin release is suppressed by sympathetic activity. Noradrenaline (NA), only released from the sympathetic nerve endings during exercise, causes lipolysis.

There is a strong interaction between the release of A and NA: a moderate level of A stimulates the release of NA, while a high level of A inhibits the release of NA; for example: emotional stress activates the adrenal medulla with a concomitant diminished activation of the neural branch of the sympathetic nervous system resulting in a lower spill over of NA from the sympathetic nerve endings into the blood.

Neuroanatomical studies have revealed that the hypothalamus can be considered as a main integration centre of the autonomic nervous system: an extended area in the hypothalamus (including the VMH, LHA and PVN) controls blood glucose level during exercise; control of plasma FFA concentrations is apparently exerted in the VMH.

Sympathetic activity is regulated at several levels: at the sympathetic nerve endings themselves, in the medulla oblongata and the brainstem and in the higher brain areas.

III

Energy expenditure

15

Energy from food

Food in the form of organic matter is produced from simple mineral compounds. The ability to perform this production process is the exclusive property of photosynthetic autotrophs. These organisms produce energy-rich organic matter with the aid of the radiant heat from the sun. The potential energy contained in the organic matter produced by plants forms the basis for life of all heterotrophs, including man. The energy-rich organic matter is either used directly by herbivores or indirectly by carnivores. Man, as an omnivore, usually consumes a mixture of vegetable and animal food. (In addition to autotrophs, there are the chemotrophs, producing organic matter from mineral compounds without using solar energy. Instead, most of them draw their energy from other organic compounds. Their contribution to the global production is minute though.)

It is estimated that about 1 per cent of solar energy is fixed in chemical energy.

The energy fixed in organic matter by autotrophs and which is potentially available for the next trophic level, is not completely utilized. Usually, a substantial part is not consumed. Beef cattle raised on grassland consume less than half of the total primary production. Of the consumed part some is eliminated, chiefly as feces and as products of nitrogen metabolism in urine. The remaining part is available for vital processes, production and reproduction. The energy fixed at each trophic level which is then available to the next level, amounts on average to 10 per cent of the energy consumed. For example, a cow can fix 15-20 per cent of the energy consumed in milk and a hen can fix about 5 per cent in eggs.

This chapter focuses on the way in which humans use food energy, what the energy is needed for and how energy intake and energy expenditure are balanced.

1   CONVERSION OF MACRONUTRIENTS TO ENERGY

Being a heterotroph, humans derive their energy from organic compounds and these can be classified as carbohydrates, lipids and proteins. (A fourth source of food energy which can sometimes make up a significant part of the energy intake is alcohol. As a fuel it is broken down to CO_2 and H_2O or converted to body fat.) Carbohydrates, lipids and proteins are usually called macronutrients. Micronutrients are minerals, vitamins and trace elements.

Carbohydrates are sugars, starches and cellulose. They are made up of carbon and water, i.e. the basic structure is $(CH_2O)_n$ in which $n > 2$. The best-known carbohydrate is the (simple) sugar glucose: $(CH_2O)_6$ or $C_6H_{12}O_6$. Examples of *lipids* are fats, oils, steroids and waxes. Like carbohydrates, they are made up of carbon, hydrogen and oxygen. However, the main building blocks consist of glycerol and fatty acids, both containing far less oxygen than carbohydrates. Consequently, lipids have more carbon-hydrogen bonds and thus contain more than twice as much energy per unit weight as carbohydrates. They are insoluble in water. *Proteins* are long chains of amino acids containing carbon, hydrogen, oxygen, nitrogen and, usually, sulphur. Proteins primarily have a structural (muscles, membranes) and a regulatory function in the body (enzymes, hormones). When used as a fuel, they are broken down to CO_2 and H_2O. The nitrogen is excreted in NH_3, urea or other organic compounds in the urine. Thus, proteins provide the body with an amount of energy per unit weight that is comparable to that of carbohydrates.

The first steps in the use of food as energy source are capture and ingestion. Next, the food is digested to make it available to the body (see Figure 15.1). The *carbohydrates*, such as sugars and starches, are hydrolysed to simple sugars or monosaccharides, mainly glucose and fructose.

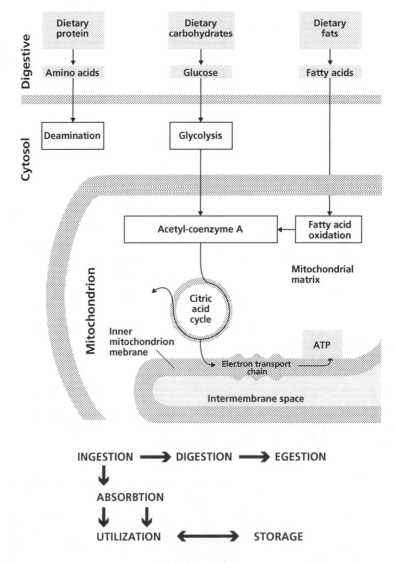

FIGURE 15.1
The breakdown of glucose, fatty acids and amino acids for energy production (redrawn from: Rhoades and Pflanzer, 1989)

The monosaccharides are absorbed, i.e. pass across the intestinal wall into the blood stream. Simple sugars can provide energy to all cells. There they are either broken down or polymerized to glycogen. Glycogen is a form in which energy can be stored in the liver and muscles.

The *lipids* are mainly ingested in the form of fats consisting of triglycerides: three fatty acids chemically bound to glycerol. The triglycerides are broken down to diglycerides, monoglycerides, glycerol and free fatty acids. The monoglycerides, glycerol and free fatty acids dissolve in bile-salt micels, so that more than 90% of the lipids are absorbed in the epithelial cells of the intestines. Short-chain fatty acids can be absorbed directly into the blood. The remaining part is packed into globules known as chylomicrons, and enters the blood indirectly through the lymph vessel system. Most of the chylomicrons are removed from the blood by fat cells and stored in adipose tissue. From there, the fat can be mobilized when needed.

The *proteins* are digested by hydrolysis of the peptide bonds between the amino acids. In that way, they are broken down to polypeptides and the to tripeptides, dipeptides and, finally, free amino acids. The latter three, tripeptides, dipeptides and free amino acids, can be absorbed in the intestinal epithelial cells where further hydrolyzation to amino acids takes place. The free amino acids are absorbed into the blood and transported for further use to the liver.

The energy in the macronutrients is released during the breakage of chemical bonds. The study of the exchange of energy is known as *thermodynamics*. The first law of thermodynamics, the law of conservation of energy, states that *energy can neither be created nor destroyed*. The second law of thermodynamics, the one that is concerned with the conversion of all forms of energy to heat, states that *processes involving energy transformations will not occur spontaneously unless there is a degradation of energy from a non-random to a random form*. According to these laws the potential energy in chemical bonds can be released for work but not all of the stored energy can be used. Some of it will always be lost because of the tendency towards disorder.

The cell temporarily stores the energy released from the breakdown of nutrient molecules in high-energy phosphate compounds like the macromolecule adenosine triphosphate (ATP), linking the energy producing and energy demanding processes. ATP is a nucleotide consisting of a ribose ring, adenine, and three phosphate groups. When a phosphate group is removed a large amount of energy is released and, conversely, adding a phosphate group uses energy. The energy change for ATP breakdown to adenosine diphosphate (ADP) and a phosphate group under cell conditions of 37 °C is about 50 kJ per mol. ATP is the main high-energy phosphate compound in the cell. Other examples are phosphocreatine and glucose-6-phosphate.

The molecules resulting from the breakdown of carbohydrate, lipids, and protein can be used for energy metabolism or can be converted and/or stored (see Figure 15.1). The breakdown of glucose, fatty acids, and amino acids for the production of energy (ATP) takes place in a series of reactions ending in the citric acid cycle. The conversion and storage processes offer the possibility to store glucose as glycogen or fat, to store lipids as fat or convert the glycerol component to glucose, and convert amino acids to glucose (Figure 15.2).

Summarizing:

Carbohydrates can yield energy, glucose, nonessential amino acids in the presence of nitrogen, and fat.

Lipids can yield energy, a minimal amount of glucose through glycerol, and fat.

Proteins can yield: energy (if needed), glucose when carbohydrate is unavailable, amino acids, and fat.

glucose $\rightleftharpoons$ 2 pyruvate $\longrightarrow$ 2 acetyl CoA $\longrightarrow$ citric acid cycle

triglycerides $\rightleftharpoons$ glycerol + 3 fatty acids, glycerol $\longrightarrow$ pyruvate, fatty acids $\longrightarrow$ acetyl CoA

protein $\rightleftharpoons$ amino acids $\longrightarrow$ pyruvate

The energy content of food can be determined by combustion. The energy contained in the organic bonds that hold its carbon and hydrogen atoms together is released in the form of heat. The amount of heat that is released can be measured. This direct measurement of the energy that is stored in the food components' chemical bonds is known as *direct calorimetry*. As the chemical bonds are broken, the carbon and hydrogen atoms combine with oxygen to form carbon dioxide and water respectively. Measuring the amount of oxygen consumed and carbon dioxide produced gives an indirect measure of the amount of energy released, named *indirect calorimetry*. Another way to arrive at the energy value of foods is to compute the energy content of the amounts of carbohydrate, protein, and lipid (and alcohol, if present) in it.

In the next sections an outline will be given of how food intake is measured and which part of the ingested energy is absorbed and available for utilization or storage. Most information on the extraction of energy from food is obtained from laboratory observations.

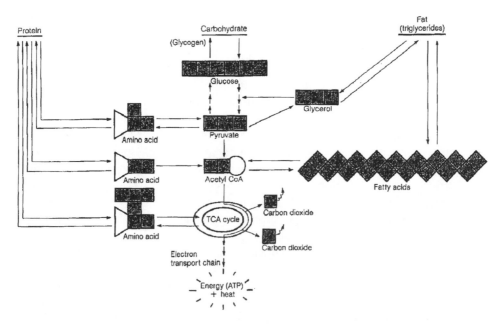

FIGURE 15.2

Conversion processes to produce or store energy from protein, carbohydrate and fat

(From: Whitney and Rolfes, 1990)

Where the processes of ingestion, absorption and egestion in laboratory and normal living conditions are comparable, the type and amount of food consumed are likely to differ. This is the main problem in individual energy balance studies in humans.

2   MEASUREMENT OF FOOD CONSUMPTION

Measurement of the habitual food consumed by humans is one of the hardest tasks in energy balance studies. The two basic problems are the accurate determination of a subject's customary food intake, and the conversion of this information to nutrient and energy intake. Any technique applied to measure food intake should not in any way interfere with the subject's dietary habits and thus alter the parameter being measured. Another problem is the time factor: how long should the food intake be measured before the information obtained can be said to truly reflect habitual food intake, i.e. the food an individual normally consumes to provide the energy and nutrient requirements for his regular everyday activities?. The standard methods of determining food intake are:

1   Indirect determination based on data on group consumption, inspection of family budgets, larder inventories, data on agricultural production etc.
2   Estimation by recall of food consumed over the last day, week, month, or even longer.
3   Measurement and recording of food intake as eaten.

Here, the focus is on methods measuring food intake at an individual level, i.e. dietary recall (method 2) and dietary record (method 3). The latter method requires the subjects to record types and amounts of all foods consumed over a given time interval. The foods are weighed or recorded in household measures like cups and spoons. The latter information is translated to weight or volume by measuring the actual 'tools' used or by adopting standard values from reference tables. Method 2, dietary recall, uses the subjects' report of intake over the previous 24h period (24h-recall) or the report of customary intake over the previous week up to the past year(s) (diet history). Here, the same methods are used to quantify the reported intake from the information on portion size. Additionally, food models, volume models, and photographs can help to recall the amounts consumed. Alternatives to the measurement of food intake with a dietary record or a dietary recall are the double portion technique or supplying subjects with their daily food and measuring the quantity and quality of the food in the laboratory. In the *double portion technique* the subjects have to collect from every food item consumed an equivalent amount for later analysis. A mixed sample is usually made per 24h interval. When the food is supplied, the subjects have to be carefully instructed to consume only the food provided by the experiment and return any left-overs.

The information on the quality of food consumption will be closer to the real life situation in the method of dietary recall than when food is supplied. On the other hand, the quantitative method (supplying food) is superior to the method of dietary recall. Generally, the subjects' energy intake and expenditure should be in balance. This is very difficult, if not impossible, to check (see Chapter 17). Moreover, sources of error may arise from reporting errors due to poor memory, inaccurate estimation of amounts, and wishful thinking. The latter is typical of overweight people who systematically 'underreport' their intake (see Chapter 16) and underweight people like anorexics who probably 'overreport'.

Finally, if the respondents are required to write down and even weigh or measure what they eat, they may alter their usual dietary habits either to make recording easier or to hide their habits.

The length of time that food intake should be measured over to determine habitual intake is a subject of discussion. If an individual lives by a regular activity pattern, e.g. 5 days work and 2 days leisure at the weekend, it would seem reasonable to suppose that his social and dietary habits are determined by this pattern of activity. Measurement of food intake should include samples of workday and weekend intake, and it would of course be preferable to measure intake for the whole week. The length of the observation period is primarily determined by the level of day-to-day variability and the level of accuracy desired.

Basiotis *et al.* (1987) performed a study entitled 'Number of days of food intake records required to estimate individual and group nutrient intakes with defined confidence'. Twenty nine apparently healthy individuals, 16 females and 13 males aged from 20 to 55, kept daily food intake records for 365 consecutive days while consuming their customary diets. They were students and scientists, trained to record the kinds of foods they ate and report the portion sizes. Daily intakes of energy, the macronutrients carbohydrate, protein and fat, and many micronutrients were calculated with standard reference tables. Each individual's average intake of nutrients and standard deviation over the year were assumed to reflect his or her 'usual' intake and day to day variability. From these, the number of days of dietary records needed for estimated individual and group intake to be within 10% of usual intake was calculated. The results indicated that the number of days of food intake records needed to predict the usual nutrient intake of an individual, varied considerably among individuals for the same nutrient and within individuals for different nutrients. Individual estimates of food intake required the fewest days, on average 31, and the longest time was needed for micronutrients, for example, on average 433 days for vitamin A. The time needed to estimate the mean nutrient intake for the group was considerably lower, ranging from 3 for food energy to 41 for vitamin A. Even fewer days would be needed for larger groups. To achieve a level of precision of 10 per cent in a single individual a relatively large number of days of food records, over 4 weeks, is required. Thus, measuring food intake at the level of the individual is a tedious job and therefore likely to interfere with the subject's dietary habits as mentioned above.

Acheson et al. (1980) evaluated three methods of determining individual food intake: dietary record, dietary recall and the double portion technique with bomb calorimetry. Twelve subjects, all males spending 1 year on an Antarctic base, participated in the study. The period of investigation of different individuals varied between 6 and 12 months. The tedious job of weighing and recording food intake and collecting duplicate food samples requires a high degree of motivation but, in the confined area of a resident base camp, the subject is easier to encourage. During the study, the food intake of each subject was determined for at least 1 week of each month. During that week the subjects weighed and recorded all food consumed. Once during the week the subjects were asked to write down everything they could remember eating during the previous 24 hours. The dietary record method and the use of food composition tables underestimated the energy intake by on average 7 per cent when compared to the analysis of duplicate meals by bomb calorimetry. Errors of over 20 per cent were found in the energy intake determined by dietary recall.

In 62 out of 68 occasions the recall underestimated the real food consumption. The authors suggested that the discrepancy might have been smaller if the subjects had been interviewed by a skilled and persistent interviewer. Errors of over 20 per cent are unacceptable in an energy balance study.

A way to improve the recall procedure is to cross-check. A trained interviewer asks about the food eaten in the recent past and cross-checks this information against data on food purchases. With a skilled interviewer and a cooperative subject one assumes the standard deviation of the estimate to be 10 per cent but the error is probably greater. The best an interviewer can do is to find out what people think they eat and this is often far from the reality (see Chapter 17). Garrow (1981) concludes that however intensively the eating habits of people are studied, it is impossible to predict their energy intake over a period of a week with an accuracy much better than 10 per cent.

3   FOOD DIGESTION

In a healthy person the digestion of food is very effective. Only in young infants does special care have to be taken when introducing new foods to the diet. The complex process of digestion and absorption can only be optimally effective if the gastro-intestinal tract and accessory organs are completely developed and fully functioning. This means that the alimentary tract with its mucosal lining and endocrine cells must operate efficiently, and the accessory organs such as pancreas, liver and gall bladder with their digestive secretions, must be physiologically mature. Usually, babies are not fed solid food until they are about 4 months old. Before this, their diet consists either of human milk or infant formula 'milk'. In infant formulas special attention is given to fats because of the inability to produce sufficient enzymes and bile salts for effective digestion at that age. To digest solid food during the first months, the infant still lacks a well-coordinated gastric motility. Usually, by the age of about 2 years the gastro-intestinal tract has reached maturity, so that in normal cases, digestion becomes an effective process.

It has been suggested that the efficiency of digestion decreases with age, due to a diminished functioning of the gastro-intestinal tract and the accessory organs. This could then partially explain the nutritional deficiencies often seen in the elderly. Although the absorption of some micronutrients like calcium has been shown to be decreased in the elderly, the overall digestion of the macronutrients carbohydrate, protein and fat is assumed to be unaffected by age.

The main variable in the digestive process is the composition of the diet. Humans usually consume a mixed diet derived from vegetable and animal sources. The digestibility of these foods is often high, because many of the products are refined. One of the few indigestible components is dietary fiber, a polysaccharide. It is a component of the plant cell wall that resists digestion by secretions of the human alimentary tract. Foods with a high fiber content include whole-grain breads and cereals, fruits, vegetables, legumes and nuts. The daily intake of fiber thus varies with the intake of such foods and usually averages 10-30 gram per day or 2-6 per cent of the gross energy intake. This forms the main part of the fecal energy loss.

It is often assumed that digestive efficiency is not only a function of the composition of the diet, but also of the level of intake. Van Es *et al.* (1984) carried out an extensive study on energy metabolism in humans measuring complete energy balances at three levels of food intake. Seventeen volunteers, 8 women and 9 men aged 18-64 years, took part. The energy balance was measured at three intake levels of about 0.5, 1.0 and 1.5 times the usual intake. The diet was a habitual diet, consisting of 15-20 foods, with an approximate energy composition of 42 per cent from fat, 13 per cent from protein and 45 per cent from carbohydrate. In principle, the diets at each of the three levels of energy intake were of similar composition except that the same amounts of vegetables were used each day at all levels to supply adequate bulk. The energy balance was measured during days 5-8 by measuring food intake and the total amount of feces. The energy in food and feces was measured with a bomb calorimeter. The intake level appeared to have no significant effect on the digestibility of the food. The average digestibility of the food at the three intake levels was 93, 94 and 94 per cent from low to high, respectively.

Norgan and Durnin (1980) overfed 6 healthy, normal-weight men aged 21-24 years for 6 weeks at 1.4 times their usual intake, and also measured digestibility. They gave the subjects free choice of normal foods. Therefore, the percentages from the macronutrients changed during the overfeeding period. In a control period of 2 weeks the protein/carbohydrate/fat ratio of the diet was 14/33/46 energy% and in the overfeeding period 12/43/38 energy% with 7 energy% alcohol in both situations. Overfeeding thus resulted in a relatively higher fat intake and a lower carbohydrate intake. The overfeeding did not influence the digestive efficiency which was on average 96 per cent in the control period and 95 per cent afterwards.

Despite these findings of a lack of relationship between the digestive efficiency and the level of food intake some people insist that they gain weight more easily than others because of a difference in digestive efficiency. Schulz (1983) performed a detailed study on the energy balance in so-called 'large eaters' and 'small eaters'. It showed a nearly twofold difference in energy intake corrected for differences in body mass. The methods were basically the same as those of Van Es *et al.* (1984), with the only difference that the subjects chose their own food and recorded everything they ate. The digestive efficiency did not differ between both groups and averaged 93 per cent. This provided further evidenc for the constancy of digestive efficiency at different intake levels, not only in an experimental situation but also at various levels of habitual intake.

All studies referred to above showed the digestive efficiency in man to be very high. Only 4-7 per cent of the energy ingested with food is lost in the feces. This figure is very high compared to that for herbivores. A goose purely feeding on grass, only digests about 30 per cent of the energy ingested with the food, being nearly unable to digest cellulose. Ruminants can digest more than 60-70 per cent of a pure grass diet. Figures as high as 90 per cent are only found in carnivorous animals feeding on large prey. Man as an omnivorous organism reaches this high level of digestive efficiency through the refinement of foods before consumption.

4   METABOLIZABLE ENERGY

As mentioned above, the ingested energy *(gross energy, GE)* is not fully available to the energy metabolism, since part of it is lost in the feces. Yet another part is lost in the urine. *Metabolizable energy (ME)* then is the gross energy minus energy in feces and urine.

The energy loss in the urine is mainly in the form of the chemical energy contained in urea, $CO(NH_2)_2$. This is an excretory product of the incomplete catabolism of the absorbed protein. When amino acids are degraded to generate energy, or used to make glucose or fat, the nitrogen-containing amino group is removed. This reaction is called *deamination*. The product is ammonia, which is chemically identical to the ammonia in cleaning solutions. It is a strong smelling poison. A small amount of ammonia is always produced by deamination reactions in the liver. Some of it is used by liver enzymes in the synthesis of non-essential amino acids. What cannot be used is combined with carbon dioxide to form urea, a much less toxic compound. Urea is released from the liver cells into the blood, where it circulates until it passes through the kidneys. One of the functions of the kidneys is to remove urea from the blood for excretion in the urine. Urea is the body's principle vehicle for excreting unused nitrogen. The heat of combustion or *GE* of a gram of protein is 22-23 kJ, but since 5.5 kJ of this is lost as urea, the metabolizable energy of a gram of digested protein is only 17 kJ.

The energy loss in the urine is mainly a function of protein catabolism. When a subject is in energy balance, i.e. energy intake equals energy expenditure, there is usually a nutrient balance as well. In that case, protein catabolism equals net protein intake. In a normal, mixed diet protein contributes 10-15 energy%. Knowing that about one fourth of the protein energy is lost as urea, the urinary energy loss is then 3-4 per cent of the daily energy intake. In situations of over- and under-feeding the energy loss in the urine is different. Especially during under-feeding protein catabolism is decreased until most of the energy reserves of the body in the form of fat are used up. Thereafter urinary energy losses increase dramatically as body protein is broken down to cover daily energy needs.

In the studies on digestibility mentioned above, *ME* was measured as well. Van Es *et al.* (1984) found *ME* to be 88-91 per cent of *GE*. Norgan and Durnin (1980) found values for *ME* of 90-94 per cent of *GE*. These figures were 3-5 per cent lower than the digested energy, a difference that is very close to the calculated urinary energy loss in urea. This justifies the assumption of urinary urea being the main route of urinary energy loss.

5   CALCULATING ENERGY FROM FOOD COMPOSITION

Already at the end of the 19th century Atwater carried out experiments in which he fed subjects for 3-8 days on mixed diets, measured the energy intake and analyzed the losses in feces and urine. His work forms the basis for the food composition tables used nowadays to calculate energy intake. From these experiments the so-called *Atwater factors* were derived (Table 15.1). Knowing the macronutrient composition of foods from chemical analysis, the metabolizable energy can be calculated by multiplying the figures for carbohydrate, protein, fat and alcohol content with the Atwater factors. This is the way the energy value of foods as shown on packages is calculated.

Calculating metabolizable energy from nutrient composition is a practical way of estimating energy intake. The method is sufficiently accurate for most applications despite some variation in the composition of different carbohydrates, proteins and fats.

TABLE 15.1

Gross energy and metabolizable energy of the macronutrients as measured by Atwater

nutrient	gross energy	fecal loss	urinary loss	Atwater factor
	(kJ/g)	(%)	(%)	(kJ/g)
carbohydrate	15-17	1	0	16
protein	22-23	8	23	16
fat	38-39	5	0	37
alcohol	29-30	0	1	29

Usually, subjects consume mixed diets with macronutrients from several sources. On the other hand, for several reasons which we have just discussed, it is impossible to estimate the energy intake any more accurately than 10 per cent. So, the small variation in energy values for macronutrients does not significantly influence the accuracy of the results.

For research purposes, it is often necessary to know the energy content of foods more accurately. In energy-balance studies subjects are fed fixed amounts of diets of a known composition and the feces and urine are collected over 24-hour intervals as in the studies by Norgan and Durnin (1980) and Van Es *et al.* (1984). Samples of food, feces and urine are combusted in a bomb calorimeter.

Comparison of the measured and calculated figures of energy intake reveals some differences, depending on the specific experimental conditions. Norgan and Durnin (1980) compared their measured values with calculated values using the Atwater factors. The calculated energy intake was always higher than the measured energy intake (Table 15.2). On average, the difference was 2 per cent (range 1-4 per cent) when the subjects ate at a level that maintained body weight. In the case of overfeeding the differences were, with an average value of 7 per cent (range 7-9 per cent), much higher although still within the suggested range of error for this method.

TABLE 15.2
Metabolizable energy intake measured from gross energy content of the diet and energy loss in feces and urine (ME_m) compared to energy intake calculated from food tables (ME_c) over a 7 day period feeding at energy balance and after 5 weeks overfeeding
(After Norgan & Durnin, 1980)

subject	control period			overfeeding period		
	ME_m (MJ/d)	ME_c (MJ/d)	Δ (%)	ME_m (MJ/d)	ME_c (MJ/d)	Δ (%)
1	12.0	12.5	4	16.5	17.7	7
2	15.0	15.1	1	21.1	22.7	8
3	12.3	12.6	2	17.3	18.9	9
4	12.8	12.9	1	13.7	14.7	7
5	11.4	12.0	5	17.7	18.9	7
6	12.0	12.1	1	16.4	17.6	7
mean	12.6	12.9	2	17.1	18.4	7

6 SUMMARY

In order to use food as a source of energy, macronutrients have to be converted to energy. Measurement of food consumption yields the amounts of carbohydrates, lipids, proteins, alcohol, minerals, vitamins and trace elements ingested.

Ingested energy yields metabolizable energy (ME) which is gross energy (GE) minus energy in feces and urine. ME is 88-94 per cent of GE.

Knowing the macronutrient composition of foods from chemical analysis, the metabolizable energy can be calculated by multiplying the figures for carbohydrate, protein, fat and alcohol content with the Atwater factors. This is the way the energy value of foods as shown on packages is calculated.

16
Energy expenditure

Metabolizable energy is energy available for energy production in the form of heat, external work, growth, offspring, milk, etc. This chapter will focus mainly on heat production.

Heat production in an animal is first of all a function of its body temperature. Two major groups of animals can be distinguished with regard to body temperature: homeotherms and poikilotherms. The body temperature of a poikilotherm follows the temperature of the environment. The homeotherm maintains an almost constant body temperature over a wide range of environmental temperatures. Moreover, this temperature is nearly the same for all homeotherms. Thus, polar bears and penguins have the same body temperature, which is also equal to that of desert rats or birds in a tropical jungle. A third group of animals, including hedgehogs and marmots, stand in between. They are homeothermic during the warm or moderate season of the year, but in wintertime, when obtaining food and maintaining a constant body temperature become difficult, they disappear into a sheltered place and hibernate, lowering their body temperature to conserve energy.

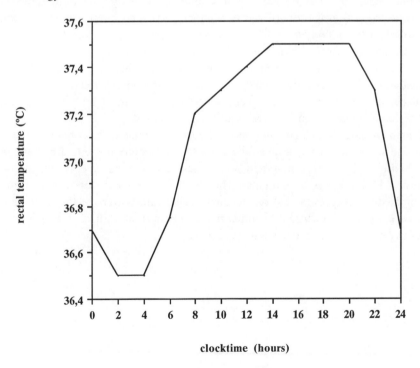

FIGURE 16.1
Body temperature in a resting subject throughout the day

The body temperature of homeotherms does not remain strictly constant. It changes regularly, even under normal conditions, during the course of the day, as can be seen in Figure 16.1.

It has been suggested that two types of humans can be distinguished as far as their patterns of temperature fluctuations during a day are concerned: early risers and late risers. The early risers have a relatively high body temperature in the morning and are bright and cheerful before breakfast. The larger group, however, are those who have difficulty in getting up in the morning and are bad-tempered and better left alone, at least until after the first cup of coffee or tea. Their body temperature is low in the morning but high in the evening, when they are wide awake while the early risers are already tired and sleepy.

Body temperature is regulated centrally through autonomic temperature control. Two brain centres are involved, a heat production centre in the posterior hypothalamus and a heat loss centre in the anterior hypothalamus. The centres receive their information from peripheral (skin) and central (blood) thermoreceptors. Autonomic responses are sweating and vasodilatation to increase heat loss and, to reduce heat loss, vasoconstriction and mechanisms to increase metabolism, such as shivering.

Heat production has been found to vary among individuals with body size and composition, age and gender. Furthermore, there are individual variations caused by changes in behavior such as food intake and physical activity.

The metabolic requirements of a homeotherm are mainly a function of the body's surface area. The physiological interpretation is that heat is lost via the body *surface*. On the other hand, heat is produced and stored in the body *volume*. The heat loss and therefore the energy requirement to maintain the body temperature, decreases with increasing body size. The body surface is proportional to the body mass to the power of 2/3, while heat production is proportional to body mass, i.e. the surface to volume ratio decreases with increasing body mass. This is shown in Figure 16.2 in which the metabolic rate is plotted against body size in animals ranging in body size from a 5 g shrew to a 1300 kg elephant. The heat production per unit body mass in the elephant is more than 50 times lower than that in the shrew. Since small animals not only have a relatively high heat production but also a relatively small energy depot in the form of body fat, it is very important for small homeotherms to have regular access to food.

The length of time an animal can withstand fasting is related to its body size. One problem at the small end of the scale is nocturnal survival. Small homeothermic animals carefully choose their microhabitat and sometimes reduce their energy expenditure through torpor. Torpor is a state of regulated homeothermy at a lower body temperature. Hummingbirds that are prevented from feeding during rainshowers in the day, subsequently show episodes of nocturnal hypothermia. In humans, newborn babies are fed at short intervals and feeding during the night is necessary.

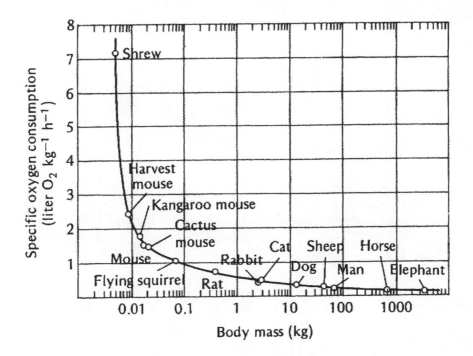

FIGURE 16.2
Metabolic rates of mammals plotted against body weight
(From: Schmidt-Nielsen, 1972)

1   COMPONENTS OF DAILY ENERGY EXPENDITURE

Daily energy expenditure consists of four components, the sleeping metabolic rate (*SMR*), the energy cost of arousal, the thermic effect of food or diet-induced energy expenditure (*DEE*), and the energy cost of physical activity (*AEE*). Sometimes the daily energy expenditure is divided into three components, taking sleeping metabolic rate and the energy cost of arousal together as energy expenditure for maintenance or basal metabolic rate (*BMR*). *BMR* is usually the main component of the average daily metabolic rate (*ADMR*). Figure 16.3 shows *ADMR* and its constituents in a 'standard' subject, a moderately active adult male weighing 70 kg with 15 per cent body fat. Nowadays there are methods that allow accurate assessment of the total energy expenditure and its components.

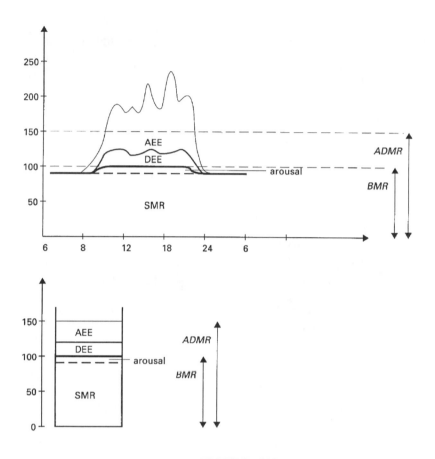

FIGURE 16.3

Average Daily Metabolic Rate (ADMR) and its constituents Basal Metabolic Rate (BMR), Diet Induced Energy Expenditure (DEE), and Activity Induced Energy Expenditure (AEE)

Data are for a moderately active adult male weighing 70 kg, with 15 per cent body fat.

2   MEASURING BY DIRECT AND INDIRECT CALORIMETRY

Living can be regarded as a combustion process. The metabolism of an organism is a process of energy production by the combustion of fuel in the form of carbohydrate, protein, fat or alcohol. In this process oxygen is consumed and carbon dioxide produced. Measuring energy expenditure means measuring heat production or heat loss, and this is known as *direct calorimetry*. The measurement of heat production by measuring oxygen consumption and/or carbon dioxide production is called *indirect calorimetry*.

The early calorimeters for the measurement of energy expenditure were direct calorimeters. At the end of the 18th century Lavoisier constructed one of the first calorimeters, measuring the energy expenditure in a guinea pig. The animal was placed in a wire cage which occupied the center of the apparatus. The surrounding space was filled with chunks of ice. As the ice was melted by the animal's body heat, the water was collected in a container, and weighed.

FIGURE 16.4
Lavoisier's calorimeter
The animal melts the ice in the inner jacket and a mixture of ice and water prevents heat flow to the surrounding.
(From: Kleiber, 1961).

The ice cavity was surrounded by a space filled with snow to maintain a constant temperature. Thus, no heat could dissipate from the surroundings to the inner ice jacket. Figure 16.4 shows Lavoisier's calorimeter schematically.

Nowadays heat loss is measured in a calorimeter by removing the heat with a cooling stream of air or water, or by measuring the heat flow through the wall. In the first case heat conduction through the wall of the calorimeter is prevented and the loss of heat is calculated from the product of the measured temperature difference between inflow and outflow and the flowing rate of the cooling medium. The second method measures the heat loss from the difference in temperature across the wall, without preventing heat loss through that wall. This method is known as *gradient layer calorimetry*.

In indirect calorimetry the heat production is calculated from chemical processes. Knowing, for example, that the oxidation of 1 mol glucose requires 6 mol oxygen and produces 6 mol water, 6 mol carbon dioxide and 2.8 MJ heat, the heat production can then be calculated from the oxygen consumption or carbon dioxide production. The energy equivalent of oxygen and carbon dioxide varies with the nutrient oxidized (Table 16.1).

TABLE 16.1

Gaseous exchange and heat production of metabolized nutrients (*a*) and resulting energy equivalents of oxygen and carbon dioxide (*b*)

(a) *nutrient*	*consumption oxygen* (l/g)	*production carbon dioxide* (l/g)	*heat* (kJ/g)
carbohydrate	0.829	0.829	17.5
protein	0.967	0.775	18.1
fat	2.019	1.427	39.6

(b) *nutrient*	*oxygen* (kJ/l)	*carbon dioxide* (kJ/l)
carbohydrate	21.1	21.1
protein	18.7	23.4
fat	19.6	27.8

Brouwer (1957) drew up simple formulae for calculating the heat production and the quantities of carbohydrate (C), protein (P) and fat (F) oxidized from oxygen consumption, carbon dioxide production and urine-nitrogen loss. The principle of his calculation consists of three equations with the three measured variables:

oxygen consumption = $0.829\,C + 0.967\,P + 2.019\,F$
carbon dioxide production = $0.829\,C + 0.775\,P + 1.427\,F$
heat production = $21.1\,C + 18.7\,P + 19.6\,F$

Protein oxidation (g) is calculated as $6.25 \times$ urine-nitrogen (g), and subsequently oxygen consumption and carbon dioxide production can be corrected for protein oxidation to enable the calculation of carbohydrate and fat oxidation.

The general formula for the calculation of energy production (E) derived from these figures is:

$E = 16.20$ oxygen consumption $+ 5.00$ carbon dioxide production $- 0.95\,P$

In this formula the contribution of P to E, the so-called protein correction, is only small. In the case of a normal protein oxidation of 10-15 per cent of the daily energy production, the protein correction for the calculation of E is about 1 per cent. Usually, only urine nitrogen is measured when information on the contribution of C, P, and F to energy production is required. For calculating the energy production the protein correction is often neglected.

3   MEASURING AVERAGE DAILY ENERGY EXPENDITURE WITH DOUBLY-LABELED WATER

The doubly-labeled water method is an innovative variant on indirect calorimetry and has only recently been validated for human use. The principle of the method is as follows. After a loading dose of water labeled with the stable isotopes ^2H and ^{18}O, ^2H is eliminated as water, while ^{18}O is eliminated as both water and carbon dioxide. The difference between the two elimination rates is therefore a measure of carbon dioxide production (Figure 16.5.) The deuterium (^2H) equilibrates, or is distributed evenly throughout the body's water pool, whereas the ^{18}O equilibrates in both the water and the bicarbonate pool. The bicarbonate pool consists largely of dissolved carbon dioxide which is an end product of metabolism and passes into the blood stream and from there into the lungs to be excreted. The rate constants for the removal of the two isotopes from the body are measured by mass spectrometry of samples of body fluid, blood, saliva or urine.

The method was developed after the discovery in 1949 that the oxygen atoms in the body water and bicarbonate pools are in equilibration. It was initially intented for studying the energy metabolism of small animals living in the wild. To that end, the animals were captured, administered the dose of labeled water, and released. After a certain interval they were then recaptured to assess the rate at which the isotopes had disappeared from their bodies. One of the first such studies involved measuring the energy cost of a 500-kilometer flight by trained racing pigeons. It was not until 1982 that the method was used in humans. The reason was that ^{18}O-water is very expensive and a human does, of course, need a much larger dose than a bird. In the first years after the initial discovery it would have cost about $ 5,000 to carry out one single measurement in an adult. The isotope is not substantially cheaper now, but the isotope ratio mass spectrometers have become so sensitive these days that the method now requires much smaller amounts of the isotope. Presently, the method is frequently used in humans in several research centers.

The doubly-labeled water method is safe to use in humans since the water is labeled with *stable* isotopes, ^{18}O and ^2H, at low abundances. Both ^{18}O and ^2H are naturally occurring isotopes which are already present in the body prior to the administration of doubly-labeled water. As such, tracer studies do not depend on measuring the actual isotope concentration, but rather on concentrations in excess of the natural abundance or background isotope concentration. The nominal natural abundances of ^{18}O and ^2H are 2000 and 150 ppm respectively. Typical doses of doubly-labeled water only produce excess isotope abundances of 200-300 and 100-150 ppm for ^{18}O and ^2H, respectively.

This method can be used to measure the carbon dioxide production and hence the energy production in free-living subjects for periods of several days to weeks. The optimal observation period is 1-3 biological half-lives of the isotopes. The biological half-life is a function of the level of energy expenditure. Table 16.2 shows the results of measurements in the Human Biology Department of the State University of Limburg, The Netherlands, in very young, middle-aged and elderly sedentary and highly active subjects. The minimum observation interval is 1×2.6 days or 3.4 days i.e. about 3 days in highly active subjects or prematures, respectively. The former were participants of the well-known bicycle race, the 'Tour de France'. The maximum interval is 3×9.8 days or about 4 weeks in elderly (sedentary) subjects.

TABLE 16.2
Biological half-life of ^{18}O ($t_{1/2}$ ^{18}O) in different subject categories

subjects		age (y)	n	$t_{1/2}$ ^{18}O	SEM (d)
premature		0	8	3.4	0.1
children		11	10	5.2	0.1
adults	sedentary	20-40	5	7.4	0.6
	active	20-40	9	4.4	0.1
	highly active	20-40	4	2.6	0.1
elderly		65-80	5	9.8	0.3

An observation starts with collecting a baseline sample. Next, a weighed isotope dose is administered, usually a mixture of 10 per cent ^{18}O and 5 per cent ^{2}H in, for a 70 kg adult, 100-150 ml water. The isotopes then equilibrate with the body water and after some time the initial sample is collected. The equilibration time is, depending on body size and metabolic rate, 4-8 hours for adults. During equilibration the subject is usually not allowed to consume any food or drink. After the initial sample has been collected the subject resumes his/her routine activities according to the instructions of the experimenter, during which time he/she is requested to collect body water samples (blood, saliva or urine) at regular intervals until the end of the observation period.

Validation studies in four laboratories resulted in an accuracy of 1-3 per cent and a precision of 2-8 per cent, when comparing the method with respirometry. The method has now been applied in subjects of a wide age range and with different activity levels, from premature infants to elderly persons and from hospitalized patients to participants in a cycling race. The method needs high-precision isotope ratio mass spectrometry, working at low levels of isotope enrichment, for the financial reasons mentioned above.

There is still some discussion on the ideal sampling protocol, i.e. the multi-point versus the two-point method. At the State University of Limburg a combination of both is preferred i.e. two independent samples at the start, in the middle, and at the end of the observation period. Thus, an independent comparison can be made in one run, calculating carbon dioxide production from the first and second samples over the first and second half of the observation interval.

The doubly-labeled water method gives precise and accurate information on carbon dioxide production. Converting the carbon dioxide production to energy expenditure requires information on the energy equivalent of CO_2. This can be calculated with additional information on the substrate mixture that is oxidized as discussed above. One option is to calculate the energy equivalent from the macronutrient composition of the diet. In energy balance, substrate intake and substrate utilization are assumed to be identical. Alternatively, substrate utilization can be measured over a representative interval in a respiration chamber (see below). In conclusion, the doubly-labeled water method provides an excellent method for measuring energy expenditure in unrestrained humans in their normal surroundings over a period of 1-4 weeks.

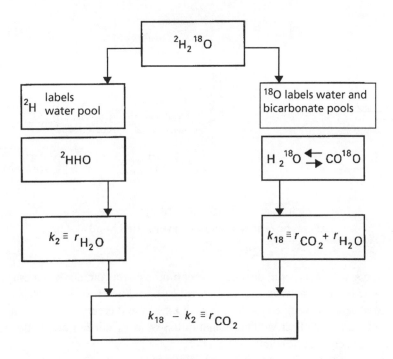

FIGURE 16.5

Principle of the doubly-labeled water ($^2H_2$ ^{18}O) method, measuring carbon dioxide production (r_{CO_2}) from the elimination rates of ^{18}O and ^2H after loading with $^2H_2$ ^{18}O
(From: Prentice, 1986)

4   MEASURING THE INDIVIDUAL COMPONENTS OF DAILY ENERGY EXPENDITURE WITH A VENTILATED HOOD AND IN A RESPIRATION CHAMBER

The ventilated hood and the respiration chamber are instruments for the continuous measurement of oxygen consumption and carbon dioxide production. They allow accurate determination of the energy production in subjects restricted to a laboratory environment. Measurements with a ventilated hood are usually performed over intervals of 1/2 to several hours to determine a subject's resting metabolic rate (*RMR*) or diet-induced energy expenditure (*DEE*). Measurements with a respiration chamber may last several hours to several days and allow determination of a subject's *RMR*, *DEE*, and energy expenditure for (standardized) physical activity (*AEE*).

4.1   VENTILATED-HOOD SYSTEM

A typical example of a *ventilated-hood* system is an open canopy. The subject lies with his head enclosed in a transparent plastic canopy, sealed off by plastic straps around the neck (Figure 16.6). Air is sucked through the canopy with a pump and blown into a mixing chamber where a sample is taken for analysis. The measurements involved are those of the air flow and of the oxygen and carbon dioxide concentrations of the air flowing in and out. The most common device to measure the airflow is a dry gasmeter comparable to that used to measure domestic calor gas consumption.

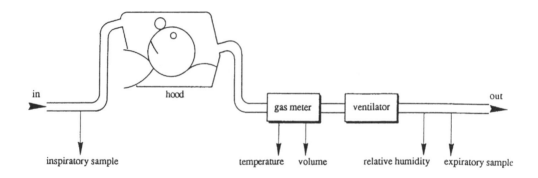

FIGURE 16.6
Schematic representation of a ventilated hood system
(From : Fredrix, 1990)

The oxygen and carbon dioxide concentrations are commonly measured with a paramagnetic oxygen analyzer and an infrared carbon dioxide analyzer respectively. The air flow is adjusted to keep the differences in oxygen and carbon dioxide concentrations between inlet and outlet air within a range of 0.5-1.0 per cent. For adults this means air-flow rates between 25 and 50 l/min.

Measuring *RMR* with a ventilated-hood system implies that the subject is at rest. Intensive physical exercise during the hours preceding the measurement should be prevented. The subjects are usually measured for 15-30 min after at least 15-30 min bed rest, i.e. the measurement lasts 30-60 min. To exclude *DEE*, the measurement takes place at least 12 hours after the last meal. In practice this means measuring *RMR* in the early morning after an overnight fast. Ideally, the subjects stay overnight in the laboratory to make sure they do not take any food and have no vigorous exercise in the hours preceding the measurement. Measuring *DEE* implies the consumption of a (standardized) meal and keeping the subjects in a supine position for many hours. *RMR* increases after a meal and does not return to pre-meal levels until at least 6 hours afterwards. The implications for the measurement of *DEE* will be discussed below.

4.2 RESPIRATION CHAMBER

A *respiration chamber* is an air-tight room which is ventilated with fresh air. Basically, the difference with the ventilated hood is a matter of size. In a respiration chamber the subject is fully surrounded instead of enclosing the head only, allowing physical activity depending on the size of the chamber. The air-flow rate and the difference in oxygen and carbon dioxide concentration between inlet and outlet air are measured in the same way. The flow rate to keep the differences for oxygen and carbon dioxide concentrations between inlet and outlet air within the range of 0.5-1.0 per cent is slightly higher as the subjects never lie down during the full length of an observation interval. In a sedentary adult a typical flow rate is 40-60 l/min, while for an exercising subject the flow needs to be increased to over 100 l/min. The latter situation requires a compromise for the flow rate if the measurements are to be continued over 24 hours, thus including both active and inactive intervals.

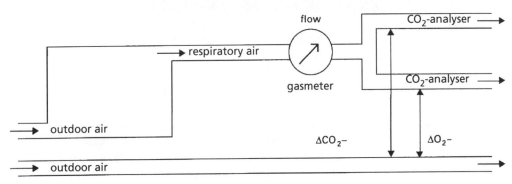

FIGURE 16.7
The interior of a respiration chamber as used at the University of Limburg, The Netherlands

During exercise bouts the 1 per cent carbon dioxide level should not be exceeded for long; during resting, like an overnight sleep, the level should not be too far below the optimal measuring range of 0.5-1.0 per cent. Changing the flow rate during an observation interval reduces the accuracy of the measurements due to the response time of the system.

A normal-size respiration chamber has a volume of 10-30 m^3 and is equipped with a bed, toilet, wash basin and communication facilities like telephone, radio and television. Basically, it is a small hotel room (Figure 16.7). The room temperature is set by the experimenter. Food and drink are delivered through an air lock according to the experimental design. Physical activity is often monitored with a Doppler radar system to know when and how often the subject is physically active. A respiration chamber can also be equipped with a cycle ergometer or a treadmill to perform standardized work loads.

The respiration chamber has a much longer response time than the ventilated hood. Although the flow rate in both systems is comparable, the volume of a respiration chamber is more than 20 times that of a ventilated hood. Consequently, the minimum length of an observation period is in the order of 5-10 hours. The shortest observation in the 14 m³ chamber is usually to measure the sleeping metabolic rate (*SMR*). The routine is usually as follows. The subject enters the chamber between 18.00 and 19.00 h and goes to sleep at 23.00 h. The *SMR* is measured during the interval from 3.00-6.00 h, i.e. 8 hours after closing the chamber door. Then, the oxygen and carbon dioxide concentration differences between the inlet and outlet air are measured within the optimal measuring range of the analyzers.

5   MAINTENANCE

The energy expenditure of a resting subject in the post-absorptive state in the region of thermoneutrality is known as the basal metabolic rate (*BMR*). *BMR* does, of course, depend on body size. The following general equation holds:

$$BMR = a . BM^b$$

in which *BMR* is expressed in watts (J/s), *BM* stands for body mass (in kg) and *a* and *b* are constants. Over a surprisingly wide range, covering not only one species but all homeotherms, quite a satisfactory approximation is obtained for *a* (3.4) and *b* (0.75). As the basal metabolic rate has diurnal rhythmicity, being lower during sleep than during arousal (Figure 16.3), the time factor should also be taken into account. In humans, *BMR* is usually measured in the awake state.

BMR is necessary for such vital processes as breathing, circulating blood, brain function, maintaining muscle tone, preserving electrochemical gradients, and replacing proteins and other macromolecules. *BMR* is not an irreducible minimum of the metabolic rate as will become clear in Chapter 19. A subject can economize on the upkeep of electrochemical gradients and on the turnover of macromolecules and other processes. The general formulae to calculate *BMR* from body dimensions are therefore valid only for strictly defined conditions.

There are several frequently used predictive equations for *BMR* in humans using subject characteristics such as *BM*, gender, age, height, and body composition. A classic equation is the one by Harris and Benedict dating back to 1919 and, although it has been criticized, this is still in use. In their experiments, Harris and Benedict measured 136 men and 103 women in the age range of 18-65 years. The resulting equations were:

women: $BMR = 2.74 + 0.774 H + 0.040 BM - 0.020 A$
men $BMR = 0.28 + 2.093 H + 0.058 BM - 0.028 A$

in which *BMR* is in MJ/day, *H* is height in m, *BM* is body mass in kg, and *A* is age in years.

Nowadays there is wide agreement that *BMR* is best predicted from the fat-free body mass (*FFM*) as this represents the active cell mass (for determination of *FFM* see Chapter 17). Using *FFM* as the only independent variable, *BMR* can be estimated with an error of about 10 per cent in different categories of subjects such as women, men, normal weight, obese, etc. The error in the prediction of *BMR* can be further reduced by adding the fat mass (*FM*) as an independent variable. Finally, there is a third variable in women, the phase of the menstrual cycle, which explains a significant part of the variation in *BMR*. Using these parameters, Meijer *et al.* (1992) studied the relationship between sleeping metabolic rate (*SMR*) measured from 3.00 - 6.00 h in a respiration chamber and body composition in 47 healthy adult subjects (23 men and 24 women). The resulting equation was:

$$SMR = 0.73 + 0.101\ FFM + 0.023\ FM + 0.364\ MC$$

in which *SMR* is in MJ/day, *FFM* and *FM* are in kg and the phase of the menstrual cycle (MC) 0 for men and pre-ovulation women, and 1 for post-ovulating women. The equation shows clearly that the contribution of *FFM*, the active cell mass, to *SMR* is far higher than that of *FM*. Secondly, pre-ovulating women have a comparable *SMR* to males, based on their *FFM* and *FM*. The *SMR* in post-ovulating women is increased by about 8 per cent. Women are known to have a higher energy intake during the post-ovulation period as well. The changes are thought to be due to increased levels of progesterone during this period.

SMR is very close to *BMR*, depending on the measuring protocol for *SMR*, the difference being the energy cost of arousal. Goldberg *et al.* (1988) measured the energy expenditure in 40 adult subjects overnight in a respiration chamber after an evening meal at 19.00 h. The subjects prepared for bed at 22.30 h and lights were out at 23.00 h. The overnight metabolic rate (*OMR*) was defined as the average metabolic rate from 23.30 - 8.00 h, and *SMR* as the lowest continuous 60-minute period during this time. *BMR* was measured during 1 hour (8.00-9.00) immediately upon waking when the subjects were still at complete rest, 13 hours post-absorptive and at thermoneutrality. The mean ratio *OMR/BMR* was 0.95 and *SMR/BMR* was 0.88. These data suggest that *OMR* is 5 per cent lower than *BMR* and *SMR* is 12 per cent lower than *BMR*. However, *SMR* was defined as the lowest 60-minute period during sleep while often a 3-hour interval during sleep is used, including more disturbed sleep. Thus *SMR* is 5-10 per cent lower than *BMR*, i. e. the energy cost of arousal is 5-10 per cent of *BMR* (see also Fredrix *et al.*, 1990).

6   FOOD PROCESSING

Energy expenditure increases after the ingestion of food, even when the subject does not change his physical activity. The increase in resting metabolic rate is due to the processing of food, the so-called diet-induced energy expenditure (*DEE*). The processing of food consists of several elements (Figure 16.8). The consumed food is digested and the digestion products - or dietary metabolites - are absorbed. Part of these metabolites are used at once for oxidation; the remainder is converted to reserve fuel for storage.

In the intestines, carbohydrate, fat and protein are broken down to hexoses, fatty acids and monoglycerides, and amino acids respectively. These metabolites are then absorbed. The hexoses are oxidized or converted to glycogen or fat for storage. The fatty acids and monoglycerides are oxidized directly, or the triglycerides are resynthesized for storage.

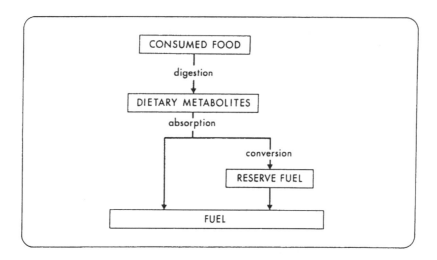

FIGURE 16.8
Diagram of the processing of consumed food
(From: Westerterp, 1976)

The amino acids are deaminated and oxidized either directly or via resynthesis and hydrolysis of protein (Figure 16.9). The conversion of amino acids to glucose or fat normally plays a minor role, given the protein content of the diet to be moderate and the subject to be in energy balance.

DEE can be calculated theoretically with the aid of *theoretical estimates* of the energy costs of the individual elements of food processing. Synthesis of digestive enzymes costs about 4 per cent of the energy content of the food supplied. The cost of bond breakage during digestion is only small: 0.6, 0.1, and 0.6 per cent of the energy content for the breakage of the glucoside, glyceride fatty acid, and peptide bonds respectively. The absorption cost for hexoses (by active transport) amounts to 2.6 per cent of the energy content. Absorption of at least some amino acids is also by active transport, and may take as much energy as the absorption of hexoses. Absorption of fatty acids and monoglycerides, mainly as micels that penetrate the cell membrane, is an energy independent process. The cost of conversion for storage depends on the metabolite, form of storage, and amount stored. Storage of hexoses in the form of glycogen takes about 5 per cent and as fat 20-25 per cent; conversion of dietary triglycerides to reserve fat costs about 5 per cent of the energy content.

The amount of energy stored is more difficult to estimate because under *ad libitum* conditions an unknown fraction of the dietary metabolites passes through storage before it is oxidized. Humans are periodic eaters and continuous metabolizers. Hence, it is necessary to store part of the food energy before use. This aspect will be discussed further in Chapter 21. On average, *DEE* appears to take about 10 per cent of the energy content of the ingested nutrients (see below).

The first *systematic measurement* of *DEE* was carried out at the turn of the 19th century by means of indirect calorimetry. The increase in oxygen consumption was measured in a dog under fasting conditions and during feeding with meat.

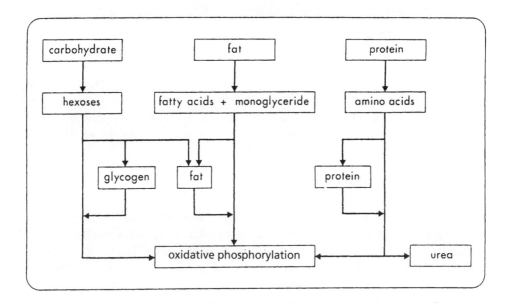

FIGURE 16.9
Diagram of the processing of separate energy containing food components: carbohydrate, fat, and protein
(From: Westerterp, 1976)

The phenomenon was termed 'specific dynamic action', which later gave way to the terms 'thermic effect of food', 'diet-induced thermogenesis', and now 'diet-induced energy expenditure', in line with 'activity-induced energy expenditure'. *DEE* can be defined as the increase in energy expenditure above the basal fasting level divided by the energy content of the food ingested, and is commonly expressed as a percentage. The post-prandial rise in energy expenditure lasts for several hours and is often regarded as being completely terminated at approximately 10 hours after the last meal. However, there is still some argument as to when the post-absorptive state is actually reached, the estimates vary between 8 and 18 hours. In the previous section a 13-hour criterium was assumed for the post-absorptive state.

Shortly after the first measurements, *DEE* was divided into two components: firstly, an obligatory part related to the digestion, absorption, and storage of nutrients as described above and, secondly, a facultative or adaptive component reflecting the energy spent in excess of the processing requirement. The latter component is referred to as 'Luxus-konsumption'. It is often regarded as a way of disposing of any excess energy intake by an increased generation of heat.

DEE is usually measured with the ventilated-hood method after an overnight fast of 12-14 hours. The subjects stay overnight in the laboratory or are transported to the laboratory, avoiding any physical activity prior to the observation period. The respiratory gas exchange is measured in the basal resting state and then for several hours after the ingestion of a meal. Weststrate *et al.* (1989) performed extensive studies on the determination of *DEE*. First of all they focused on a number of methodological points. When measuring *DEE* in post-absorptive subjects and using a meal size of maximally 2.1 MJ for women and 2.6 MJ for men or about 25 per cent of the daily energy intake, 90 per cent of *DEE* was measured within 4 hours after the meal. This is about the maximum duration of a venti-lated-hood measurement in which subjects have to be in rest. Using a mixed test meal, i.e. with all macronutrients present, *DEE* was about 8 per cent of the ingested energy. Comparing the measured *DEE* values with the theoretical estimates mentioned above, in this situation *DEE* comprised hardly any more than the obligatory component including digestion and absorption, leaving 1-2 per cent for temporary storage. The facultative com-ponent was negligible in this situation. Based on observations as mentioned above, *DEE* in real life is assumed to be 10 per cent of *ADMR* in subjects consuming the average mixed diet and being in energy balance.

7   PHYSICAL ACTIVITY

Activity-induced energy expenditure (*AEE*) is the most facultative component of daily energy expenditure, and hence is difficult to assess. It requires a technique capable of measuring *AEE* under free-living conditions with reasonable accuracy. Until recently, in-formation on *AEE* was only available from studies in laboratory settings like a metabolic ward or a respiration chamber. In such an environment, variation in physical activity is usually very low due to its restrictive nature. The doubly-labeled water method described earlier offers the possibility to measure the average daily metabolic rate (*ADMR*) in unre-strained humans in their normal routines.

The doubly-labeled water method results in a measure of *ADMR*. *AEE* can only be es-tablished by subtracting *BMR* and *DEE*. These components may be individually estimated or determined in the laboratory. Alternatively, the activity level can be calculated as a multiple of *BMR*, i.e. *ADMR/BMR*. The latter procedure gives a good measure for the level of physical activity since *DEE* is supposed to be a fixed proportion of *ADMR*. Thus, *AEE* is the only component of *ADMR* that shows major variations as reflected in the term *ADMR/BMR*. Therefore, this quotient is known as the *physical-activity index* (*PAI*).

In daily life *PAI* ranges between 1.2 and 2.2. Its minimum value, for subjects staying in bed all day and not consuming any food, is obviously 1.0 (*ADMR* = *BMR*; *DEE* and *AEE* are both 0). As soon as someone gets up and eats up to energy balance, *PAI* becomes higher than 1.2. The maximum value is higher than 2.2, but nowadays values over 2.2 are only reached in endurance sports such as cycling races taking several days. Figure 16.10 shows a frequency distribution of *PAI* in 96 subjects, compiled from recent literature. The subjects were 66 women and 30 men aged between 19 and 48. They cover most studies in the literature up to 1991 in which data were available on subject characteristics to calculate *BMR*, and on *ADMR* as measured with doubly-labeled water.

In 1985, an expert consultation by FAO/WHO/UNO (Food and Agriculture Organization/World Health Organization/United Nations Organization) estimated the *PAI* of subjects based on observed patterns of activity and the laboratory determinations of the associated energy costs of the observed activities. *PAI* was calculated for subjects in three categories: light-activity work (office clerk), moderate-activity work (subsistence farmer), and heavy work (not specified). The calculated *PAI* values were 1.5, 1.8, and 2.1 respectively. Comparing the calculated estimates with the measurements that are available now (Figure 16.10) there appears to be a discrepancy in the lower *PAI* range. Apparently, there are many people, 31 of the 96 subjects or nearly one third, who have a very low activ-ity level. On the upper side of the activity range the measured values and the estimates be-come very close.

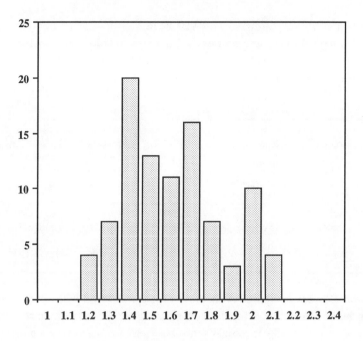

physical activity index

FIGURE 16.10
Frequency distribution of the physical activity index (average daily metabolic rate/basal metabolic rate) in 96 subjects as compiled from the recent literature

Extension of the doubly-labeled water determinations in the near future will tell whether this conclusion is based on a representative sample of the general population.

Translating *PAI* to the size of *AEE* allows conclusions on the relative contribution of physical activity to *ADMR*. The *PAI* of highly physically active subjects is 2.2. In this situation *AEE* is half of *ADMR*. In nearly everybody's life, therefore, *AEE* is smaller than *BMR*, as *BMR* is the largest component of *ADMR*. In the average subject *BMR* is 65 per cent, *DEE* 10 per cent, and *AEE* 25 per cent of *ADMR*.

8   ENERGY EXPENDITURE AS A MEASURE TO VALIDATE MEASUREMENTS OF ENERGY INTAKE

Nutrition research uses data on energy intake (*EI*) for comparing the energy balance between subjects, to detect changes in energy balance within subjects or as a basis for nutrition intervention. Recently, there is some discussion on the reliability of the results of self-reported food intake as a measure for *EI* since the introduction of doubly-labeled water as a method to measure energy expenditure (*EE*). The doubly-labeled water method allows validation of data on *EI* in subjects who are in energy balance.

There are several methods to measure the dietary intake. Of these, dietary recall, including dietary history and dietary records, are commonly used. The method of dietary record is supposed to be superior to that of dietary recall with regard to actual consumption, as it does not rely on the memory of the subject. The disadvantage is, however, that the subjects may change their feeding habits as they have to write down all they eat.

TABLE 16.3

Subject characteristics in three studies, A Comparison of lean and obese; B Activity intervention; C Feeding intervention

study	subjects		n	age (y)	height (m)	body mass (kg)	BMI (kg/m)
A	women	lean	5	28 ± 7	1.66 ± 0.04	53.8 ± 2.1	19.7 ± 1.3
		obese	6	34 ± 7	1.73 ± 0.11	94.5 ± 19.9	31.3 ± 3.3
	men	lean	6	37 ± 3	1.75 ± 0.05	74.0 ± 8.4	24.0 ± 1.5
		obese	4	37 ± 4	1.83 ± 0.06	109.2 ± 20.1	32.3 ± 4.2
B	women		8	36 ± 4	1.68 ± 0.07	67.0 ± 6.9	23.7 ± 2.0
	men		8	37 ± 3	1.78 ± 0.06	71.0 ± 5.4	22.4 ± 2.0
C	men		9	41 ± 11	1.77 ± 0.06	77.0 ± 12.9	24.5 ± 3.5

Data on *EI* were compared with those on *EE* measured with doubly-labeled water. The latter method measures *EE* under normal living conditions with an accuracy of 1-3 per cent and a precision of 2-8 per cent. When subjects are in energy balance *EI* equals *EE*, and hence, doubly-labeled water is the ideal method to validate recordings of *EI*. In the following, data on *EI* and *EE* in three different settings are compared:

1 In obese and lean subjects.
2 Energy metabolism in subjects before and after activity intervention.
3 In subjects fed according to their reported dietary intake.

The subjects' characteristics are presented in Table 16.3, grouped according to the three studies: comparison between lean and obese (A), activity intervention (B), and feeding intervention study (C). Studies A and B included both females and males, study C concerned males only. All subjects were adults aged between 22 and 62. Most of them had regular jobs or were engaged in housework. The mean body-mass index (*BMI*) in the lean groups in study A was slightly, but not significantly, lower than the mean *BMI* of subjects of the same gender in studies B and C. Obese subjects in study A had a significantly higher *BMI* than subjects of the same gender in studies B and C.

Study A was designed to compare the energy metabolism of obese men and women under free-living conditions. There was no interference with either *EI* or *EE*. In study B, the effect of an increase in physical activity on the energy balance was investigated, without interfering with energy intake. Initially, the sedentary subjects were trained during 40 weeks to run a half marathon (21.1 km). In study C, the effect of the pattern of energy intake on energy expenditure was investigated. During two weeks the subjects were fed with a fixed amount of energy, based on their recorded intake. During one week the food provided had to be taken in two big meals a day, during the other week in seven small meals.

The energy intake was measured with a 7-day dietary record. The subjects recorded all foods consumed in their diary, including brand names and cooking recipes where applicable. The diary was divided into 7 periods a day, 3 meals and 4 inter-meal periods. The foods were weighed or quantified in household measures. Volumes of repeatedly used utensils were measured with a 400 ml cup with 10 ml scaling. The subjects were instructed by a dietician before the observation period started. The diary also included a preprinted example of a full day's intake. The subjects were assisted further by telephone in the middle of the week to solve any problems. During a short session with the subject at the end of the week, the diary was examined to eliminate inconsistencies, etc. The energy content of the foods was calculated according to the Dutch food composition table.

Energy expenditure was measured over two weeks with doubly-labeled water. In studies A and B, the first week of the two-week interval coincided with the measurement of *EI*. In study C, the subjects were fed during the two-week interval according to their intake as recorded during the week preceding the interval. The subjects drank a weighed amount of an isotope mixture of ^{18}O and ^2H. The dose was calculated to create an excess of 300 ppm ^{18}O and 150 ppm ^2H. The isotope mixture was administered in the evening between 22.00 and 23.00 h, just before the subjects went to bed, after collecting a background sample. Further urine samples were collected from the second voiding in the morning on days 1, 8, and 15 after administration. The carbon dioxide production was calculated from the isotope ratios in a baseline sample. The carbon dioxide production was then converted to *EE* by using an energy equivalent of 23.7 kJ/l as calculated from the 7-day dietary records (Table 16.1).

The energy balance over the 2-week observation interval with doubly-labeled water was measured by weighing the subjects (studies A and C). The night before the start of the interval and the last night of the interval they stayed in a respiration chamber. Here, body mass was measured upon rising, after emptying the bladder and without clothes, on the same scales with an accuracy of ±0.1 kg. The energy balance of the subjects in study B was calculated from changes in body mass and body composition over the full 40-week intervention period. Here, *EI*, *EE*, and body composition were measured before and at the end of the training period.

The subjects in study A were in energy balance over the 2-week observation interval with doubly-labeled water. Body mass changes were not signicantly different from zero in all groups, i.e. 0.1 ± 1.3, –0.2 ± 1.2, 0.0 ± 1.3, and –1.1 ± 0.9 kg (mean ± SD) for lean and obese females and for lean and obese males respectively. When combining females and males, the body-mass changes were not significant either, i.e. –0.1 ± 0.6 and –0.5 ± 1.1 kg for the lean and the obese, respectively. Thus, there was a tendency towards a slight but not significant weight loss. The differences between reported *EI* and measured *EE* were –11 ± 15 and –30 ± 18 per cent of *EE*, for the lean and obese respectively. Thus, there was no significant difference between *EI* and *EE* in the lean group, but *EI* in the obese group was systematically lower than *EE* ($p < 0.01$). The difference between *EI* and *EE* was not related to body-mass changes, neither in the individual lean group and obese group, nor in the group as a whole. The difference between *EI* and *EE* was related to the degree of overweight as expressed in the value of *BMI* (Figure 16.11(*a*), $r = 0.56$, $p < 0.01$).

The subjects in study B were weight-stable before the start of the training period (B_0). During the 40-week training interval *EE* increased by 30 per cent (B_{40}). Females did not show a significant change in body mass. On average they lost 2.8 kg fat mass (FM) and gained 2.1 kg fat-free mass (FFM). Males showed an average body mass change of –1.3 kg, made up of a loss of 4.5 kg FM and an increase of 3.2 kg FFM. Before the start of the training intervention, the difference between *EI* and *EE* was –5 ± 28 per cent, a result comparable to the measurement in the lean subjects of study A. The average *EI* did not differ from the average *EE*, although individual discrepancies ranged from +22 to –52 per cent. Figure 16.11(*b*) shows the results of study A and B_0 plotted together, supporting the result already shown in Figure 16.11(*a*). Of the original 16 subjects, 13 completed the training and were measured again after 40 weeks. At that time, there was a significant difference between intake and expenditure, $EI - EE = -19 \pm 17$ per cent ($p < 0.01$). Figure 16.12 shows *EI* of the 13 subjects plotted as a funtion of their *EE*, before and at the end of the training. Before the training *EI* was higher than *EE* in 6 out of 13 and at the end of the training in only 1 out ot 13 subjects.

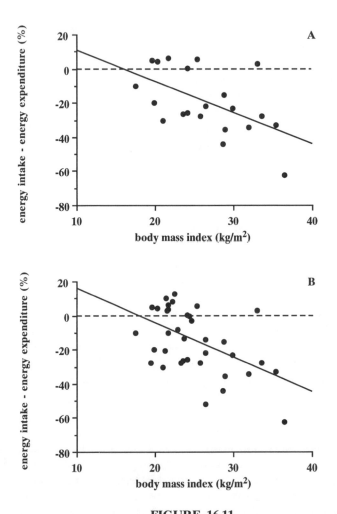

FIGURE 16.11

Reported energy intake (*EI*) minus measured energy expenditure (*EE*), expressed as a percentage of *EE* and plotted as a function of the body mass index (*BMI*) with the calculated linear regression line

(*a*): results from study A, $n = 21$; (*b*): results from study A and B_0, $n = 37$.

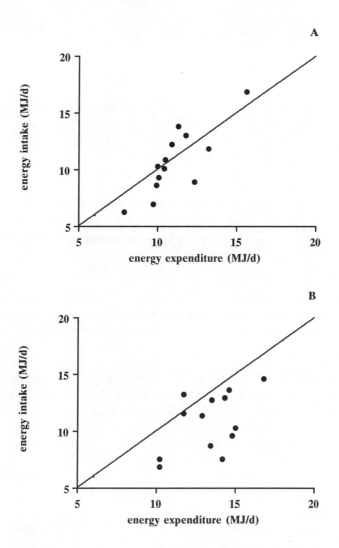

FIGURE 16.12
Reported energy intake (*EI*) plotted as a function of measured energy expenditure (*EE*)
with the line of identity

(*a*): results from 13 subjects before the start of a training program; (*b*): results from the same 13 subjects at the end of a 40-week training program.

For 2 weeks the subjects in study C were given food according to their reported *EI*. The average body mass change was −0.6 ± 0.9 kg during the 2-week interval. Individual body-mass changes ranged between +0.6 and −2.3 kg. There was a significant relationship between the discrepancy between *EI* and *EE* and the body mass change during the observation interval ($r = 0.80$, $p < 0.01$). Subjects with the highest energy deficit lost most weight (Figure 16.13).

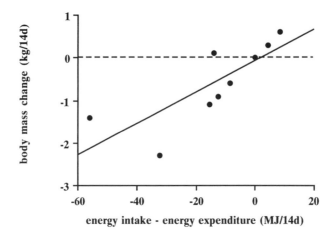

FIGURE 16.13

Body-mass change over a 14-day interval plotted as a function of energy intake (*EI*) minus energy expenditure (*EE*) over the same interval with the calculated linear regression line
EI is the energy content of the food as supplied according to reported intake and *EE* is measured (see text for further details).

In all three studies the reported *EI* tended to be lower or was even significantly lower than the measured *EE*. There are two possible reasons for this discrepancy. Firstly, the subjects were not in energy balance; secondly, they underreported their intake. In these studies it was assumed that the procedure used to calculate *EI* from the 7-day diary was accurate, as well as the calculated *EE* from the doubly-labeled water method. The two reasons mentioned above will be considered after discussing the two assumptions.

To facilitate the procedure, the subjects reported their intake mainly in household measures instead of weighing every item. This might have introduced some bias and therefore the measured volumes of the utensils of their own household were used to calculate the quantities. Using the national food-composition table with added information on local foods guaranteed the most accurate procedure to convert the quantities to energy.

The measurement of *EE* with doubly-labeled water has proven to be accurate and precise. Differences between the simultaneous measurement of carbon dioxide production with doubly-labeled water and respirometry showed the validity of the method. There is consensus regarding the calculation procedure of carbon dioxide production from the isotope elimination rates. In the present study, the conversion of carbon dioxide production to energy expenditure needs further discussion.

The energy equivalent of carbon dioxide depends on the energy substrate used, i.e. carbohydrate, protein, or fat. For the oxidation of pure carbohydrate the value is 21.1 kJ/l, of pure fat 27.5 kJ/l, and of pure alcohol 30.5 kJ/l. In practice, the body uses a fuel mixture which can be assessed from the nutrient composition of the diet, correcting for changes in body composition. This procedure reduces the potential error from this source to less than 2 per cent. Even in the situation of imbalance between *EI* and *EE* as in study C and during the activity intervention in study B, the potential error from this source is less than 2 per cent. In any case, the calculated value of *EE* increases when the body fat contributes to oxidation at a negative energy balance, as can be seen from the energy equivalents referred to above. Thus, the discrepancy between *EI* and *EE* seen in all studies is not overestimated, because the subjects were in energy balance or tended to be in a negative energy balance.

In conclusion, a self-reported dietary intake underestimates the energy needs, even when a reliable method is used to measure the intake. One can only speculate on the reasons for underreporting. Obese people often suggest that their intake is normal or even below normal and consequently suggest a low level of *EE* as the reason for overweight. Prentice *et al.* (1986) showed that *EE* in obese people is significantly higher than in lean controls. They observed a mean difference between reported *EI* and measured *EE* of 33 per cent, while the corresponding difference in the lean group was only 2 per cent, which is in agreement with the present study. The reason for underreporting by the obese might be the wish to eat less and consequently they underestimate their real intake. Lean people are not expected to have any reason for underreporting on purpose. The measured under-reporting by lean subjects was only significant in study B when *EE* was increased by 30 per cent with a training program and when the subjects had to report their intake again at the end of the program. Probably, both factors are important. The subjects may have reported with a bias to their usual, lower, intake. The fact that they had to keep a 7-day dietary record for the fourth time appeared to be a real burden according to their reactions. The results from the second and third dietary record, 8 and 20 weeks after the start of the training period have not been reported here, as *EE* was only measured in a subselection of the subjects. Finally, keeping a dietary record possibly influences the habitual intake in a negative sense. It is impossible to discern whether the small differences between *EI* and *EE* in the lean subjects are due to underreporting or to dieting. Whatever the reason may be, data on self-reported intake should be interpreted with great care, particularly in the case of the obese.

9   SUMMARY

Heat production varies with body size and composition, age and gender.

The components of daily energy expenditure are sleeping metabolic rate, arousal, diet-induced energy expenditure (DEE), and physical activity (AEE).

Measuring energy expenditure as heat production is known as direct calorimetry. The measurement of heat production by measuring oxygen consumption and carbondioxide production is known as indirect calorimetry.

The doubly-labeled water method measures average daily energy expenditure in free living individuals.

The individual components of daily energy expenditure can be measured with a ventilated hood and in a respiration chamber.

Measured energy expenditure is used as a validation of measurements of energy intake.

17

Body composition

Together, energy from food (*EI*) and energy expenditure (*EE*) result in growth (*EG*) which can be formulated as follows: $EI = EE + EG$ or $EG = EI - EE$

This chapter focuses on growth as a consequence of a discrepancy between energy intake and energy expenditure, i.e. growth in terms of a change in the body's energy content. A further restriction here is the age of the subjects under consideration. Whereas growth is usually seen as the development of infants from birth to maturity, this chapter mainly considers changes in weight and body composition in adult life, as a result of an energy deficit or surplus. The one criterion in this context for adult life is that the body, for instance the skeleton, is fully developed. Thus, most studies referred to are on subjects of at least 18 years of age.

1   BODY WEIGHT AND BODY COMPOSITION

The body weight of adults is supposed to be regulated to remain at a constant level. There are, however, hardly any longitudinal studies to confirm this. Insurance companies in the United States have set up tables for ideal weight and height associated with the lowest mortality. In real life, however, most people seem to be heavier than their ideal weight for height and, despite health education efforts, the discrepancies do not change much.

Thus, weight homeostasis is not perfect and overweight is a familiar problem in Western societies. One of the few and earliest longitudinal studies providing information on the constancy of body weight is the *Framingham Study*. This long-term study included 5209 adults, 30-59 years of age, living in the town of Framingham (USA) at the start of the study (in 1948, 1949). Every two years the subjects underwent a standard medical examination, including the measurement of body weight, for at least 20 years unless prevented by illness or death. Figure 17.1 shows the proportion of subjects changing weight over any part of the observation period by the indicated amount as calculated by James (1985). From Figure 17.1 it is clear that nearly no-one retained a constant weight but that most people lost or gained between 5 and 10 kg over any part of the 20-year period in adult life.

A weight change of 5-10 kg over an interval of several years is no indication of an imperfect system for the preservation of energy balance in terms of the size of the discrepancy between energy intake and energy expenditure. This can be shown by comparing the difference between intake and expenditure with the total energy turnover during the interval of several years. A weight change in adult organisms can be the result of a change in body water, carbohydrate, protein, or fat. The body stores glycogen and fat as energy reserves, the former in liver and muscle, and the latter in adipose tissue. Under special circumstances a weight change can be the result of a change in lean tissue, i.e. muscle wasting in patients and exercise-induced muscle increase. Table 17.1 shows the energy equivalent of the individual body components.

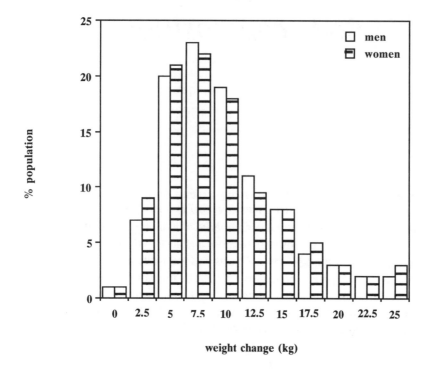

weight change (kg)

FIGURE 17.1

Weight change in a population of 5209 adults, men and women initially aged 30-59 living in the town of Framingham, observed over 20 years at two-year intervals Proportion of subjects changing weight over any part of the observation period by the indicated amount, as calculated by James (1985)

Adipose tissue obviously is the main energy store of the body. Fat not only has a more than two times higher energy equivalent than carbohydrate and protein, but it can also be stored without much extra water. Also, the storage capacity for glycogen is limited while there is no known limit to the formation of fat tissue. The energy equivalent of a change in body weight ranges from 0 to 38 MJ/kg, the minimum when it is only a change in body water and the maximum when the change represents purely fat. A reduction in body weight by 5-10 kg in, say, 5-10 years, i.e. a weight loss of 1 kg per year, is now known to represent an energy imbalance of 38 MJ per year at the most. Knowing, on the other hand, that an individual has a daily energy turnover of 8 to 12 MJ under normal living conditions, i.e. a mean energy turnover of 10 MJ per day or 3650 MJ per year, the discrepancy is less than 1 per cent. In other words, there is a perfect regulation of energy balance.

There are speculations in the literature on the existence of more than one regulatory system for energy balance. It is assumed that each macronutrient, i.e. carbohydrate, protein, and fat, is regulated individually in addition to water which has a known precise regulation. The basis for the hypothetical nutrient control system lies in the comparison of the daily nutrient intake and the size of nutrient stores in the body. To illustrate this point, the average daily intake of the above mentioned 10 MJ having the advised macronutrient energy% ratio of 55/15/30 for carbohydrate/protein/fat, can be compared to the size of the body nutrient stores in 'reference man'. The 'reference man' is an adult male subject of

1.70 m in height and with a body weight of 70 kg, of which 20 per cent is body fat. Figure 17.2 shows the nutrient intake as energy% of the nutrient stores. The carbohydrate, protein, and fat intake is 5.5, 1.5, and 3.0 MJ/day respectively. The size of the stores is 6, 120, and 520 MJ respectively. The latter figures are based on an available quantity of glycogen in liver and muscle of 250-500 g, a muscle mass of 30 kg with 7.5 kg protein which, of course, can only partly be used to cover energy shortages, and a fat mass of 14 kg being 20 per cent of the body mass. Thus, as depicted in Figure 17.2, the daily carbohydrate, protein and fat intake is 90, 1.3, and 0.6 per cent of the body stores respectively. The daily carbohydrate intake is close to the carbohydrate store in the form of glycogen. The daily protein and fat intake are a fraction of the amount present, mainly in muscle and fat tissue respectively. Animal studies indicate that there are nutrient-specific hungers to facilitate not only sufficient energy intake but also sufficient intake of individual nutrients. In terms of energy balance at least a sufficient amount of carbohydrate should be eaten. The human body can not function without carbohydrate as an energy substrate. Protein is not essential for energy production and fat is usually available in unlimited amounts.

Body weight is a fairly reliable indicator of the state of energy balance. Excess energy intake or expenditure generally causes weight gain or weight loss respectively. However, information on body mass alone is not sufficient for valid conclusions on the energy metabolism of a subject. This is evident from the great difference in energy densities of the major body components (Table 17.1). Two examples can illustrate this point. The first example is the short-term effect of a change in energy balance in daily practice.

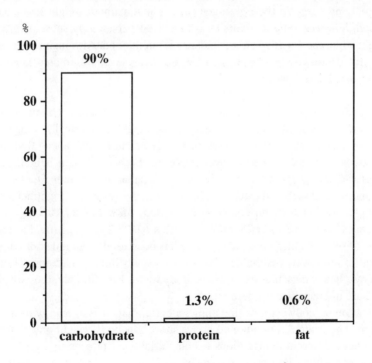

FIGURE 17.2
Daily nutrient intake as a percentage of the quantities of each nutrient in body stores

TABLE 17.1

The composition and the energy equivalent for the major body components

component	water (%)	dry matter (%)	energy density (MJ/kg)	
			dry	wet
water	100	0	0	0
glycogen	75	25	16	4
lean tissue	73	27	16	4
fat tissue	15	85	38	32

Many people are or bring themselves in a negative energy balance during certain periods, their energy intake being lower than their energy expenditure. Initially, the discrepancy between intake and expenditure will be covered by using the energy store of the body in the form of glycogen. Carbohydrate in the form of glucose is the first nutrient to be in short supply as we have seen above. Subsequently, the body will cover the deficit between intake and expenditure mainly by breaking down body fat. The depletion of the glycogen pool of the body causes a rapid decrease of body weight while weight changes due to fat mobilization are much slower because of the 8 times difference in energy density (Table 17.1). Thus, people adopting a low-energy or carbohydrate-free diet initially lose weight rapidly by using up their glycogen pool. The maximum weight loss from glycogen mobilization, however, only amounts to 1-2 kg. That person may subsequently even gain body weight while the negative energy balance is maintained. Although the energy store of the body still decreases, this is masked by the increase in water associated with some repletion of the glycogen stores.

The second example is the diagnosis of overweight or obesity. Overweight or obesity is often based on a classification using weight and height. Emperically, weight divided by the square of height turns out to correct weight for height in such a way that small people are not underweight and tall people overweight just because of their height. This was formulated, at the end of the 19th century, by the Belgian astronomer Quetelet, who proposed an index to classify subjects for obesity with the formula weight:height², in which weight is in kg and height in m. The *Quetelet* or *Body Mass Index (BMI)* in normal-weight subjects ranges between 20 and 25. Subjects with a *BMI* > 25 are classified as having a certain degree of overweight. Subsequently, *BMI* has been used as an index of adiposis or fatness and this is not always justifiable. Overweight usually indicates adiposity but could also be caused by a heavy musculature or excess body water. For example, football players with a *BMI* > 25 are not usually considered too fat.

Information on body composition allows a better diagnosis of energy balance than just body weight. The examples mentioned above show how body weight and changes in body weight can have different explanations needing additional information on body *composition*. Body composition in terms of energy balance contains, first of all, information on fat mass (*FM*) and fat-free mass (*FFM*). The latter is sometimes further subdivided in, for instance, water mass or total body water, protein mass and mineral or bone mass, neglecting the small compartments such as glycogen. Obviously, it is difficult to measure body composition. Body weight can easily be measured with reasonable accuracy, even on bathroom scales. Body height is a measure that needs to be taken only once in adult life and will not change much afterwards. Thus, *BMI* can be calculated without much trouble. There are no such easy methods to measure body composition.

2   MEASUREMENT

In vivo, body composition can only be measured indirectly. These days, there is a great variety of methods based on different assumptions and each with its limitations. They all stem from the chemical analysis of six adult cadavers of subjects with a normal body condition until death, carried out at the end of the 1940s and early 1950s. The general model for body composition is the two-compartment model, i.e. *FM* and *FFM*. *FM* is assumed to be triglyceride without water and potassium, with a density of 0.90 g/ml. *FFM* is assumed to have a water content of 73 per cent and a potassium content of 60-70 mmol/kg, with a density of 1.10 g/ml. There are four generally accepted methods for the measurement of body composition based on one or more of the assumptions mentioned above:

1 Densitometry.
2 Total body water.
3 Total body potassium.
4 Anthropometry.

In this section, these four well-established methods will be discussed before multi-compartment models and methods for *in vivo* determination, which have been developed more recently, are described.

2.1   DENSITOMETRY

Densitometry assumes a constant chemical composition of *FM* and *FFM* resulting in a density of 0.90 and 1.10 respectively, as mentioned above. The method requires the measurement of body weight and body volume. The most widely used technique for measuring body volume is according to Archimedes' principle: the volume of an object submerged in water equals the volume of water displaced by the object. The difference between weight in air and weight under water, corrected for the density of the water at the temperature at the time of measurement, is the body volume. The body volume thus obtained should be corrected for lung volume, ideally by measuring simultaneously the residual lung volume during submersion (Figure 17.3). Densitometry has gained widespread application and is the 'gold standard' for body-composition measurement with other techniques.

The theoretical error of densitometry for predicting *FM* and *FFM* is 3-4 per cent and is associated with the uncertainty of the density and chemical composition of *FFM*. The main variables are the water content and bone density. In practice, additional error sources are variability in gastro-intestinal gas volume and residual lung volume, the latter especially when the lung volume is not measured during but before or after submersion. An error of 0.1 l in one of the two is roughly equivalent to a 1 per cent error in *FM* and *FFM*. Usually, the overall accuracy of densitometry for body composition is 1-2 per cent.

Body volume measurement by submersion in water is not always applicable for adults, for example patients and the elderly. Several laboratories are working on alternative methods to measure body volume, for example in air instead of in water.

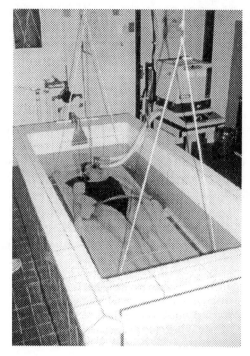

FIGURE 17.3
Measurement of body composition with densitometry
The weight of the subject is measured during submersion with simultaneous measurement of lung volume. Therefore the subject lies down on a stretcher suspended from a scale while connected to a lung volume apparatus.

These methods are still in a developmental stage and the first results have only just appeared in literature. The advantage would not only be the applicability but also the required time for an observation. Under-water weighing of a trained subject takes half an hour while measuring body volume in an air tank can be done in 5-10 min.

2.2   TOTAL BODY WATER

Total body water (*TBW*) is a measure for body composition assuming a fixed hydration of *FFM*, usually 73 per cent. Measuring *BM* and *TBW* allows the calculation of *FFM* as *TBW*/0.73 and *FM* as *BM* - *FFM*. *TBW* is measured with isotope dilution, mainly using isotopes of hydrogen and oxygen, ^3H, ^2H, and ^{18}O. The method is based on the assumption that these isotopes have the same distribution volume as water. A subject is given an accurately measured oral or intravenous dose of labeled water, followed by an equilibration period of at least 2 hours, and subsequent sampling of the body fluid. Dose, equilibration time, and sampling medium depend on the isotope, the dosing route and the facilities for sample analysis. Tritium (^3H) is a radio-isotope which is measured with liquid scintillation counting. Deuterium (^2H) and ^{18}O are both stable isotopes. ^2H can be measured in higher concentrations with infrared absorption. Both isotopes are measured in low concentrations with isotope ratio mass spectrometry (IRMS). Body fluids suitable for sampling are saliva, blood, and urine. The minimal length of the equilibration period is 2 hours, which is achieved when administering intravenously and taking blood as the sampling medium. The less invasive oral administration route and sampling urine need a minimal equilibration time of 4-6 hours. Calculation of *TBW* is based on the following relationship:

$$C_1 \cdot V_1 = C_2 \cdot V_2$$

in which C_1 and V_1 are the concentration of the tracer and the volume of the dose, respectively, and C_2 and V_2 the concentration of the tracer and the volume of the body water.

In practice, using a non-invasive method with stable isotopes at low concentrations, subjects receive a dose of labeled water in the post-absorptive state after collecting a background sample of saliva or urine. Background levels for ^{2}H, and ^{18}O are around 150 and 2000 ppm, respectively. The minimal excess enrichment to be reached is around 100 ppm. After equilibration, lasting 4-6 hours, a final saliva or urine sample is collected. For urine this should be a sample from at least a second voiding after administration of the labeled water.

The use of ^{18}O as a tracer is preferred over that of ^{3}H and ^{2}H since the dilution space of ^{18}O is very close to *TBW*. On average, the dilution space of the hydrogen isotopes is 4 per cent larger, whereas that of ^{18}O is only 1 per cent larger than *TBW*, due to the exchange of the label with non-aqueous substances in the body. On the other hand, the cost of ^{18}O is 100 times higher than the other labels.

The results from body-composition measurements with isotope dilution usually show good agreement with densitometry. This is illustrated by a recent study by Van der Kooy *et al.* in which both techniques are applied simultaneously in an intervention study. Eighty-four subjects, 42 women and 42 men aged 35-45, participated in a weight-reduction program. The initial *BMI* in all subjects was higher than 27. They were allowed an energy deficit of 4.2 MJ/day for 13 weeks, losing on average 13 kg BM. Their body composition was measured before and after weight loss. Figure 17.4 shows the values for body composition as obtained from densitometry and ^{2}H dilution. *FFM* calculated from isotope dilution, assuming the ^{2}H-dilution space to be 1.04 × *TBW*, is plotted as a function of *FFM* measured with densitometry. There is a very high correlation between the results of the two methods in both situations, before and after weight loss ($r = 0.98$, $p < 0.001$). The slopes of the regression lines do not differ significantly from unity, indicating that on average the results of both methods are the same. The regression lines calculated separately for the observations before and after the dietary treatment and the resulting weight loss, show no difference either. However, while the two methods give on average the same results, discrepancies for individual observations can be as high as 1.9 kg. Whatever the reason for the discrepancies between the results from the different methods for the measurement of body composition, the absolute accuracy of densitometry and isotope dilution is in the same range of 1-2 per cent.

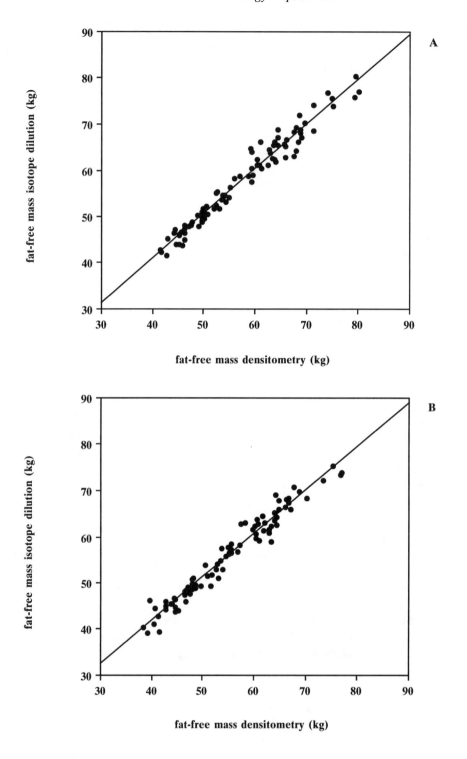

FIGURE 17.4
Fat-free mass (*FFM*) measured with isotope dilution plotted as a function of *FFM* measured with densitometry in 84 subjects (*A*) before and (*B*) after weight loss

2.3   TOTAL BODY POTASSIUM

Potassium is a cation which is distributed mainly intracellularly, and which is not present in stored triglyceride. A fixed fraction of 0.012 per cent of potassium is the radio-isotope ^{40}K. These two phenomena allow estimation of *FFM* from the external counting of ^{40}K. Chemical analysis of the earlier mentioned six adult cadavers resulted in values of 2.50 and 2.66 g potassium per kg *FFM* in women and men respectively.

Quantitation of total body potassium (*TBK*) requires a whole-body counter, consisting of a shielded room containing a γ-ray counter, partially or completely surrounding the subject. The background radiation from cosmic and terrestrial sources can be adequately reduced with the shielded room. A further significant reduction of contamination can be reached by showering, including hair cleaning, beforehand, and wearing clean clothing during whole-body counting.

Provided an accurately calibrated and well-shielded system for whole-body counting is used, the *TBK* method permits an estimation of *FFM* within 3 per cent accuracy. The accuracy and precision is comparable to that of densitometry and isotope dilution. However, the cost of the equipment is very high, higher than that for the sophisticated IRMS equipment for *TBW* determination.

2.4   ANTHROPOMETRY

The quickest and cheapest method to measure body composition is measuring *skinfold thickness*. The assumptions forming the basis for this method are: (a) the thickness of the subcutaneous *FM* reflects a constant proportion of the total *FM*; and (b) the average thickness of skinfolds at selected sites reflects the subcutaneous *FM*.

Skinfolds are usually measured at four sites: triceps, biceps, subscapula, and iliac crest. The sites are measured at least three times and the sum of the individual measurements is then averaged. Durnin and Womersley (1974) developed a regresssion equation to predict body fat from the sum of four skinfolds, measuring skinfold thickness and body fat (with densitometry) in 481 men and women aged 16-72 years. The equation uses a logarithmic transformation of skinfold thickness, as body fat is not linearly related to skinfold thickness, age and gender of the subject.

Skinfold thickness is measured with a caliper. The skin is grasped between thumb and forefinger, gently shaken to exclude underlying muscle, and pulled away to allow the jaws of the caliper to take over. The caliper is calibrated to exert a constant pressure. The measurement requires the skill to grasp the skin in the right way at the right site. Other sources of variation are inter-individual differences in compressibility of the subcutaneous tissue. Some people have firm subcutaneous tissue and in others it is flabby and easily deformed. The method is not applicable in extremely fat subjects in whom the subcutaneous fat layer is too thick to allow the proper grasping of a skinfold.

The errors estimating *FM* and *FFM* from *BM* and skinfold thickness are reported to be between 3 and 9 per cent, strongly depending on the experience of the examiner, and generally higher than the methods mentioned earlier. On the other hand, it is sometimes one of the few possibilities to gather information on body composition as is illustrated by the results of a study on subjects of contrasting weight.

TABLE 17.2
Body mass index and body composition measured with three techniques

Subject	Sex	BMI (kg/m²)	isotope dilution FFM (kg)	fat (%)	densitometry FFM (kg)	fat (%)	skinfold FFM (kg)	fat (%)
1	F	15.0	32,7	22	32.7	22	35.8	14
2	F	14.2	26.7	33	28.3	29	36.8	8
3	F	16.0	33.4	17	29.2	27	36.8	8
mean		15.1	30.9	24	30.0	26	36.5	10
4	M	20.8	66.2	8	66.6	8	67.1	7
5	M	26.1	67.9	13	66.9	14	67.4	14
6	M	21.3	51.0	12	50.6	13	50.7	13
7	F	21.6	43.2	26	44.7	23	44.7	23
8	F	23.3	45.1	32	44.5	32	44.9	32
9	F	21.5	48.3	31	50.4	28	49.0	30
mean		22.4	53.6	20	54.0	20	54.0	20
10	M	42.9	86.8	41	93.0	37		
11	M	61.7	86.9	56	87.1	60		
12	M	43.2	81.6	42	86.5	38		
13	F	47.8	68.2	54	71.0	52		
14	F	45.1	56.3	57	54.8	58		
15	F	45.1	51.8	56	48.2	61		
mean		47.6	71.9	51	73.4	51		

The subjects were 9 females and 6 males, aged between 19 and 48. Three subjects (nrs. 1-3, Table 17.2) were low-weight, as shown by their *BMI* which ranged between 14 and 16, showing symptoms of anorexia nervosa. All three were females. Six subjects (nrs. 4-9) had a normal weight, ans six subjects (nrs. 10-15) were of extremely high weight and waiting for surgical treatment (vertical banded gastroplasty) (Westerterp *et al.*, 1991b).

Body composition was measured with three techniques: isotope dilution, hydro-densitometry with simultaneous lung-volume measurement (He-dilution) and skinfold thickness. Each technique is based on some assumption and is therefore not absolutely accurate. The assumptions were:

isotope dilution: fat free mass = body water/0.73

$$\text{body water} = 0.5 \times \left(\frac{^2\text{H dilution space}}{1.04} + \frac{^{18}\text{O dilution space}}{1.01} \right)$$

densitometry: average density of lean body mass = 1.10
                     average density of body fat = 0.90

skinfold thickness: summing the biceps, triceps, supra-iliac and subscapular sites and
                     using the tables by Durnin and Wommersley

Skinfold thickness could not be measured in the extreme high-weight group. There was a pronounced discrepancy between the results of the three measuring techniques in the low-weight group. The data obtained from skinfold measurement had to be chosen as being valid since isotope dilution and densitometry both resulted in the unrealistically high figures of 17-33 per cent body fat for female subjects with a *BMI* of 14.2-16.0, possibly because of a lower than normal hydration of *FFM*. In the normal-weight subjects the percentage body fat measured in the same subject with the three techniques differed 3 per cent at the most and was on average the same. In the high-weight group only isotope dilution and densitometry could be used, giving comparable results as in the normal-weight subjects. Thus, anthropometry was the only method applicable in low-weight subjects while not applicable in the extreme high-weight group.

2.5   OTHER METHODS FOR MEASURING BODY COMPOSITION

Recently, there has been a tendency to move from the two-compartment model of *FM* and *FFM*, to a four-compartment model for body composition, subdividing *FFM* into *TBW*, protein mass (*PM*) and bone or mineral mass (*MM*). In addition, there are new methods to quantify the individual body components. Unfortunately, none of the recent developments in the measurement of body composition have resulted in a more direct *in vivo* method. Of the new methods, each has its own hypotheses and often needs expensive technology. Examples are the measurement of total body nitrogen and total body calcium with neutron activation analysis and magnetic resonance imaging. The four-compartment model is mostly used to calculate changes in body composition, measuring the three variables body mass, body volume, and total body water, with the 'traditional' methods and assuming no changes in one of the four compartments, for example, *MM*. Both developments will be discussed.

The most widely applied new technique for measuring body composition is that using the electrical conductance of the body. The conductivity of the body is assumed to be a reflection of *FFM* as this contains virtually all the water and conducting electrolytes in the body. The two techniques for measuring the electrical conductance for determining body composition are total body electrical conductivity (*TOBEC*) and body impedance (*BI*). A *TOBEC* instrument consists of a solenoid coil that creates an oscillating field and induces a current in any material placed within the coil. The difference between the impedance of the empty coil and the coil with the subject inside it is a measure for body composition. A *BI* instrument creates a current through the body and measures the impedance with contact electrodes positioned on hands and feet. In both *TOBEC* and *BI*, corrections are necessary for conductor length and configuration. To this end, *TOBEC* intruments are usually calibrated with a phantom of known composition. *BI* results have been validated with simultaneous *TBW* measurements using isotope dilution. The results of both techniques are good as long as the appropriate equations are applied for calculating the body composition from body impedance for the population under study. Most laboratories use their own equations because of differences in equipment and methodology. The method has not been validated for subjects undergoing changes in body composition. In such cases, even the assumptions of traditional techniques may not hold and a combination of techniques is advised as described below.

The other new development involves the use of a multi-compartment model. Body composition is usually based on body mass and one additional measurement. The traditionally added measurements are body volume (densitometry), total body water (isotope dilution), and skinfold thickness (anthropometry). Each method carries with it a degree of uncertainty which allows detection of changes in body composition in the order of magnitude of minimally 1.5 kg *FM* and *FFM* if the underlying assumptions are valid for the situation prior to and after the change. A combination of independent measurements (body mass, body volume, and total body water) can reduce this measurement bias. Two studies will be described here in which these measurements were performed simultaneously before and after the subjects changed their energy metabolism with possible consequences for body weight and body composition.

The first study is an *activity intervention*, in which sedentary subjects were trained for 40 weeks to run a half marathon (about 21 km). The second study is a *feeding intervention*, in which morbidly obese subjects underwent vertical banded gastroplasty and were measured before the intervention and one year afterwards. In both studies densitometry overestimated the loss of *FM* (Table 17.3) by about 10 per cent, showing the importance of the combination of densitometry with the measurement of total body water. Apparently, the assumptions of densitometry and isotope dilution did not have the same validity before and after the change in weight or body composition. Possible explanations for the systematic discrepancies are a decrease in hydration of *FFM* or an independent decrease in *TBW* with the loss of *FM*.

The claimed precision of the measurements of changes in body composition using body mass, body volume and total body water is 1.0 kg for *FM* and 0.7 kg for *PM*. The precision will never reach the level of 0.001 kg for *BM* with integrating electronic balances. On the other hand, the precision of the average bathroom scales for the measurement of *BM* is usually not better than 1.0 kg. Putting the scales elsewhere in the room may well result in a difference of 0.5-1.0 kg. Also, everybody is familiar with the discrepancies between body weight measurements using different scales. Thus, the essential starting point of the measurement of body composition and changes in body composition is an accurate measurement of *BM*. For comparative studies, the subjects should be measured with minimal clothing, minimal intestinal contents (post-absorptive) and an empty bladder.

TABLE 17.3

Changes in body composition measured with hydrodensitometry only and with hydroden-sitometry in combination with isotope dilution in two studies

A   From prior to until 40 weeks after the start of a training period to run a half marathon in 8 males and 5 females.

B   From prior to until 54 weeks after vertical banded gastroplasty in 4 males and 1 female.

	FM* (kg)	FFM* (kg)	FM† (kg)	body water† (kg)	PM† (kg)
A					
males					
1	−1.3	+1.2	−0.6	−0.2	+0.8
2	−3.6	+1.3	−3.8	+1.2	+0.4
3	−3.3	+3.3	−1.9	+0.1	+1.8
4	−3.4	+3.6	−3.4	+2.2	+1.4
5	−6.3	+2.0	−5.2	−0.3	+1.2
6	−6.1	+5.8	−5.7	+3.0	+2.4
7	−7.1	+5.0	−6.1	+1.8	+2.2
8	−4.8	+3.4	−3.7	+0.5	+1.8
mean	−4.5	+3.2	−3.8	+1.0	+1.5
SD	1.9	1.6	1.9	1.2	0.7
females					
1	−3.6	+1.9	−2.3	−0.6	+1.2
2	+0.3	+2.1	+0.0	+1.7	+0.7
3	−1.9	+0.7	−2.0	+0.6	+0.2
4	−2.2	+4.0	−1.8	+1.9	+1.7
5	−4.0	+1.6	−2.5	−0.8	+1.3
mean	−2.3	2.1	−1.7	+0.6	+1.0
SD	1.7	1.2	1.0	1.2	0.6
B					
males					
1	−77.8	−3.1	−70.4	−8.4	+4.1
2	−25.7	−13.0	−24.7	−9.5	−4.5
3	−20.8	−11.4	−20.2	−7.9	−4.1
4	−48.1	−14.9	−44.6	−14.2	−4.2
female					
1	−58.7	−2.3	−55.8	−5.5	+0.3
mean	−46.2	−8.9	−43.1	−9.1	−1.7
SD	23.6	5.8	21.0	3.2	3.8

*Calculated from body mass and body volume, †calculated from body mass, body volume, and total body water.

3   GENDER DIFFERENCES

At birth, small differences exist between boys and girls. On average, boys are slightly taller with a higher weight and more muscle mass than girls, while the latter have more body fat. These differences are small and probably reflect influences of sex hormones in the prenatal period. Body composition, in terms of body-fat percentage, is the same in both sexes until puberty when girls gain relatively more *FM* than boys, to reach a higher body-fat percentage as adult (Figure 17.5). The greater differences appear with sexual maturation. The *FFM* increases quickly during adolescence to reach a maximum at about age 17 in females and age 20 in males. *FM* of females increases throughout adolescence while in males *FM* shows a decline around age 15 followed by an increase as in females. It is hypothesized that body fat has a regulatory effect on the reproductive ability of human females (Chapter 22). An adult woman needs a minimum of about 17 per cent body fat to have normal menstrual cycles. The 'normal' body composition at age 20 is 10-15 per cent body fat in men and 20-25 per cent body fat in women. The adult years show an increase in body fat percentage in both sexes.

Differences in body composition between the sexes are a reflection of the influence of sex hormones. There is evidence that women defend a higher *FM* than men as illustrated by an exercise intervention study (Westerterp *et al.*, 1992). In a long-term intervention under normal daily living conditions, Westerterp and associates studied the effect of an increase in physical activity on energy balance and body composition, without interfering with the energy intake. The subjects were sedentary males and females whose weight ranged from normal to slightly overweight. Out of 370 respondents, 16 females and 16 males were selected who did not participate in any sport such as running or jogging, and who were not active in any other sport for more than one hour per week (Table 17.4).

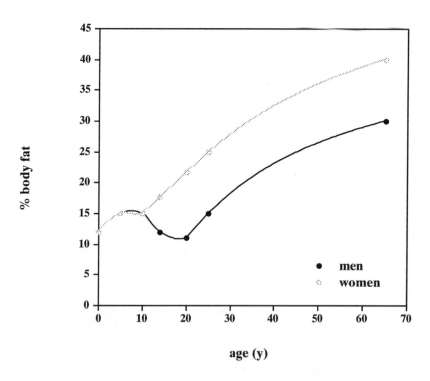

age (y)

FIGURE 17.5
Body composition as a function of age in men and women
• men; o women

TABLE 17.4

Subject characteristics of males and females, both in sequence of ascending %body fat as determined by densitometry

	subject no.	age	height	BM	body fat
males					
	1	33	1.73	63.2	15.1
	5	34	1.88	75.6	15.1
	6	40	1.76	69.7	18.1
	7	39	1.75	59.4	19.4
	2	32	1.86	79.6	19.9
	3	40	1.79	66.3	20.2
	8	35	1.80	72.0	21.8
	*	40	1.80	75.7	24.1
	9	33	1.79	69.4	24.7
	10	39	1.85	74.2	25.4
	*	39	1.80	79.5	26.4
	11	40	1.73	77.0	26.5
	*	36	1.78	77.3	26.8
	12	37	1.69	70.3	26.8
	*	35	1.80	79.3	30.5
	4	41	1.73	68.0	31.5
females					
	13	35	1.72	63.3	24.8
	14	32	1.63	54.0	25.5
	19	41	1.65	64.4	27.0
	20	32	1.57	52.6	27.7
	15	36	1.76	66.8	28.1
	16	35	1.67	60.1	29.6
	21	32	1.81	70.4	30.6
	22	38	1.73	68.5	32.6
	17	38	1.65	61.0	33.2
	23	41	1.68	65.8	35.0
	*	40	1.79	83.3	35.6
	18	32	1.58	61.1	36.0
	*	28	1.68	68.8	37.0
	*	35	1.67	73.7	38.4
	*	31	1.64	65.6	43.3
	*	39	1.68	74.6	45.4

* subjects withdrawn from the study.

Further criteria were applied to create a comparable group for both sexes with regard to age (28-41 yr) and body-mass index (19.4-26.4 kg/m²). The training program was aimed at running a half marathon competition after 44 weeks. In between, subjects ran a 10 km and 15 km race after 10 and 24 weeks, respectively. The training consisted of four training sessions per week, increasing the running time to 10-30 min, 20-60 min and 30-90 min per session after 8, 20 and 40 weeks, respectively. Measurements included those of energy intake, energy expenditure and body composition. Energy intake (*EI*) was measured with a 7-day dietary record. Energy expenditure was measured at rest during an overnight stay in a respiration chamber in which sleeping metabolic rate (*SMR*) and average daily metabolic rate (*ADMR*) were measured over two-week intervals with doubly-labeled water. Body composition was measured with hydrodensitometry and isotope dilution. Measurements were performed at four time intervals: before the start of the training and 8, 20 and 40 weeks afterwards, in schedule with the training program and the competitions. This was done in such a way that the subjects were still on their normal training regimen, and not yet engaged in the final preparations for the competition. The protocol of the observations was identical for each of the four intervals. During the study, 9 subjects withdrew (Table 17.4), as they were unable to keep up with the training program. They all dropped out within 20 weeks after the start of the training, two females and 1 male because of lack of time to join the training, 3 males and 2 females because of injuries and finally 1 female had to slow down the training, finishing by running a 15 km contest. Thus, 11 females and 12 males finished in the half marathon competition after 44 weeks of preparation, from a completely untrained starting situation. The subjects who withdrew all had an initial body-fat percentage higher than the group mean for their gender. Their *EI* did not show significant changes. However, there was a clear tendency towards an increase in energy intake in the females from the second interval onwards. At the same time the *EI* of the males, after an initial drop, remained completely unchanged from the second interval until the end of the study. *ADMR* increased from interval 1 to 2 and remained at the same level until interval 4, or even showed a tendency to decrease, while the training intensity further increased. When looking at the physical-activity index ($PAI = ADMR/SMR$), the subjects started at a low level of physical activity. The initial *PAI* was 1.7 (± 0.2) and 1.6 (± 0.1) in males and females respectively, but increased to values between 1.9 and 2.4 during interval 2. Males showed a slight tendency to a higher increase in *PAI* with training than females. Body mass (*BM*) did not show much change (Table 17.5).

In females the change from a mean value of 62.5 kg ($SD = 5.6$) during interval 1 to 61.8 kg ($SD = 5.3$) during interval 4 was not significant. In males the change in *BM* reached significance during interval 4 ($p < 0.01$), going down from an initial mean value of 70.4 kg ($SD = 5.8$) to 69.1 kg ($SD = 5.9$), i.e. an average loss of 1.3 kg. However, the changes in body composition were more pronounced. Based on hydrodensitometry only, there was a near-linear decrease in *FM* and a corresponding tendency to an increase in *FFM* in both sexes (Table 17.4). Females lost on average 2.8 kg *FM* ($SD = 0.5$) and males 4.5 kg ($SD = 0.6$), from the start until interval 4, 40 weeks afterwards. The increase in *FFM* during the corresponding interval was 2.1 kg ($SD = 0.4$) in females and 3.2 kg ($SD = 0.5$) in males. Based on changes in *BM*, body volume and total body water, females lost on average 1.7 kg *FM* ($SD = 1.0$; $p < 0.05$) and gained 1.0 kg *PM* (0.6) ($p < 0.05$), while males lost 3.8 kg *FM* ($SD = 1.9$; $p<0.001$) and gained 1.5 kg *PM* ($SD = 0.7$; $p < 0.01$). In males, the change in *FM* was related to the initial percentage body fat of the subject ($n = 12$, $r = 0.92$, $p < 0.001$). Subjects with a higher initial percentage body fat lost more fat than subjects which were leaner at the start. In females, such a relationship did not occur (Figure 17.6).

The energy intake did not show a significant change in response to the training program, while energy expenditure increased. Thus, the energy balance was negative in the males during intervals 2 ($p < 0.05$), 3 ($p < 0.05$) and 4 ($p < 0.01$). In the females the differences between *EI* and *EE* did not reach significance during any of the observation intervals, probably due to the persistent upward trend of *EI* after the start of the training program. The changes in body mass were not impressive. Only males showed a significant drop of on average 1.3 kg ($p < 0.01$), when comparing the initial mass with *BM* during interval 4.

TABLE 17.5

Body mass (BM), fat mass (FM), and fat-free mass (FFM) before (0) and 8, 20, and 40 weeks after the start of the training period

week	BM (kg)				FM (kg)				FFM (kg)			
	0	8	20	40	0	8	20	40	0	8	20	40
males												
1	63.2	65.0	64.5	63.0	9.5	10.7	10.8	7.9	53.7	54.3	53.7	55.1
2	79.6	80.6	80.6	79.0	15.9	15.1	14.9	13.2	63.7	65.5	65.7	65.8
3	66.3	67.7	66.6	64.6	13.4	13.2	11.2	8.4	52.9	54.5	55.4	56.2
4	68.0	66.8	67.0	65.7	21.4	17.3	14.8	13.1	46.6	49.5	52.2	52.6
5	75.6	76.6	77.3	75.5	11.4	12.3	11.0	10.1	64.2	64.3	66.3	65.4
6	69.7	71.3	70.4	67.4	12.6	13.6	12.7	9.0	57.1	57.7	57.7	58.4
7	59.4	59.1	59.7	59.4	11.5	11.1	11.1	8.2	47.9	48.0	48.6	51.2
8	72.0	73.7	72.0	72.2	15.7	14.0	10.5	12.3	56.3	59.7	61.5	59.9
9	69.4	69.3	68.3	65.1	17.1	19.4	14.5	10.8	52.3	49.9	53.8	54.3
10	74.2	73.4	71.9	73.9	18.8	15.0	15.3	12.8	55.4	58.4	56.6	61.1
11	77.0	77.5	74.2	74.9	20.4	17.4	11.6	13.3	56.6	60.1	62.6	61.6
12	70.3	70.3	67.9	68.9	18.8	18.5	16.4	14.0	51.5	51.8	51.5	54.9
mean	70.4	70.9	70.0	69.1**	15.6	14.8	12.9**	11.1**	54.8	56.1**	57.1**	58.0**
SD	5.8	5.9	5.7	5.9	3.9	2.9	2.1	2.3	5.3	5.8	5.7	4.8
females												
13	63.3	66.4	63.4	61.3	15.7	15.2	15.0	10.8	47.6	51.2	48.4	50.5
14	54.0	56.3	55.1	54.4	13.8	15.0	15.0	12.5	40.2	41.3	40.1	41.9
15	66.8	67.5	66.5	65.9	18.8	18.5	17.6	17.3	48.0	49.0	48.9	48.6
16	60.1	62.4	61.5	60.7	17.8	18.0	17.2	14.7	42.3	44.4	44.3	46.0
17	61.0	59.3	60.3	60.6	20.2	18.1	18.1	17.5	40.8	41.2	42.2	43.1
18	61.1	60.3	58.4	56.5	22.0	21.5	21.3	16.4	39.1	38.8	37.1	40.1
19	64.4	63.5	63.3	62.7	17.4	16.3	14.6	13.8	47.0	47.2	48.7	48.9
20	52.6	54.6	54.7	55.0	14.6	14.8	14.0	14.9	38.0	39.8	40.7	40.1
21	70.4	70.8	71.4	69.2	21.5	20.8	20.5	19.6	48.9	50.0	50.9	49.6
22	68.5	68.9	69.4	70.3	22.4	20.9	22.4	20.1	46.1	48.0	47.0	50.2
23	65.8	65.6	63.0	63.4	23.0	22.3	21.0	19.0	42.8	43.3	42.0	44.4
mean	62.5	63.2	62.4	61.8	18.8	18.3*	17.9*	16.0**	43.7	44.9**	44.6*	45.8**
SD	5.6	5.2	5.3	5.3	3.3	2.8	3.0	3.0	3.9	4.4	4.5	4.0

* $p < 0.05$	difference with pre-training value (Wilcoxon signed-rank)
** $p < 0.01$	difference with pre-training value (Wilcoxon signed-rank)

Changes in body composition, on the other hand, were quite pronounced. Based on the results of hydrodensitometry, males lost significantly more *FM* and gained more *FFM* than females ($p < 0.05$). The magnitude of the changes in *FM* and *FFM* calculated with hydrodensitometry is only an overestimation when comparing the results with the figures obtained from a combination of density and *TBW*. Then, the change in *FM* is –3.8 (±0.5) and –1.7 (±0.5) in males and females, respectively. The change in *FFM*, calculated as the sum of the changes in *PM* and *TBW*, then becomes +2.5 (±0.6) and +1.6 (±0.6), respectively.

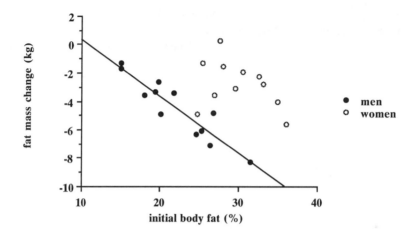

FIGURE 17.6

**Fat-mass change from before until 40 weeks after the start of the training period
plotted against initial percentage body fat with the calculated linear regression line
for men ($p < 0.001$)**

The overall effect of the increase in physical activity on energy balance and body composition is in line with earlier observations. Females show a better defense of their energy balance than males, and their loss of *FM* is significantly less than that in males. Whether the difference between the sexes is based on a decreased intake in males or an increased intake in females cannot be stated, although there is a very clear tendency towards the latter.

The exercise was not within the capability of the subjects with a higher *FM*. The absolute changes in *BM* and *FM* support the earlier view that exercise is not the first-choice therapy to reduce weight. Especially females do not lose much *FM*, even when a high exercise level can be maintained.

4   AGE

Body composition changes with increasing age, the increasing fat mass steadily replacing the fat-free mass. Thus, a healthy subject at the age of 25 has a body composition which is different from that of a healthy subject aged 65, even while of the same sex, height and body weight (see Table 17.6). The principal cause of loss of *FFM* is diminishing skeletal muscle. In the young male adult, the skeletal muscle mass may account for about 45 per cent of his body weight while for a man aged 65 that percentage would only be 27 per cent. Despite the loss of muscle mass, there is on average a slight increase in total body weight with increasing age. The relative stability of the body weight in aging individuals is due to an increase in body fat. It is not yet known whether the change in body composition with age can be influenced by life-style factors such as energy intake and physical activity.

TABLE 17.6

The body composition of reference humans at ages 25 and 65 for men and 25 and 70 for women

	man		woman	
	25	65	25	70
TBW (%)	60	53	54	45
PM (%)	19	12	15	11
MM (%)	6	5	6	4
FFM (%)	85	70	75	60
FM (%)	15	30	25	40

5   SUMMARY

Growth is a consequence of a discrepancy between energy intake and expenditure i.e. a change in the body's energy content.

Growth is usually measured as weight change.

A weight change in adult organisms can be the result of a change in body water, carbohydrate, protein, or fat. So information on body composition allows a better estimate of energy balance or growth than just body weight.

Body composition depends on gender and changes with age, and might be influenced by life-style factors.

The four accepted methods for measurements of body composition are densitometry, measurement of total body water, of total body potassium, and anthropometry.

18

Limits of energy expenditure

The main component of the daily energy turnover (i.e. average daily metabolic rate, *ADMR*) in the average subject is the energy spent on maintenance processes, usually called the basal metabolic rate (*BMR*). This is the energy expenditure of the ongoing processes in the body in the resting state, when no food is digested and no energy is needed for temperature regulation, i.e. in the post-absorptive state in a thermoneutral environment. *BMR* is usually expressed as a function of body mass to allow comparison between subjects and even between species. The remaining components of *ADMR* are the diet-induced energy expenditure (*DEE*) and the energy spent on (physical) activity (*AEE*). The *DEE* fraction of energy intake is about 10 per cent depending on the macronutrient composition of the food consumed. *AEE* is the most variable component of the daily energy turnover, ranging between an average value of 25-35 per cent up to 75 per cent in extreme situations during strenuous sustained physical exercise. These days it is hard to find people working at such a level in daily life, but endurance sports are a way of taking people up to their physical limits. This chapter focuses on energy turnover in relation to physical performance during sports activities in which body size and energy supply are often limiting factors. Table 18.1 shows some examples of energy intake by endurance, strength and team sport athletes.

Body *size* is a limiting factor in sports like gymnastics, body building and judo, where a low body mass or fat mass is preferable for the performance, or where the competitors are grouped into specific weight classes. In these types of sport, the subjects often try to maintain a negative energy balance to lose weight. The other extreme are endurance sports, like cycling and the triathlon, where the only way to success is to maintain energy balance at the upper limit of human performance in terms of energy turnover. Both extremes, the consequences of a limiting energy intake on energy turnover and the upper limit of energy turnover during endurance exercise, will be discussed. Finally, two problems confronting the average athlete will be dealt with: firstly, the changes in energy metabolism during the preparation period when the physical activity is gradually increased; secondly, how to regulate the energy intake while the energy expenditure changes from day to day.

TABLE 18.1
Energy intake data of varying endurance, strength, and team sport athletes

Type of sport[1]	Sex	Energy intake (kJ/kg/day)	
		Mean	Range
Endurance			
Tour de France	M	347	286-388
Tour de l'Avenir	M	316	247-378
Triathlon	M	272	246-295
Cycling, amateur	M	253	207-314
Marathon skating	M	222	175-294
Swimming	M	221	119-300
Rowing	M	189	167-225
Running	M	193	127-311
Rowing	F	186	140-200
Cycling, amateur	F	164	115-215
Running	F	168	123-218
Sub-top swimming	F	200	92-338
Strength			
Body building	M	157	106-183
Judo	M	157	77-210
Weight lifting	M	167	99-203
Judo	M	177	60-325
Top gymnastics	F	158	91-216
Sub-top gymnastics	F	206	113-334
Body building	F	110	91-133
Team sports			
Water polo	M	194	92-299
Soccer	M	192	118-287
Hockey	M	181	167-217
Volleyball	F	140	101-229
Hockey	F	145	91-199
Handball	F	142	78-271

[1] Subjects were top athletes including European, World and Olympic medal winners. Data were obtained from a 4- or 7-day food diary. Recently, it has become possible to compare the recorded intake with measured expenditure using the doubly-labeled water method. The results from this comparison often show the recorded intake to be a lower estimate of energy requirements.
(From Erp-Baart *et al.* 1989).

1   LOWER LIMIT

The energy intake by, especially female, gymnasts and ballet dancers is often extremely low. The energy expenditure in the average sedentary subject ranges between 1.4 and 1.6 *BMR*, while the reported energy intake in top gymnasts is usually below this level, despite the fact that they train for 3-4 hours a day. This can probably be explained by two factors: underreporting (see Chapter 16.8) and the urge to limit energy intake, the former being a result of the latter. Whether the intake data are realistic or not, in some sports the athletes often limit their energy intake to reduce their body mass and fat mass which has consequences for their energy turnover and, possibly, performance. The lower limit of *ADMR* is set by the sum of *BMR*, *DEE* and a minimum *AEE*. Reducing intake to lose body mass directly reduces *DEE* and indirectly *BMR*, probably while it is difficult to mobilize energy for physical activity, and this will result in a decrease in *AEE* as well.

Subjects on a weight-reducing diet show a reduction in their metabolic rate, not only through loss of metabolically active tissue but also per unit active tissue (see also Chapter 19.1). The efficiency of their energy turnover improves. This phenomenon is more pronounced when the energy-restrictive periods are repeated. This is practiced in some sports where weight control takes place before the match, with a regain afterwards. The result is weight cycling and in this situation it becomes increasingly difficult to lose weight while regain is facilitated.

Top female athletes often encounter another problem when they restrict their energy intake, namely menstrual dysfunction (see also Chapters 17.3 and 22). A high percentage of top athletes experience secondary amenorrhea. Although energy restriction is probably not the only reason, food intake in amenorrheic athletes is lower than in eumenorrhoeic athletes. Relative undernutrition occurs through a high energy expenditure during intense exercise and a decreased energy intake. One of the important consequences of amenorrhea is a decrease in bone density which predisposes a subject to stress fractures. Resumption of the menstrual function does not result in a complete regain of lost bone mass.

There are no studies on exercise performance at a lowered energy intake but there is evidence for a reduction in spontaneous activity. The body not only defends itself against a negative energy balance through a reduction in energy expenditure at rest (*BMR* and *DEE*). People in a negative energy balance become slower, i.e. they reduce their *AEE*, and this can not be the goal of an athlete.

The long-term effect of a negative energy balance is not only loss of body fat but also loss of fat-free mass, i.e. muscle mass. Wrestlers who want to lose weight by reducing their energy intake, show a diminished protein nutritional status. The practice of starvation in order to achieve a rapid weight loss in the days before competition in weight-class sports is well-known. Several methods are used including an increase in exercise instead of tapering off the training intensity. Sometimes food restriction is combined with deprivation or total abstention of fluid intake. Besides excessive loss of lean body mass, depletion of glycogen stores - the most important energy substrate for high-intensity performance - will occur. Therefore, the net result is a diminished body reserve for athletic events which more than offsets any advantages of competing in a lower weight class.

2   UPPER LIMIT

The limiting factors for human performance depend on the type of exercise. Nutrition is an important determinant during the preparation phase and, for endurance exercise, also during the exercise itself. Athletes can only sustain a high level of exercise when they manage to maintain energy balance. Here, only the latter situation, the limit of energy expenditure that can be maintained over several days up to several weeks, will be dealt with. In such a case, energy intake should meet energy requirements.

The limits of human performance were, for a long time, only reached by labourers such as lumberjacks. They were supposed to work for 8 or more hours per day at an energy expenditure level of 8 times *BMR*. Assuming a daily sleep period of 8 hours at 1 *BMR* and the remaining hours at 2 *BMR*, their *ADMR* reached a level of 4 times *BMR*. This is nearly three times higher than the minimum level of 1.4 *BMR* in a subject of minimal activity. Nowadays, the upper limit of energy turnover is reached in endurance sports like bicycle racing and triathlon. Here, athletes work at a high level of energy expenditure for up to 8 hours per day, sometimes for several weeks.

The level of energy expenditure during endurance sports has been measured in the world's most demanding cycling race, the 'Tour de France', a race of more than 20 stages lasting about three weeks (Westerterp *et al.*, 1986). Five subjects were selected who were assumed to be capable of finishing the tournament, and four actually did. The observations started in the morning of the first stage and ended in the morning of the last stage, finishing in Paris. Figure 18.1 summarizes, on a daily basis, the distances covered and some characteristics of the route. The 22-day period was split into three parts: A, B, and C of 7, 8, and 7 days respectively.

Energy intake was measured with the food record technique. The participants noted their daily food consumption during the tournament in specially designed diaries. A trained nutritionist gave instructions on filling in the diaries, made weekly checks, and cross-checked with the information on the food supply during the race including caloric beverages. Energy-rich drinks, making up a substantial part of the daily energy intake, were provided in standard bottles and their consumption was checked. The data were converted to macronutrient intake and energy intake using the computerized Dutch nutrient data bank.

Energy reserves were calculated either from body mass and skinfold thickness or from body mass and body water values assuming the body water compartment to be 72.5 per cent of the lean body mass. Body mass was measured at the start and end of periods A, B, and C. Subjects were weighed after voiding to the nearest 0.1 kg directly before and/or after bedtime, at 20.00-21.00 h and 06.00-09.00 h respectively. Body water was calculated from the $H_2^{18}O$ dilution space. The isotope was administered the night before the start of periods A, B, and C after a baseline blood sample was taken and body mass was measured. The next morning at rising, the initial blood sample of the observation period was taken and the final sample was taken at the end of the interval, 7 or 8 days later. Total body water was calculated from the isotope dose and the excess isotope concentrations in the initial and final samples by extrapolating back to time 0.

Energy expenditure was measured with doubly-labeled water as described before (Chapter 16.3). The administration and sampling schedules are presented in Figure 18.1. Energy expenditure was calculated from carbon dioxide production using the energy equivalent calculated from the measured macronutrient intake (Table 16.1).

This procedure was justified by the fact that the subjects were in nutrient balance, showing no systematic changes in body mass and body composition during the observation intervals (see below).

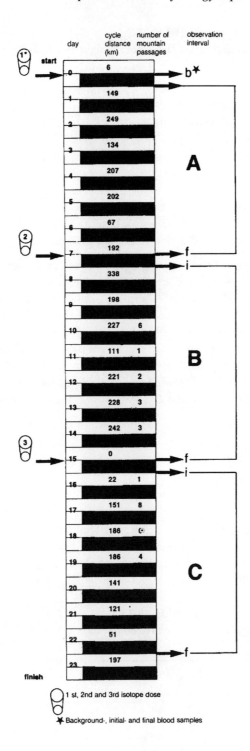

FIGURE 18.1
Characteristics of a cycling race and application of the doubly-labeled water technique
(From: Westerterp *et al.*, 1986)

TABLE 18.2

Body mass of four subjects during a cycling race, at night and early in the morning

Subject No.	Day 0 night (kg)	Day 7 night (kg)	Day 8 morning (kg)	Day 15 night (kg)	Day 16 morning (kg)	Day 23 morning (kg)
1	71.6	72.1	71.0	73.2	71.4	70.9
2	75.2	74.1	73.1	73.1	71.4	73.8
3	61.4	61.7	61.0	61.8	60.1	59.9
4	68.7	68.6	67.4	68.7	68.1	66.6
mean	69.2	69.1	68.1	69.2	67.7	67.8

All four subjects who finished the race managed to cover their energy expenditure with their energy intake in view of their unchanged body-energy reserves. Table 18.2 shows the body mass at night and in the early morning. Both evening and morning values for the average body mass of the four subjects were the same throughout the 22-day race. There were no differences in the average daily values for night or morning masses in the course of the race. The variations in individual values were probably due to differences in intestinal contents. Combining body mass figures with measurements of skinfold thickness as determined before the race, or water mass figures calculated from isotope dilution as determined at the start of periods A, B, and C, allows calculation of the body fat reserves (Table 18.3).

Values are expressed as percentage of total body mass. Before-race values were calculated from body mass and skinfold thickness. Values on days 1, 8, and 16 were calculated from body mass and water mass.

The values for the average body-water mass of all four subjects was the same at the start of periods A and C. The value for period B was systematically lower in all four subjects. This comparison of the body fat compartments is based on the body-water mass in periods A and C, where the interval between the isotope dose and the initial sample was 12.0 and 11.7 hours respectively; the interval for period B was 10.3 hours.

TABLE 18.3

Fat mass of the four subjects before the race and at days 1, 8, and 16 of the race

Subject No.	Day 0	Day 1	Day 8	Day 16
1	13.2	11.4	15.7	11.6
2	11.1	11.9	13.5	11.6
3	9.8	7.8	9.9	5.8
4	12.1	7.7	9.6	8.1
mean	11.6	9.7	11.2	9.3

TABLE 18.4

Energy intake (*EI*) and energy expenditure (*EE*) of four subjects in three periods of a cycling race

Subject No.	A		B		C	
	EI	*EE*	*EI*	*EE*	*EI*	*EE*
1	24.8	30.5	27.1	34.2	23.9	38.0
2	26.1	29.9	26.8	35.2	24.5	37.0
3	22.3	28.7	25.2	38.4	21.5	33.4
4	24.9	28.4	26.0	36.3	23.0	34.2
mean	24.5	29.4	26.3	36.0	23.2	35.7

Values are expressed in MJ per day.

The shorter equilibration period coincided with a higher isotope level in the blood, possibly indicating another equilibration status.

The figures for energy intake and energy expenditure are presented in Table 18.4. Those for expenditure are 13-35 per cent higher than then the figures for intake. This discrepancy systematically increases from the first to the third period and is most likely due to methodological difficulties. Measuring food intake with the food record technique is a delicate task (see Chapter 15.1). The doubly-labeled water technique for measuring energy expenditure, however, proved to be valid, even under the given conditions (Westerterp *et al.*, 1988). Whatever the right value for energy turnover is, the subjects operated during a period of three weeks at a level of 3.5-5.5 times *BMR* (Table 18.5).

It is obvious that there is a limit to the performance of an organism, set by energy intake and energy mobilization. Free-living birds have a maximum working capacity on a sustained basis of four to five times their *BMR* (Westerterp & Bryant, 1984). Kirkwood (1983) suggested, based on data from a wide range of homeotherms under energy demanding conditions like rapid growth, heavy lactation, strenuous work, and living in a cold environment, that this is an interspecies limit. In absolute figures, using the allometric comparison based on body size (a constant times body mass to the of power 0.72; $a./aB^{0.72}$), individual energy expenditure over 3 weeks was 1.6-1.8 MJ/day/kg$^{0.72}$. This is in accordance with the interspecies limit of 1.7 MJ/day/kg$^{0.72}$.

TABLE 18.5

***ADMR* presented as a multiple of *BMR* in three intervals of a cycling race**

Subject No.	A	B	C
1	3.5 - 4.3	3.9 - 4.9	3.4 - 5.4
2	3.7 - 4.2	3.8 - 5.0	3.5 - 5.2
3	3.6 - 4.6	4.1 - 6.2	3.5 - 5.4
4	3.6 - 4.1	3.7 - 5.2	3.3 - 4.9
mean	3.6 - 4.3	3.9 - 5.3	3.4 - 5.2

The lower estimates are from food intake, the higher estimates from isotope turnover (Table 18.4). *ADMR* =average daily metabolic rate; *BMR* = basal metabolic rate.

It is not known what sets the upper limit of energy turnover, energy intake or energy expenditure, i.e. the capacity to eat and process food or the capacity to supply energy through oxidation. The power output of the human muscle is higher on carbohydrate than on fat and protein and the body store of carbohydrate energy in the form of glycogen can cover energy expenditure for 1-2 hours only. Thus, the upper limit of power output during high intensity (> 60 per cent of the aerobic power) endurance exercise can be increased when one can eat while one works, which is common practice in cycling races.

It is clear that the power output is directly related to the energy turnover of the body. Brouns *et al.* (1989) did a simulation study in a respiration chamber, measuring the influence of diet manipulation on performance in subjects who cycled for two days reaching an energy expenditure level of 26 MJ/day. The subjects received *ad libitum* a conventional solid diet with a high carbohydrate content (CHO; 60 energy%) supplemented with water, or the same diet supplemented with a 20 per cent enriched carbohydrate liquid (total CHO 80 energy%). They were able to maintain energy balance during the exercise days on the carbohydrate-supplemented diet, while on the water-supplemented solid diet the energy intake per day appeared to be 5-10 MJ too low. Under such circumstances the athlete must drink and eat to maintain the body CHO stores during the exercise as well as to replete the stores within 18 hours for the exercise performance the next day. From measurements on oxygen consumption and carbon dioxide production, an oxidation rate of nearly 13 g CHO/kg/day was calculated. The actual intake in the case of CHO supplement was 16 g CHO/kg/day. This suggests that also with respect to CHO intake there is a ceiling in the energy intake above which other factors play a role which have to do with fatigue. A higher intake of fat as an alternative, with its higher energy density, is not appropriate in top sports performance because the muscles need carbohydrate instead of fat for maximum power output. From these experiments it has become clear that energy intake cannot meet energy expenditure when physically highly strenuous labour is performed even when conventional high-carbohydrate foods are consumed, unless special liquid formulas are supplemented.

People working intensively for many hours do not have the time to consume and process food outside the exercise period to balance their energy expenditure. Therefore, the strategy of intake of large quantities of liquid CHO-rich formulas is one of the appropriate nutritional answers to optimize performance during extreme sustained exercise.

3   CHANGES IN ENERGY EXPENDITURE DURING TRAINING

Novice athletes starting on a training program change their physical activity with subsequent consequences for their energy turnover, depending on the intensity and the duration of the training sessions. Again, the most profound effects are to be expected in those taking up endurance training. A recent study on novice athletes preparing for a half marathon (Westerterp *et al.*, 1991a, also referred to in Chapter 17.3) provided some data on the consequences.

Energy expenditure, as measured with doubly-labeled water in 4 men and 3 women before and 8, 20, and 40 weeks after the start of the training period, showed an initial increase of on average 30 per cent, and remained stable afterwards (Figure 18.2).

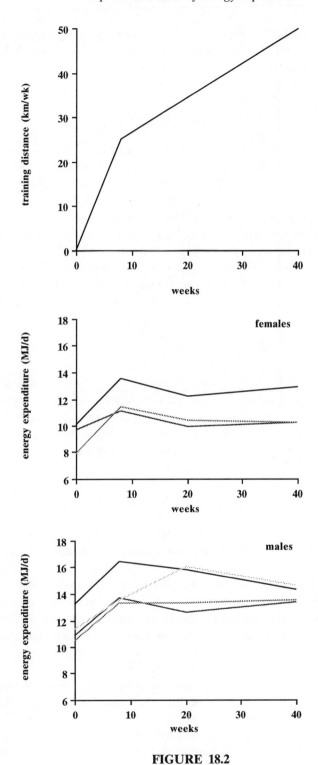

FIGURE 18.2
Training distance and energy expenditure in novice athletes, 4 men and 3 women, from
before until the end of a training period in preparation for the a half marathon
(From: Westerterp *et al.*, 1991)

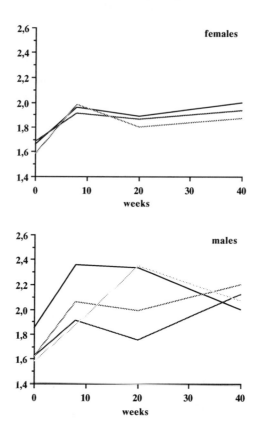

FIGURE 18.3

Physical activity index (*PAI = ADMR/BMR*) in novice athletes, 4 men and 3 women, from before until the end of a training period in preparation for a half marathon
(From: Westerterp *et al.*, 1991a)

The training 'volume' of the subjects doubled after the energy turnover had stabilized at 1.3 times the pre-training level. To explain this result, it was suggested that already in the beginning of the training period, after 8 weeks, the subjects had reached the natural limit of their daily energy turnover. After that, they could then increase their training volume without an increase in energy expenditure because of an increase in efficiency as a mere result of their training. Thus, training (in this case running) improves energy efficiency.

The subjects in this study were genuine novice athletes. At the start of the training period they were sedentary, as can be seen from their physical activity index (*PAI = ADMR/BMR*, Figure 18.3). The initial *PAI* value was 1.66 ± 0.07 (mean ± *SD*), which is close to 1.54, the value for sedentary subjects (WHO, 1985). The *PAI* after 8 weeks of training was 2.03 ± 0.18, close 2.14, the value for subjects engaged in physically hard work.

One indication for the increase in the efficiency of energy turnover after the start of the training, apart from changes during the training itself, is a change in metabolic rate at rest. Although there are indications of an increase in resting metabolic rate directly after an exercise bout, in the long term subjects showed a decrease in their resting energy expenditure. In males the resting metabolic rate, as measured in a respiration chamber, was 4 per cent lower ($p < 0.05$) after 40 weeks of training. In females the same tendency was observed, although here the variation between individuals was higher and the change was probably delayed even longer.

The decrease in resting metabolic rate is surprising for another reason. Resting metabolic rate is closely related to lean body mass. All subjects showed an increase in their lean body mass during the training period, females gained 2.1 ± 0.4 kg ($p < 0.01$) and males gained 3.2 ± 0.5 kg ($p < 0.01$). The decrease in resting metabolic rate, combined with the increase in lean body mass, especially in the males who gained significantly more lean body mass than the females ($p < 0.05$), is further evidence of an increase in energy efficiency during the training period.

Energy intake does not seem to be a reliable measure for energy turnover during the training period. Before the training started, there was a nonsignificant underreporting of intake as measured with a 7-day dietary record compared to simultaneous measurement of energy expenditure with doubly-labeled water (-12 ± 17 per cent). After 40 weeks of training the discrepancy was (-19 ± 17 per cent). Thus, the subjects did not show a significant change in reported intake between the measurement before and during the training when energy expenditure had increased by as much as 30 per cent.

Despite the lack of change in *quantity* of energy consumed as measured by self-report, there are systematic changes in the *quality* of food consumption in the course of a training program. In the half marathon study, there were no changes in the quality of the reported food consumption. However, Janssen *et al.* (1989) reported an increase in the carbohydrate energy% from 47 to 55 and 48 to 52 in females and males respectively, over a period of 18 months while preparing for the event. The increase in carbohydrate consumption was at the expense of dietary fat. Thus, the increase in physical activity caused a favourable change in food habits with regard to the nutritional guidelines for good health. There were no indications for a gender difference in substrate utilization as suggested in the literature, such as females demonstrating greater lipid utilization and less carbohydrate and protein metabolism than males.

4   REGULATION OF ENERGY BALANCE DURING A CHANGE IN ENERGY TURNOVER

Humans do not balance energy intake and energy expenditure on a daily basis, as smaller animals do. Apparently, humans can afford to rely on their body reserves while smaller species show signs of energy shortage sooner, as expressed in a lowered body temperature and reduced physical activity. Smaller species have a higher energy expenditure per kg body mass as well as a relatively smaller body energy reserve. Thus, a mouse cannot survive 3 days without food, while a normal adult human being can survive more than 30 days.

Of course, humans maintain a perfect energy balance in the long term as shown by a constant body weight in adult life. Energy intake strongly correlates with energy expenditure on a weekly basis. Discrepancies on a daily basis between intake and expenditure are especially large when days with a high energy expenditure are alternated with quieter intervals. Military cadets did not show an increase in energy intake on days with a higher energy expenditure when they joined a drill competition. The corresponding increase in energy intake came about two days afterwards (Edholm, 1955). A more recent review, based on 69 subjects in 6 energy-balance studies also concluded that there is no precise short-term control mechanism.

Even during intensive sustained exercise like the earlier mentioned Tour de France, in which it is essential for success to match energy intake with energy expenditure, there are day to day discrepancies. Despite the fact that cycle racers eat while they work, they do not always manage to cover their energy needs on days of the highest energy expenditure, i.e. on those days with the longest stages. However, the overall match of energy intake and energy expenditure is striking compared to studies on a lower exercise level, indicating that at higher levels of exercise the risk of deregulation of the energy balance is smaller.

There may be two mechanisms involved: firstly, the physiological regulation system, and secondly, the fact that highly trained athletes have learned how to eat the maximum amount of food during physically hard work. The digestive system is a limiting factor and it is difficult to compensate for a negative energy balance acquired on one day later on. In a study of food intake and energy expenditure in four amateur cyclists it was shown that with increasing energy expenditure two cyclists had great difficulties keeping in energy balance. Therefore one of the factors endurance athletes have to adapt to is the consumption of CHO-rich liquid formulas in order to compete at the top level.

5   SUMMARY

Energy expenditure demonstrates lower as well as upper limits.

A low energy intake limits energy turnover and hence a reduction in BMR, DEE and AEE occurs.

The upper limit of energy turnover is reached during endurance exercise. Training changes energy turnover.

After the natural limit of daily energy turnover is reached, the training volume can be increased without an increase in energy expenditure, because of an increase in efficiency. During this change in energy turnover also regulation of energy balance is adapted.

19

Balance between energy intake and energy expenditure

Adult humans maintain a balance between their energy intake and energy expenditure. The energy store of their body does not fluctuate much, as the constancy of body weight and body composition shows (Chapter 17.1). This can be achieved by control of either energy intake or expenditure. Humans however, do not balance energy intake and energy expenditure on a daily basis (Chapter 18.4). On days of high energy expenditure the energy intake is usually normal or even below normal. The corresponding increase in energy intake does not take place until a few days later. Man can change energy intake by at least a factor three to adapt it to energy expenditure. Under sedentary living conditions the energy balance is maintained at about 1.5 times *BMR* while during sustained exercise levels of 4.5 times *BMR* are reached (Chapter 18.2).

The possibility of humans adapting energy expenditure to energy intake is often questioned. It is even stated by some that the control of energy expenditure is contrary to what the body weight would require in such cases as hyperactivity in anorexics and hypoactivity in obesity. This chapter presents evidence for energy expenditure adapting to energy intake and vice versa. Adaptations in both directions will be shown by means of studies in which the energy *intake* is either decreased or increased, and studies in which the energy *expenditure* is decreased or increased. Finally, two practical questions will be answered with regard to energy balance: Why is it impossible to maintain energy balance at high altitude? and why does alcohol consumption have no effect on energy balance?

1   ENERGY EXPENDITURE DURING STARVATION AND DIETING

A decrease in energy intake has consequences for energy expenditure. The body weight does not remain at equilibrium level as during *ad libitum* conditions, and this may cause a drop in energy expenditure. Some metabolic savings will result from the mere fact that one handles and digests less food. The question to be answered here is whether there is an additional genuine adaptation of energy expenditure to a decreasing energy intake. Does an underfed person spend less energy on a given function (e.g. protein turnover, muscular effort for a certain behavior) than an *ad libitum* eating conspecific of the same weight and body composition would do? If there is such a phenomenon, evolution has provided the individual human with a special capability of eking out provisions in order to increase the chances of tiding over temporary food scarcity, a condition quite common historically, and still so today in several parts of the world. On the other hand, many well-fed people in the Western world struggle to maintain a normal body weight as most of them are overweight or even manifestly overweight or obese. To rid themselves of their excess body mass, mainly in the form of body fat, they think they have to to spend more energy, even on an energy-restricted diet. Energy expenditure has been studied from both points of view: how to survive on a lowered energy intake and how to maximize the effect of an energy-restricted diet to lose weight. Recently, a new element has been added in the form of a suggested increase in lifespan by energy restriction. In the following, examples of all three aspects are used to show the evidence for and mechanisms of changes in energy expenditure in dieting and starvation.

1.1   STARVATION

Keys *et al.* (1950) carried out a classical experiment on the effect of semi-starvation in normal-weight men, the so-called Minnesota Experiment. In 1944, during World War II, people in America realized there was mass famine in the occupied areas of Europe. It was recognized that with the end of the hostilities in sight, relief feeding would be necessary. This created the need for a controlled experiment to determine the changes in man induced by semi-starvation. The subjects were volunteers recruited from camps with conscientious objectors who were under the control of the Selective Service System for the duration of the war. Thirty-two men, aged 25.5±3.5, completed the experiment. The subjects stayed in the laboratory from November 19, 1944, until October 20, 1945. The period included a 12-week baseline period, 24 weeks of semi-starvation, and the first 12 weeks of rehabilitation. During the study, the subjects were assigned to specific tasks, such as general maintenance, laboratory assistance, shop duties and desk work requiring about 15 hours per week. They also participated in an educational program which took about 25 hours per week. In addition, per week they walked 35 km out-of-doors and for half an hour on a treadmill.

The diet in the baseline period was normal with regard to the variety of the food items. It provided 14.6 MJ/day with a carbohydrate/protein/fat energy% ratio of 55/13/32. The quantity was adjusted to the individual need of the subject to maintain his ideal weight. The diet during the semi-starvation period was designed to represent the types of foods available in European famine areas: whole-wheat bread, potatoes, cereals, considerable amounts of turnips and cabbage, and token amounts of meat and dairy products. The average daily intake was 6.6 MJ with a carbohydrate/protein/fat energy% ratio of 71/12/17. It was served in two meals per day at 8.30 and 17.00.

The body weight of the subjects decreased from a mean value of 69.4 (*BMI* 21.4) to 52.6 kg (*BMI* 16.3). The total weight loss was about 25 per cent of the initial body weight. The weight loss decreased progressively and nearly reached a plateau value towards the end of the semi-starvation period (Figure 19.1). Thus, the subjects adjusted their energy expenditure to reach a situation of energy balance after 24 weeks at 45 per cent of *ad libitum* energy intake. The adaptation was accompanied by a loss of active tissue. On average, body fat loss made up about one third of the weight change, i.e. there was a relatively high loss of lean body mass over this prolonged period of energy deficiency in normal-weight subjects, decreasing from a mean value of 9.8 kg to 3.0 kg. Table 19.1 shows the energy saved at the end of the 24-week period. Of the mean total of 8.0 MJ/day the main part stemmed from reduced activity costs (*AEE*). The energy saved can be split into a 'passive' component and an 'active' or real adaptation component. Together, the decrease in active tissue mass for maintenance (*BMR*), the reduced processing costs by lowering food intake (*DEE*) and the reduced body weight for physical activity (*AEE*) explained 3.6 of the 8.0 MJ/day saved. The remaining 4.6 MJ/day are a consequence of real adaptation by a reduction in tissue metabolism (*BMR*) and physical activity (*AEE*).

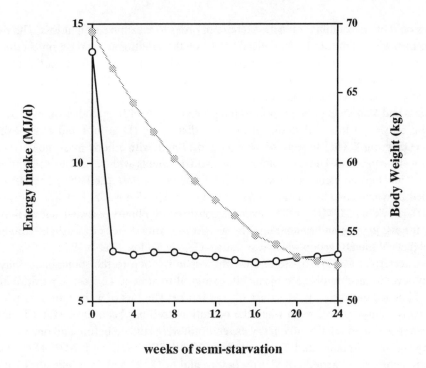

FIGURE 19.1
Mean daily energy intake (o) and mean body weight (•) of 32 men during 24 weeks of semi-starvation
(Source: Keys *et al.*, 1950)

TABLE 19.1
Energy saved by 24 weeks semi-starvation in the Minnesota Experiment

	(MJ/d)	Energy saved (% of total)	
BMR	2.6	32	65% for a decreased active-tissue mass
			35% for a lower tissue metabolism
DEE	0.8	10	
AEE	4.7	58	40% for a reduced body weight
			60% for a reduced physical activity
total	8.0	100	

1.2   DIETING

Maximizing an energy deficit on an energy-restricted diet to lose weight is thought to be feasible by combining energy restriction with an exercise program. Such a program is thought to curtail the reduction in tissue metabolism that normally results from energy restriction and at least maintain or increase the costs of physical activity. There are many studies on diets in combination with exercise in order to maximize weight loss. The results are inconsistent, some find a beneficial effect of the additional exercise on dieting and some report no difference or even an adverse effect. Both outcomes, the beneficial and the possible adverse effects of exercise in combination with a restriction of energy intake will be illustrated with the following examples.

Saris and Van Dale (1989) reported the pooled results of four studies on the effects of exercise on weight loss. All studies included a diet group (D group) and a diet-exercise group (DE group). The subjects, 45 women and 17 men with a body mass index (*BMI*) of 29-40, were equally assigned to one of the two treatments with regard to sex and over-weight. The D group included 25 women and 9 men, mean *BMI* 32.6 kg/m^2. The DE group included 20 women and 8 men, mean *BMI* 33.1 kg/m^2. The diet was a 'very low calorie diet' (VLCD), providing 3.0 MJ/day. The exercise consisted of 5 hours per week supervised aer-obics, fitness, jogging and swimming. The weight loss after 5 weeks was slightly higher in the subjects of the DE group which was due to a higher fat loss (Table 19.2).

Westerterp *et al.* (1991b) studied weight loss in relation to the spontaneous physical activity level in morbid obese subjects after surgical treatment, i.e. gastric partitioning by vertical banded gastroplasty, to reduce energy intake. The loss of weight after gastric par-titioning is highly variable even when the operation itself has been successful. Changes in body composition in relation to energy expenditure were studied before and one year after surgery in one woman and 5 men, aged 20 to 38 years, and *BMI* 42-62 kg/m^2. Measurements were carried out shortly before and 6, 12, 27 and 54 weeks after surgery. Body composition was measured by hydrostatic weighing and isotope dilution, and the en-ergy expenditure over a 24-hour period (24 EE) and during complete rest (sleeping metabolic rate, *SMR*) in a respiration chamber. In 5 of the 6 subjects the energy expendi-ture was measured with doubly-labeled water under daily living conditions during 2-week intervals (average daily metabolic rate, *ADMR*).

Weight loss and fat loss during the full observation period until 54 weeks after surgery was 54 ± 8 and 43 ± 9 kg (mean ± SEM), respectively. There was a more than threefold dif-ference in fat loss between the subject losing most (70 kg) and the one losing least (20 kg). The fat loss in this group of 6 subjects was not related to the initial fat mass but there was a significant negative relationship between fat loss and loss of protein mass.

TABLE 19.2
**Weight loss and fat loss after 5 weeks on an energy-restricted diet with
and without additional exercise**

treatment	n	BM_0 (kg)	BM_5 (kg)	ΔBM (%)	FM_0 (MJ/d)	FM_5 (MJ/d)	ΔFM (%)
diet	34	90.4	80.8	−10.5	34.5	27.2	−21.4
diet+exercise	28	93.9	82.9	−11.6*	36.4	27.4	−25.0**

BM_0, FM_0, BM_5, FM_5: body mass and fat mass before and at 5 weeks on the diet respectively
*$p < 0.05$, **$p < 0.01$ for difference with diet only.
(From: Saris & Van Dale, 1989)

There was a strong negative relationship between fat loss and activity level of the subjects expressed as the quotient *ADMR/SMR*.

These results are in contrast with what might be expected. Reduction of fat mass after successful gastric partitioning is highest in the subjects with the lowest physical activity level (before and) after surgery; these subjects lose less fat-free mass. One explanation is that the subjects who lowered their intake most were not able to maintain or increase their pre-operative activity level. Alternatively, the reason may be the lower tendency to eat in the subjects with lower energy expenditure after the reduction in their stomach size, i.e. they showed a proportionally more negative energy balance. They managed to save more fat-free mass at a higher loss of fat mass showing a relatively higher use of fat as energy substrate.

2   ENERGY EXPENDITURE DURING OVERFEEDING

Energy expenditure in overfeeding has been studied in several laboratories, mainly to increase our understanding of the causes and consequences of obesity. To that end, lean people can be persuaded to gain weight to increase the understanding of 'spontaneous' obesity. One of the first studies in this field was by Sims *et al.* (1973). In their first attempt, these researchers overfed university students who volunteered for the experiment. The outcome was that the students had great difficulty in adding even 10 per cent to their normal weight while continuing their normal curricular and extra-curricular activities. Gaining weight turned out to be a full-time job for normal young people. However, they received support to carry on with the project at Vermont State Prison. The warden there believed the project might have rehabilitative dividends, and so a special kitchen and dining facilities were installed. All 22 volunteers achieved a weight gain of 20 per cent or more. At the end of the overfeeding interval, which lasted 25 weeks, the subjects were fed at energy balance to measure the energy needs to maintain their higher body weight. Unfortunately, no direct measurements on energy expenditure were performed. Only energy intake, changes in body weight and body composition, and changes in physical activity were measured. The latter was crudely monitored by means of a pedometer (a device to count step frequency). The main conclusion was that there is a wide variation in the ability to convert food into body mass. One extreme was a man who initially weighed 61 kg and reached a final weight of 69 kg after consuming on average 80 per cent above his intake that was initially sufficient to maintain his body weight over the 25 weeks (Figure 19.2). The other extreme was a man with an initial weight of 69 kg who reached a final weight of 87 kg, i.e he gained twice as much, after 26 weeks of eating on average 35 per cent above the initial intake to maintain his body weight. The first subject could not maintain his body weight of 69 kg at an intake level equalling the level at which the second subject managed to maintain his higher body weight of 87 kg. The pedometer did not show the first subject to be more active than the second to explain the difference. During the overfeeding period physical activity, as measured with the pedometer, tended to decrease. The men were reported to have less initiative and their performance on work assignments was less satisfactory.

This first study merely showed individual differences in response to overfeeding. The data do not allow calculations of changes in energy balance and energy expenditure in greater detail. On the other hand, there are no more overfeeding studies in literature in which subjects were overfed with a comparable excess, mainly because of the length of the described study which lasted more than six months.

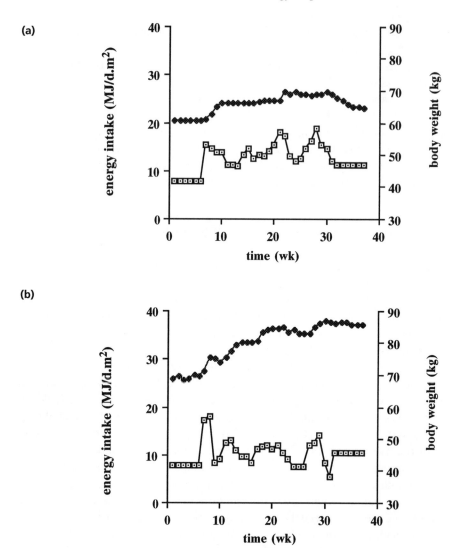

FIGURE 19.2
Energy intake (▫) and body weight (◆) in (*a*) a volunteer who gained weight with difficulty and (*b*) a volunteer who gained readily

Energy intake is expressed as MJ/m² body surface per day, body surface is calculated from body weight and body height
(From: Sims *et al.*, 1973)

The effect of overfeeding on energy balance and energy expenditure was studied in more detail in two later, shorter, studies. Norgan and Durnin (1980) overfed 6 young healthy adult males , classified as thin, average, muscular or plump respectively (Table 19.3). The subjects lived for 9 weeks in a metabolic unit where they took all their meals. Extra portions and snacks were freely available throughout the experiment.They were free to attend lectures and classes during the day and to go out in the evenings and at weekends, as long as they returned to their unit at meal times.

TABLE 19.3

Physical characteristics of the 6 men in the overfeeding study by Norgan and Durnin (1980)

subject	age (y)	height (m)	weight (kg)	body fat (%)	classification
1	21	1.68	70.6	20	plump
2	22	1.85	61.5	10	thin
3	21	1.63	59.1	21	plump
4	24	1.75	54.5	12	thin
5	21	1.78	70.5	14	muscular
6	21	1.80	65.0	12	average

After 4 days of familiarizing with the procedures, a 2-week baseline period followed with normal food intake, followed by 6 weeks of overeating. The mean energy intake in the baseline period was 12.6 MJ/day and this was increased to 17.5 MJ/day in the overfeeding period. This meant an average increase of nearly 40 per cent. The mean body weight gain was 6.0 kg, mainly in the form of fat as measured with hydrodensitometry (Table 19.4). The excess energy intake was compared with the energy gain in the form of body reserves and the results are shown in Table 19.5. The daily metabolic rate (*ADMR*, column A) during the baseline period averaged 11.4 MJ as calculated from the time spent on activities and the measured energy cost of each activity.

ADMR during the overfeeding period was assumed to increase as a result of the weight gain itself, the increased mass to be maintained (*BMR*) and carried during activities (*AEE*), and as a result of normal costs of processing food (*DEE*), and tissue gain. The average extra energy cost for *BMR* and *AEE* was assumed to be 5 per cent of *ADMR*. This assumption was based on an average body mass gain over the 6-week interval of 10 per cent divided by two, justified by the linear gain apparent from column B. The extra energy cost of *DEE* and tissue gain was assumed to be 10 per cent (column C).

TABLE 19.4

Body weight and body fat at the beginning and end of 42 day overfeeding

subject	body weight (kg)			body fat (kg)		
	initial	final	gain	initial	final	gain
1	71.60	76.50	4.90	13.9	16.5	2.6
2	62.10	70.60	8.50	6.2	11.3	5.1
3	59.75	66.80	7.05	12.8	18.6	5.8
4	54.80	59.80	5.00	6.5	9.6	3.1
5	71.25	78.25	7.00	9.6	14.1	4.5
6	64.50	68.25	3.75	8.2	9.3	1.1
mean	64.00	70.03	6.03	9.5	13.2	3.7

(From: Norgan & Durnin, 1980)

The excess intake was larger than energy gain, assuming an energy equivalent of 38.9 MJ/kg body fat gain, even when the excess intake is corrected for increased maintenance and activity costs (column B), and increased costs for food processing and tissue gain (column C). The mean discrepancy after the latter maximally corrected figure for excess intake is still 29 per cent. This means that 71 per cent of the excess energy intake is accounted for by the observed body fat gain. Of course, the subjects also gained some lean body mass which explains at least some of the discrepancy. The mean body mass gain was 6.0 kg and the mean fat gain was 3.7 kg (Table 19.4), hence, the mean increase in lean body mass was 2.7 kg. Assuming 73 per cent hydration and the dry material to be protein, resulting in an energy equivalent of 6.3 MJ/kg lean body mass gain, this represents a mean energy gain of 17 MJ or 8 per cent of the excess intake. Twenty per cent of the excess energy intake remains unexplained and is left to speculation on mechanisms increasing energy expenditure above that associated with increased body size and tissue gain. Direct measurements of energy expenditure, for example with doubly-labeled water, are required to answer this question. There are no indications for differences in energy gain between subjects related to their weight 'profile' at the start of the experiment. The plump subjects 1 and 3 gained as much fat as the thin subjects 2 and 4 (Table 19.4), while there was no difference in excess intake either (Table 19.5).

Webb and Annis (1983) overfed 9 middle-aged subjects classified as either lean or obese (Table 19.6). The subjects lived at home and came to the laboratory for a meal in the early morning, at noon and after work. They were mainly sedentary, avoiding strenuous exercise and sports activities. Instead of asking them their usual food intake they were fed for 30 days on a mixed diet of which the amounts were adjusted to maintain a constant body weight. Then, they were overfed for 30 days with an amount sufficient to achieve an expected weight gain of 5 kg.

TABLE 19.5

Comparison of excess energy intake and energy gain

subject	energy intake	energy expenditure			excess intake	energy gain	body reserves
		A	B	C	B	C	
	(MJ/d)	(MJ/d)	(MJ/d)	(MJ/d)	(MJ/wk)	(MJ/wk)	(MJ/wk)
1	16.2	11.3	11.9	12.4	181	160	101
2	21.9	12.4	13.0	13.6	374	349	198
3	17.7	11.3	11.9	12.4	244	223	226
4	15.6	10.9	11.4	12.0	176	151	121
5	17.9	12.5	13.1	13.8	202	172	175
6	15.8	10.1	10.6	11.1	218	197	43
mean	17.5	11.4	12.0	12.6	233	209	144

A: calculated from time spent on activities and energy cost of these activities
B: assuming an increase in energy expenditure of 5% as a result of weight gain
C: assuming an increase in energy expenditure of 10% as a result of weight and tissue gain
(From: Norgan & Durnin, 1980)

TABLE 19.6

Physical characteristics of the 5 women and 4 men in an overfeeding study

subject	sex	lean (L) overweight (O)	age (y)	height (m)	weight (kg)	body fat (%)
1	F	L	55	1.64	58.3	25
2	F	L	43	1.60	45.9	27
3	M	L	45	1.81	67.2	10
4	M	L	55	1.80	56.6	14
5	M	L	42	1.72	64.3	11
6	F	O	45	1.69	71.8	42
7	F	O	43	1.71	94.9	46
8	F	O	48	1.54	71.8	44
9	M	O	40	1.82	108.8	27

(From: Webb & Annis, 1983)

Four subjects received a high-protein and high-fat diet (carbohydrate/protein/fat = 30/20/50), four subjects an average diet (carbohydrate/protein/fat = 45/14/41), and four subjects a high-carbohydrate diet (carbohydrate/protein/fat = 60/10/30). Three subjects were overfed twice on a different diet (subjects 1, 3, and 9 in Table 19.7).

Thus, each diet was consumed by at least one lean and one overweight subject and one woman and one man. The net energy intake was measured by bomb calorimetry. Energy expenditure was measured in the laboratory during two 36-hour intervals in the control period and two 36-hour intervals in the overfeeding period with a suit calorimeter (direct calorimetry). On all three diets the subjects gained less body weight than the anticipated 5 kg (Table 19.7).

The weight gain was slightly smaller on the high-protein high-fat diet. Surprisingly, the overweight subjects (numbers 6-9) tended to gain less weight than the lean subjects (numbers 1-5). For comparing the excess intake with the energy gain Webb and Annis used an approach that was slightly different from that of Norgan and Durnin (Table 19.8). Energy expenditure was measured in sedentary conditions, thus including increased maintenance cost (*BMR*), cost of processing the extra food (*DEE*) and tisue gain. *ADMR* in these conditions increased from a mean value of 8.1 to 8.7 MJ/day, i.e. by 7.5 per cent. Calculating the excess intake as the difference between energy intake and sedentary energy expenditure, the mean excess is 136 MJ during the 30-day interval (Table 19.8, column A). When calculating the excess intake as the difference between energy intake and estimated energy expenditure including *AEE* (see Table 19.6 for the calculation procedure), the mean excess is 99 MJ (Table 19.8, column B).

The measured energy gain, based on the adopted energy equivalent of 38.9 MJ/kg for fat and 6.3 MJ/kg for lean body mass (see above), averaged 72 MJ. In conclusion, the excess intake is very close to the energy gain, when energy expenditure is estimated from the measured sedentary expenditure and the calculated cost of physical activity is added.

TABLE 19.7
Body weight, body fat, and lean body mass changes after 30 days overfeeding

subject	diet*	body weight (kg)			body fat (kg)	lean body mass (kg)
		initial	final	gain	gain	gain
1	1	60.2	62.0	1.8	1.3	0.5
3	1	66.9	69.8	2.9	0.7	2.2
6	1	73.8	75.1	1.3	1.0	0.3
9	1	107.7	109.2	1.5	1.4	0.1
mean				1.9	1.1	0.8
1	2	50.0	52.0	2.0	2.1	-0.1
4	2	56.6	60.1	3.5	2.7	0.8
7	2	94.0	96.3	2.4	2.1	0.3
9	2	94.0	96.3	2.3	2.1	0.2
mean				2.6	2.3	0.3
2	3	45.9	48.7	2.8	2.1	0.7
3	3	70.2	73.5	3.3	2.1	1.2
5	3	63.6	66.1	2.5	1.1	1.4
8	3	70.8	72.4	1.6	3.1	-1.5
mean				2.6	2.1	0.5
mean				2.4	1.8	0.5

*1) high-protein and high-fat diet (c/p/f ratio = 30/20/50); 2) average diet (c/p/f ratio = 45/14/41); 3) high-carbohydrate diet (c/p/f ratio = 60/10/30).
(From: Webb & Annis, 1983)

Combining the results of the two overfeeding studies it can be concluded that most of the energy added to a maintenance diet is stored. Norgan and Durnin fed their subjects 17.5 MJ/day with a mean energy expenditure of 11.4 MJ/day for 42 days, i.e. the added energy over 42 days was on average 256 MJ and the energy gain was on average 144 MJ or 56 per cent. Webb and Annis' subjects, with a mean energy expenditure of 9.2 MJ/day, were fed 13.2 MJ/day for 30 days, i.e the added energy over 30 days was on average 120 MJ and the average energy gain was 76 MJ or 63 per cent.

TABLE 19.8
Comparison of excess energy intake and energy gain

| subject | baseline | | overfeeding | | | excess intake | | energy |
| | EI | EE_s | EI | EE_s | EE_e | A | B | gain |
	(MJ/d)	(MJ/d)	(MJ/d)	(MJ/d)	(MJ/d)	(MJ)	(MJ)	(MJ)
1	7.0	7.1	11.7	8.2	8.1	105	107	54
3	11.2	10.8	14.7	10.9	11.3	115	101	41
6	7.3	7.3	11.3	7.6	7.5	113	113	41
9	11.4	8.5	15.7	8.3	11.2	222	135	55
mean	9.2	8.4	13.3	8.7	9.5	139	114	48
1	7.5	7.6	11.7	7.8	7.7	117	120	81
4	8.0	7.8	12.0	8.3	8.6	112	104	110
7	7.5	8.2	11.9	9.5	8.7	72	96	84
9	11.0	10.0	14.8	11.3	12.3	107	76	60
mean	8.5	8.4	12.6	9.2	9.3	102	99	84
2	7.6	6.7	11.2	6.9	7.8	128	100	86
3	13.0	8.0	17.1	9.4	15.3	230	53	89
5	10.0	7.7	14.0	8.2	10.7	174	100	52
8	9.1	7.6	12.4	8.0	9.7	133	83	111
mean	9.9	7.5	13.7	8.1	10.9	166	84	85
mean	9.2	8.1	13.2	8.7	9.9	136	99	72

$*EE_s$ = expenditure measured under sedentary conditions
EE_e = energy expenditure estimated as EI baseline x (EE_s:overfeeding/EE_s: baseline)
A = excess intake calculated as (EI-EE_s)
B = excess intake calculated as (EI-EE_e)
(From: Webb & Annis, 1983)

3   ENERGY INTAKE AFTER A DECREASE IN ENERGY EXPENDITURE

There is little information on the consequences of a decrease in energy expenditure for energy intake and the resulting energy balance. It is often suggested that humans eat according to their needs to maintain energy balance, just to prevent a negative energy balance. In evolution, there is no evidence for selection against or even in favour of mechanisms to promote a positive energy balance. Thus, there might not be an adaptive mechanism to reduce energy intake after a decrease in energy expenditure.

It is common knowledge that people gain weight when they decrease their energy expenditure by reducing their physical activity. Well-known examples are athletes who stop their training and people quitting from manual work. A more complicated example is weight gain in people who stop smoking. Traditionally, the weight gain after the cessation of smoking is ascribed to an increase in energy intake. However, there are indications that smokers do have a higher energy expenditure than non-smokers. Stopping with smoking would thus mean a decrease in energy expenditure, which would explain why ex-smokers gain weight even when their energy intake remains the same.

Unfortunately there are no systematic studies on energy intake and energy balance in subjects before and after a reduction in energy expenditure.

TABLE 19.9
Energy intake before (0), and 8, 20, and 40 weeks after the start of a training period

		energy intake (MJ/d)		
	0	8	20	40
males				
1	12.7	12.6	11.5	12.2
2	9.8	10.2	10.0	10.0
3	11.6	11.6	11.4	9.4
4	9.8	9.0	9.3	9.2
5	13.0	11.9	11.6	10.3
6	12.3	10.1	11.2	8.7
7	8.9	9.7	9.5	9.6
8	16.9	15.0	12.9	14.6
9	10.9	11.9	10.5	12.7
10	11.9	12.1	12.5	12.9
11	13.8	11.9	13.4	13.6
12	10.1	8.3	8.5	7.5
mean	11.8	11.2	11.0	10.9
SD	2.2	1.8	1.5	2.2
females				
13	9.9	10.0	8.7	10.0
14	9.5	6.5	9.1	7.5
15	9.1	9.4	9.5	9.7
16	9.0	9.2	9.3	9.5
17	6.0	7.5	6.5	9.0
18	9.1	8.4	8.0	8.0
19	8.6	9.0	8.6	13.2
20	6.3	7.4	8.1	6.9
21	10.3	10.1	12.0	11.6
22	9.3	10.6	9.8	11.4
23	7.0	5.6	7.6	7.5
mean	8.6	8.5	8.8	9.5
SD	1.5	1.6	1.4	2.0

(From: Westerterp *et al.*, 1992)

4   ENERGY INTAKE AFTER AN INCREASE IN ENERGY EXPENDITURE

The consequences of an increase in energy expenditure for energy intake and energy balance have been measured by raising the *EEA*. Military cadets increase their energy intake in response to a drill competition (Chapter 18.4). Novice athletes starting a training program manage to maintain energy balance after a mean increase of their energy expenditure with 30 per cent (Chapter 18.3). The latter example will be described in more detail with regard to changes in energy intake.

The particulars of the athletes, training program and observational protocol have already been described in Chapter 17.3. The energy intake (EI) did not show significant changes (Table 19.9). However, the females showed a clear tendency towards an increase in EI from the second interval onwards while the males, after an initial drop, kept their EI completely unchanged from the second interval until the end of the study. $ADMR$ increased in the period between weeks 8 and 10 and remained at the same level up to week 40 (Figure 18.2), on average 30 per cent higher than before the start of the training. Body mass (BM) did not show much change, indicating the subjects were in energy balance. The discrepancy between the observed increase in energy expenditure without a significant increase in EI clearly is an artefact. It is another example of the difficuties in measuring food consumption (Chapter 15.1). The conclusion must be that it is possible to maintain energy balance after an increase in energy expenditure by increasing the energy intake. Usually, there is, at least initially, a negative energy balance. In the long term an increased level of physical activity affects the body composition as described in Chapter 17.3.

5   ENERGY BALANCE AT HIGH ALTITUDE

It seems to be difficult, if not impossible, to maintain energy balance above 4500 m and the ensuing weight loss is a phenomenon well-known to high-altitude climbers. Until now there is no clear evidence whether the negative energy balance is mainly the result of a lowered energy intake or of an increased energy expenditure as well. This section presents the results of a study on energy metabolism in climbers at high altitude, measuring simultaneously the energy intake, energy expenditure, and changes in body composition.

The subjects were two women and three men, all members of an expedition to reach the summit of Mount Everest. Two subjects were observed while preparing for the expedition. This included a 4-day stay in a field laboratory (Vallot Observatory, 4260 m) on Mont Blanc in the French Alps, with daily climbing activities between 3500 and 4800 m, and a subsequent 4-day stay in a hypobaric chamber. Here they took daily bicycle ergometer exercise, simulating daytime ascents of 5600-7000 m on Mount Everest. The five subjects were also observed during a 7-10 day period while making the final preparations for the summit attempt, climbing between 5300 and 8000 m (Figure 19.3). Three subjects reached the summit (8872 m) within two days after the observation interval, for one subject the ascent to the summit was included in the observation interval. Measurements of energy metabolism comprised energy and water intake, energy expenditure, water loss, and changes in body composition over the observation interval.

The energy intake (EI) was measured with a dietary record. The subjects recorded their food and fluid intake in a diary in household measures, including brand names and cooking recipes where appropriate. Where possible, the items were weighed as well. After the expedition, a trained nutritionist examined the records to clarify and eliminate inconsistencies with the subject. Energy expenditure was measured under both resting pre-absorptive conditions (resting metabolic rate, RMR) and field conditions (average daily metabolic rate, $ADMR$). RMR was measured in the morning immediately after waking up, in the Alps just before the observation period and on Mount Everest just after. Body composition was calculated from body mass and skinfold thickness measured at the start and at the end of the observation interval of EI and $ADMR$.

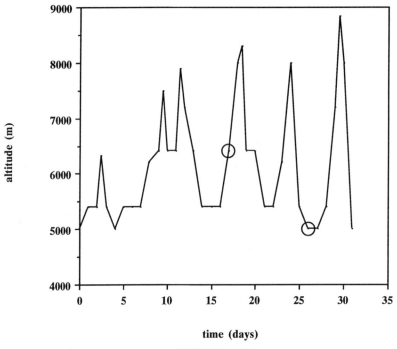

time (days)

FIGURE 19.3
**Altitude profile of two subjects (nos 1 and 2, Table 19.10) during
the expedition on Mount Everest**
The start and end of the observation interval are encircled.(From: Westerterp *et al.,* 1992)

All subjects consumed a diet which was very close to the current nutritional advice of 55 energy% carbohydrate, 10-15 energy% protein and 30-35 energy% fat. *EI* of subjects 1 and 2 was 9 and 13 per cent respectively lower on Everest than during the preparation in the Alps (Table 19.10). There was a tendency of *EI* to decrease during the observation interval while the nutrient composition remained the same. The fluid intake was also lower on Mount Everest than during the preparation in the Alps. *ADMR* was higher than energy intake in all subjects in both situations (Table 19.10). Calculating the activity level of the subjects from the quotient *ADMR/RMR*, the value ranged between 2.0 and 2.9 based on measured *RMR* values, and between 2.2 and 2.5 based on calculated *RMR* values. There was no indication of a difference in energy expenditure during the preparation in the Alps and the expedition on Mount Everest.

All subjects, apart from subject 1 in the Alps, lost weight. For practical reasons, the figure on change in fatty mass *(FM)* in one subject on Mount Everest was not reliable. Being the only person present in the camp at the start of the observation interval who had experience with measuring skinfold thickness, she could not measure here own skinfolds. Hence, her data on *FM* have been omitted. The remaining figures on the changes in *FM* are all negative, i.e. subjects lost body fat in both situations, with one subject even gaining 0.6 kg body mass in the Alps. When comparing *EI* and *ADMR*, all subjects were in negative energy balance in the Alps and on Mount Everest. The differences between *EI* and *ADMR* ranged from 2.6 to 8.9 MJ/day. The energy deficit was not related to the change in *BM* during the observation period. However, a significant relationship appeared between the energy deficit and the change in *FM* ($r = 0.86$, $p < 0.05$).

TABLE 19.10

Total body water, fractional elimination rates from body water of excess ^{18}O and ^{2}H, fluid intake, fluid output, energy intake and average daily metabolic rate

	Subject	TBW* (l)	Fluid in (l/d)	Fluid out (l/d)	EI (MJ/d)	ADMR (MJ/d)
Alps	1	29.0	2.9	3.9	9.4	12.1
	2	40.4	2.7	3.4	10.7	17.2
mean		34.7	2.8	3.7	10.1	14.7
SD		8.1	0.1	0.4	0.9	3.6
Everest	1	28.5	2.7	3.8	8.2	12.3
	2	39.9	2.2	2.5	9.7	14.9
	3	30.8	1.8	2.9	5.8	11.7
	4	41.8	2.4	4.0	7.0	15.6
	5	32.7	2.0	3.4	6.9	13.5
mean		34.7	2.2	3.3	7.5	13.6
SD		5.8	0.3	0.6	1.5	1.7

* *TBW* = mean total body water as calculated from initial isotope-dilution space and body-mass change (see text)

Fluid in = fluid intake

Fluid out = fluid output

EI = energy intake

ADMR = average daily metabolic rate

As expected, the subjects lost weight during the observation period on Mount Everest. The weight loss amounted to an average of 4 per cent of the starting weight and consisted of two thirds *FM* loss and of one third *FMM* loss. The values are not different from changes in body composition observed in studies including an energy restricted diet, and other studies at high altitude. Energy intake at high altitude was low, especially on Mount Everest, as has been reported before. No measurements were performed in the same subjects at low altitudes but the decrease in *EI* in subjects 1 and 2 in the Alps compared to the higher Everest situation is indicative. The energy intake was lower than *EI* measured with the same method in a comparable group of sedentary subjects at sea level, on average 9.1 and 11.8 MJ/day in women and men, respectively. Of course, there is a tendency to underreport when self-reporting. However, the figures on *EI* in this study are less liable to be the result of underreporting as the subjects were closely supervised in the Alps. On Mount Everest all the food was carried by the subjects themselves which gave them optimal opportunity to keep track of their intake.

The activity level of the subjects, as reflected in the *ADMR/RMR* ratios, was surprisingly high and comparable to the level of highly active subjects at sea level, such as women and men engaged in endurance exercise, for example long-distance running. *ADMR/RMR* values normally range from 1.5 to 1.8 for modestly to moderately active subjects. Here, the activity level was higher in all subjects. The discrepancy between *EI* and *ADMR* was related to the loss of *FM*. The mean energy equivalent of weight loss was 31 MJ/kg, which was, within the range of 28-32 MJ/kg, caused by fat for two thirds and the remainder by water loss only or protein loss only, respectively. Thus, the energy balance equation fits, i.e. $EI = EE - \Delta BM$.

In conclusion, the results from this study show the problems of maintaining energy balance at high altitude. The energy intake is lower than normal while the energy expenditure is equivalent to values which are only found in endurance athletes at sea level.

6   ALCOHOL INTAKE AND ENERGY UTILIZATION

In epidemiological studies alcohol consumers have been shown to have a higher daily energy intake, while on the other hand, they appear to be less obese than abstainers. The mechanism behind these apparently contradictory phenomena needs further investigation.

Alcohol forms a significant component of the diet in many countries. Nationwide studies indicate an average per capita consumption of about 10-30 g/day or about 3-9 per cent of the daily energy intake. Alcohol supplements rather than displaces the food-supplied daily energy intake. Even in heavy drinkers the average energy intake, excluding alcohol, is the same as in non-drinkers, i.e. drinkers have a higher energy intake than non-drinkers and the difference in energy intake between drinkers and non-drinkers seems to be equivalent to the energy value of the alcohol consumed (Colditz *et al.*, 1991; Figure 19.4). On the other hand, alcohol does not seem to be fattening as is traditionally believed. Surprisingly, in studies on the relationship between alcohol consumption and body weight, it was found that drinkers are less obese than abstainers (Table 19.11). This must mean that either the data on food intake and alcohol consumption are in error or the consumption of alcohol raises energy expenditure. With regard to the latter, one study showed an elevated basal metabolic rate in men consuming one or more drinks per day as compared to abstainers and light drinkers. However, the increased energy expenditure only partially offset the alcohol energy at lower levels of alcohol intake. It did not substantially offset the large energy surplus seen at higher levels of alcohol consumption.

The studies on alcohol consumption and energy balance mentioned above were all cross-sectional, i.e. they compared the energy balance and/or body weight of subjects differing in alcohol intake. This type of study is not conclusive with respect to the consequences of alcohol consumption for energy balance and body weight regulation. Surprisingly, there are hardly any longitudinal and experimental studies on the implications of alcohol consumption for energy metabolism. It is suggested that during alcohol ingestion ethanol becomes the major fuel of hepatic metabolism without any mechanism of storage or feedback control, i.e. the energy is wasted.

There are a few studies on the effect of alcohol on the resting metabolic rate. All studies showed an increased energy expenditure after alcohol consumption. However, the increased energy expenditure only partly offset the alcohol energy at lower levels of alcohol intake and did not substantially offset the large energy surplus seen at higher levels of consumption. One reason may have been the duration of the observation period of 1.5-4 hours. Rosenberg and Durnin (1978; Figure 19.5) saw a not significantly higher energy expenditure during the first 1.5 hour of the post-prandial period after a meal with alcohol while the higher energy expenditure was significant for the final 1.5 hour ($p < 0.01$).

Nowadays, energy expenditure and its components can be measured during 24-hour intervals in a respiration chamber. The doubly-labeled water method even allows assessment of daily energy expenditure under daily living conditions over 1-3 weeks.

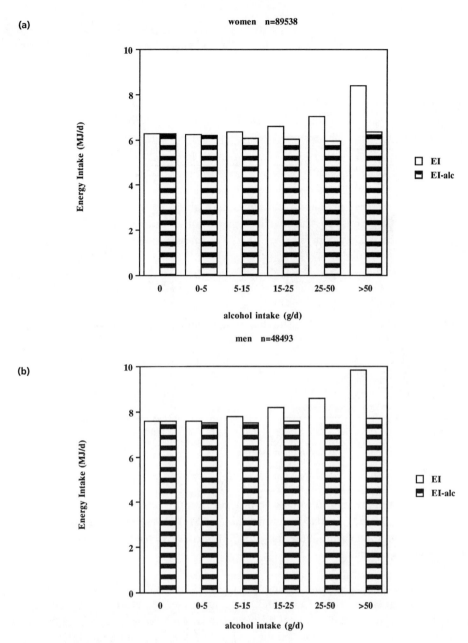

(a)

women   n=89538

(b)

men   n=48493

FIGURE 19.4

Daily energy intake with and without alcohol consumption in relation to daily alcohol intake (*a*) in 89 538 women and (*b*) in 48 493 men

(From: Colditz, 1991)

With these instruments records of daily energy intake can be validated to see whether alcohol really supplements the daily energy intake, and simultaneously measure the average daily metabolic rate and its components (basal metabolic rate, diet induced thermogenesis and energy expenditure for physical activity) during intervals with and without alcohol 'supplementation'.

TABLE 19.11

Body mass index in drinkers and abstainers

	women		*men*	
	abstainer	*drinker*	*abstainer*	*drinker*
	(n=1458)	*(n=814)*	*(n=2056)*	*(n=281)*
BMI (kg/m²)	23.2	22.2**	24.6	24.1*

* $p < 0.05$; ** $p < 0.001$

(From: Le Marchand *et al.*, 1989)

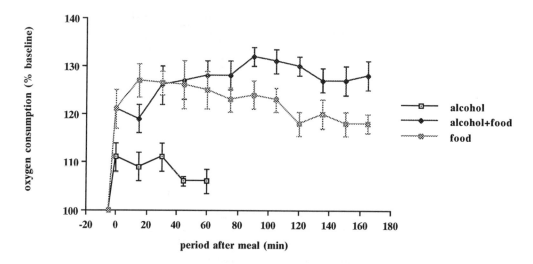

FIGURE 19.5

Mean oxygen consumption in 10 women before and after the consumption of only alcohol (230 ml red wine, 0.6 MJ), food (ham sandwiches, 2.5 MJ) plus alcohol (230 ml red wine, 0.6 MJ), and food (ham sandwiches, 2.5 MJ) plus 230 ml fruit drink (0.6 MJ)
(From: Rosenberg & Durnin, 1978).

The 3-9 per cent alcohol supplementation to the daily food intake mentioned above has drastic consequences if the food intake and energy expenditure remain unaltered. Converted to the equivalent of body fat this would mean an increase of 2.5-7.5 kg body fat per year. Obviously very few people gain weight so fast. On the other hand, intake and/or expenditure changes of this magnitude can be detected with existing instruments. Thus, the mechanism behind the relationshop between alcohol consumption and maintenance of energy balance is open to research.

7   SUMMARY

Usually a balance between energy intake and energy expenditure is maintained, and body weight and body composition are constant. This is achieved by control of energy intake and expenditure. Energy expenditure can be adapted to intake and vice versa.

During starvation and dieting, decreased energy expenditure caused by a decrease in active tissue mass for maintenance (BMR), reduced processing costs by lowering food intake (DEE), and reduced body weight for physical activity (AEE). Further adaptation comes from a reduction in tissue metabolism (BMR) and in physical activity itself (AEE).

During overfeeding 56-63 per cent of the energy added to a maintenance diet is stored. In addition ADMR increases as a result of the weight gain itself, the increased mass to be maintained (BMR) and carried during activities (AEE), as a result of costs of processing food (DEE) and tissue gain.

It is possible to maintain energy balance after increase in energy expenditure. Initially there is a negative energy balance. Moreover in the long term an increased level of physical activity, which is usually the cause of increased expenditure, affects body composition. For instance at of high altitude maintaining energy balance is a problem: a lower than normal energy intake and energy expenditure equivalent to values found in endurance athletes at sea level occur together.

20
Nutrient utilization and energy balance

Humans need to consume food to remain in energy balance. Food energy is consumed in the form of carbohydrate, protein and fat, together forming the macronutrients of our diet. The distribution of food energy over these macronutrients differs between cultures and countries. There is a fairly wide range of carbohydrate/protein/fat (c/p/f) ratios within which energy balance can be maintained, as shown by nationwide studies of nutrient intake (Table 20.1). The carbohydrate intake ranges from 3 to 82 energy% and the fat intake from 6 to 54 energy%, while the protein intake is at least 11 energy%. Within a culture or country dietary changes take place with changes in food supply. There is an increasing availability of fats in Western countries. In the USA the contribution of fats to the diet has steadily increased from 32 energy% in 1910 to 43 energy% in 1985 (Figure 20.1). The shift to a higher fat diet has often been quoted as the reason for the increasing incidence of overweight, i.e. a positive energy balance. This chapter will discuss some of the reasoning concerning and evidence for the relationship between the macronutrient composition of the diet and the maintenance of energy balance.

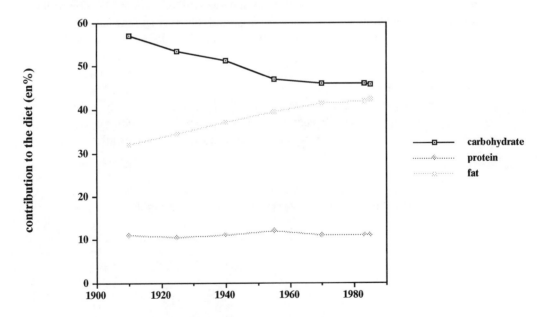

FIGURE 20.1
Distribution of dietary energy intake as carbohydrate, protein and fat in the USA between 1910 and 1985

1   DIETARY FAT/CARBOHYDRATE RATIO AND ENERGY BALANCE

Dietary fat is the main determinant in the energy density of our diet. The metabolizable energy for dietary carbohydrate, protein and fat is 16, 16 and 38 kJ/g respectively (Table 20.1). Fat also has an important function as an energy depot in the body. It can be stored with minimal additional weight. The energy density of the fat stores is approximately eight times higher than the energy density of the carbohydrate (glycogen) stores (Table 17.1). Thus, in most circumstances energy balance can be reached with the minimum bulk by consuming high-fat diets, and the energy surplus is mainly stored as body fat.

There are some studies that show a changing energy intake in people when they change to a diet with a lower or a higher energy density. Duncan *et al.* (1983) allowed subjects to eat to satiety from a diet low in energy density (3 kJ/g) and from a diet high in energy density (6.5 kJ/g). Each diet was provided for 5 days in a randomized cross-over design with a weekend in between. The subjects ate nearly twice as much the high-energy density diet (12.5 MJ/day) or, when put the other way round, twice as little after the low-energy density diet (6.5 MJ/day). Surprisingly, there was no tendency to a higher energy intake on subsequent days on the low-energy diet and to a lower energy intake on subsequent days after the high-energy density diet.

Lissner *et al.* (1987) provided subjects with three different diets, one with a low, one with a medium and one with a high energy density, by exchanging carbohydrate for fat. All subjects were given each diet for 14 days in a balanced sequence. The energy intake increased from a mean value of 8.7 MJ/day on the low-fat diet (15-20 energy% fat) to 9.8 MJ/day on the medium fat diet (30-35 energy% fat) and 11.4 MJ/day on the high fat diet (45-50 energy% fat). Again, there was no trend in intake over time in any of the subjects during any 14-day dietary treatment, indicating there was no intake compensation with respect to the type of diet during the observation period. On average, the subjects were in energy balance on the medium-fat diet, lost weight on the low-fat diet and gained weight on the high-fat diet.

Animals, such as the laboratory rat, compensate in time if the energy density of their food is changed, to achieve the same energy intake and reach the same body weight as controls. In humans, the failure to compensate for a change in the energy density of the food is suggested as a means to prevent overweight. Energy dilution with carbohydrate or fat analogs, like aspartame and sucrose polyesthers, is applied in the new market for 'light' products.

TABLE 20.1
Macronutrient ratios of typical diets in various countries

country	year	carbohydrate (en%)	protein (en%)	fat (en%)
Nigeria	1959	82	12	6
Japan	1958	77	12	11
India	1958	77	11	12
Czechoslovakia	1962	59	11	30
USA	1958	46	12	42
Greenland	1937	3	43	54

On the other hand, there are indications that we compensate for energy dilution when not all the high-energy density food items are reduced-energy versions by increasing the consumption of non-manipulated items. Subjects failed to compensate for increases in energy intake by covertly substituting part of the food items with high-energy versions.

2   DIETARY FAT/CARBOHYDRATE RATIO AND OBESITY

There are quite a few studies on the macronutrient ratio of the diet and overweight or obesity. This section will restrict itself to studies in which subjects are allowed to choose their own foods freely under ordinary living conditions. In such situations, overweight subjects appear to take diets with a higher energy density.

Jiang and Hunt (1983) asked 11 adult men to collect a double portion of all they ate during 7 days for analysis. The energy density of the diet, including drinks, ranged from 1.8 kJ/g in normal weight subjects to 3.9 kJ/g in overweight subjects. Surprisingly, they found the energy density of the diet in overweight subjects was not higher because of a higher fat content, but no alternative explanation was presented. Dreon *et al.* (1988) related diet composition (measured with 7-day food records) to body composition (measured with hydrostatic weighing) in 155 middle-aged men. The subjects with a higher percentage of body fat consumed a diet with relatively more fat and less carbohydrate. Tremblay *et al.* (1989) performed a comparable study in 244 male adults, measuring diet composition with a 3-day food record. They also found the energy% fat of the diet to be positively and the energy% carbohydrate to be negatively correlated to the fatness of the subjects. Finally, Miller *et al.* (1990) made similar observations in 216 adult subjects, 50 per cent men and 50 per cent women, and in their study too, adiposity was positively correlated to the dietary fat content and inversely related to the dietary carbohydrate consumption. In all four studies mentioned here there was no relationship between energy intake and indices for overweight or obesity.

Recently, there are indications that self-reported intake tends to be an underestimate of the real habitual intake. Reported intakes tend to be lower than energy expenditure as measured simultaneously with doubly-labeled water, especially in obese subjects and subjects with a high energy intake like those engaged in endurance exercise. It is not known whether the results of the studies referred to, mainly using self-report to measure dietary intake, were influenced by this phenomenon. On the other hand, it is almost impossible to measure the habitual intake while the subjects freely choose their own foods under ordinary living conditions with a more accurate alternative technique. The double-portion technique as used by Jiang and Hunt (1983) is one of the best, but not really feasible in large groups. Also, it interferes with daily routines when the subjects are not at home every time they eat something.

Thus, the conclusion is that fat intake may play a role in obesity, independent of energy intake.

3   DIETARY FAT/CARBOHYDRATE RATIO AND SUBSTRATE UTILIZATION

Knowing that obese subjects generally do not eat more than but different from normal-weight subjects with respect to the fat/carbohydrate ratio of their diet, the next question is why an iso-energetic diet is fattening when it contains relatively more fat and less carbohydrate. In terms of energy carbohydrate can be made from protein, and fat from carbohydrate and thus fat and carbohydrate are not essential nutrients. However, the conversion processes of proteins consume much energy and produce a good deal of waste products like ammonia and urea. Ideally, the body covers its energy needs with a mixture of fat and carbohydrate. Some tissues preferentially use carbohydrate, while fat is less bulky to consume. Additionally, man is a periodic eater and a continuous metabolizer, i.e. part of the energy intake is stored before it is used.

The storage capacities for fat and carbohydrate are very different and this may have consequences for the regulation of body weight. Flatt (1987) suggests a model with a regulated carbohydrate (glycogen) store while the fat store is a function of nutrient intake and nutrient utilization. An increase in the fat content of the diet needs an increase in fat oxidation and it is hypothesized that the latter needs an expansion of the body-fat mass, which would explain the increased adiposity in subjects consuming high-fat diets.

Several groups studied the effect of a change in nutrient intake on nutrient utilization. The latter can be measured with indirect calorimetry from oxygen consumption, carbon dioxide production and urinary nitrogen excretion. When a subject is in nutrient balance, the measured oxygen consumption and carbon dioxide production are equal to the calculated oxygen consumption and carbon dioxide production from the consumed nutrients, and the intake of nitrogen with protein equals the nitrogen output in the urine. Sometimes protein oxidation is measured with primed infusion of a ^{13}C-labeled amino acid. Measurement of the elimination of the label in the respiratory gas then provides a rapid responsive index. A first indicator for nutrient balance is the comparison of the ratio between measured carbon dioxide production (V_{CO_2}) and oxygen consumption (V_{O_2}), the so-called respiratory quotient ($RQ = V_{CO_2}/V_{O_2}$), with the ratio of the calculated carbon dioxide production and oxygen consumption when the consumed food is fully oxidized, known as the food quotient (FQ).

Studies on nutrient utilization can be divided into short-term studies, usually measuring the effects of a single meal, and long-term studies, covering at least one 24-hour cycle on a fixed diet with or without an adaptation period beforehand. The indirect calorimetry system in short-term studies is a ventilated hood, measuring subjects in a lying or sitting position. Long-term studies are performed in a respiration chamber allowing the subjects to move around on 10-20 m^2 floor space.

3.1   SHORT-TERM STUDIES ON NUTRIENT UTILIZATION

Short-term studies on nutrient utilization are usually started in the post-absorptive condition after an overnight fast. Food is consumed as breakfast after baseline measurements are taken, and observations are continued for several hours. Acheson *et al.* (1982) measured the effect of a large carbohydrate meal (500 g or 9 MJ, Table 20.2) on nutrient utilization. The high-carbohydrate load lowered the fat oxidation rate, however there were no indications of the conversion of carbohydrate to fat. Assuming complete processing of the meal in the subsequent 10-hour observation interval, the storage capacity for carbohydrate in the form of glycogen was probably sufficient to accommodate the intake surplus since only 133 g carbohydrate were oxidized. In a subsequent experiment Acheson *et al.* (1984) controlled the diet over 3-6 days preceding a comparable test. The subjects consumed either a low-carbohydrate diet, a mixed diet or a high-carbohydrate diet.

TABLE 20.2
Food intake and nutrient utilization

study	intake	c/p/f	observation time	C-oxidation		P-oxidation		F-oxidation	
(reference)	(MJ)	(en%)	(h)	(g/h)	(en%)	(g/h)	(en%)	(g/h)	(en%)
Acheson (1982)	9.0	93:5:2	10	13.3	66	2.9	15	1.7	19
Acheson (1984)	8.4	100/0/0* [1]	14	12.6	60	2.6	12	2.6	28
	8.4	100/0/0* [2]	14	17.2	75	2.4	11	1.4	14
	8.4	100/0/0* [3]	14	18.4	84	2.4	11	0.5	5
Flatt (1985)	2.0	62/27/11	9	9.4	42	3.1	14	4.3	44
	3.6	35/15/50	9	9.4.	41	2.8	12	4.6	47
	3.6	35/15/50**	9	9.4	40	3.2	14	4.7	46

* dextrin maltose solution flavored with fruit juice
** three quarters of the fat in the form of medium-chain triglycerides
[1] c/p/f diet 3-6 days preceding test: 14/11/75 en%
[2] c/p/f diet 3-6 days preceding test: 60/12/28 en%
[3] c/p/f diet 3-6 days preceding test: 80/11/9 en%
c = carbohydrate; p = protein; f = fat
en%: contribution to energy intake and energy expenditure in %.

The higher the carbohydrate content of the initial diet the higher the carbohydrate oxidation rate after the carbohydrate meal: 177±5, 241±11, and 258±9 g during the 14 hours after consumption of 500 g in the low, medium- and high-carbohydrate groups respectively.

Carbohydrate oxidation was higher than in the first experiment, possibly because of the use of a different dietary carbohydrate. Again, most of the carbohydrate surplus was stored as glycogen although there was significant lipid synthesis from carbohydrate amounting to 3.4±0.6 and 9.0±1.0 g in the medium- and high-carbohydrate group respectively. However, the limited lipogenesis combined with the extremely high carbohydrate intake was judged to be unimportant in daily life. Finally, Flatt *et al.* (1985) studied the effects of the addition of fat to a standard, mixed meal. Fat and carbohydrate oxidation appeared not to be influenced by the increased fat consumption (Table 20.2). In the post-absorptive state the main fuel of the body is fat with a rapid shift to carbohydrate on the commencement of feeding, independent of the fat content of the food consumed.

3.2   LONG-TERM STUDIES ON NUTRIENT BALANCE

Ideally, observations of nutrient balance should cover at least a 24-hour interval or a multiple of 24 hours due to the existence of a diurnal pattern of nutrient utilization (see also Chapters 6 and 21). Nowadays several laboratories have facilities like a respiration chamber to perform this kind of study. There are at least 4 studies available on the consequences of nutrient exchange for nutrient utilization as measured over at least 24 hours. The results of these studies are summarized in Table 6.3.

TABLE 20.3

Mean 24-hour food quotient (*FQ*) and respiratory quotient (*RQ*) on diets with different carbohydrate/protein/fat ratios (c/p/f) in four studies

study (reference)	c/p/f (en%)	days on diet	energy balance (MJ/d)	FQ	RQ
Hunri	44/16/40	7	−1.4±0.4	0.85	0.80±0.01*
(1982)	78/16/6	7	−1.9±0.5	0.95	0.88±0.01*
Lean	45/15/40	0	−0.1	0.85	0.82±0.01
(1988)	82/15/3	0	−0.1	0.96	0.87±0.01
Abbott	43/15/42	5-43	−0.3±0.3	0.83	0.84±0.01
(1990)	65/15/20	6-32	−0.1±0.2	0.91	0.88±0.01**
Hill	20/20/60	3	–	0.77	0.75±0.01
(1991)		7	–	0.77	0.75±0.01
	35/20/45	3	–	0.83	0.79±0.01
		7	–	0.83	0.78±0.00
	60/20/20	3	–	0.92	0.86±0.02
		7	–	0.92	0.86±0.01

*$p < 0.05$, and **$p < 0.001$ *FQ* compared with *RQ*

FQ = ratio of the calculated carbon dioxide production (V_{CO_2}) and oxygen consumption (V_{O_2}) when the consumed food is fully oxidized, the food quotient ($FQ=V_{CO_2}/V_{O_2}$)

RQ = ratio of the measured carbondioxide production (V_{CO_2}) and oxygen consumption (V_{O_2}), the respiratory quotient ($RQ=V_{CO_2}/V_{O_2}$)

Hunri *et al.* (1982) measured the nutrient utilization in subjects consuming a mixed diet followed by a high-carbohydrate low-fat diet. Both diets were consumed for 7 days, with a 2-week wash out period in between. The measurements of nutrient utilization took place on the last day of each 7-day period. Unfortunately the subjects were in negative energy balance during the 24-hour respiration chamber measurements, with intake on average 14 per cent lower than expenditure. Despite the negative energy balance, carbohydrate and protein oxidation were lower than carbohydrate and protein intake, even on the mixed diet. However, these results are still within the error range of the applied methods. The carbohydrate oxidation was more than doubled on the high-carbohydrate diet as compared to that on the low-carbohydrate diet, thus illustrating the flexibility of the body to use carbohydrate for energy metabolism. Lean and James (1988) measured the metabolic effect of an iso-energetic exchange of nutrients over one 24-hour interval after the consumption of a standard evening meal. The subjects were measured during fasting, and subsequently on a low-fat and on a high-fat diet. The mean energy intake, based on energy expenditure during 24 hours of fasting plus 5 per cent, was 1.7 per cent higher than the mean energy expenditure. The individual differences were all smaller than 10 per cent, so the subjects were in energy balance. Nutrient utilization was closer to nutrient intake on the high-fat diet than on the low-fat diet, when comparing the calculated *FQ* with the *RQ* presented in Table 20.3. Abbott *et al.* (1990) measured the energy metabolism in subjects after 5-43 days on a high-fat and on a low-fat diet in a metabolic ward. Some of the subjects started on the high-fat diet, the others on the low-fat diet at the first admission. The interval between first and second admission was more than 4 weeks. The subjects were in energy balance, since the mean difference between energy intake and energy expenditure (< 5 per cent) was within the error range of both measurements. *FQ* and *RQ* did not differ on the high-fat diet, indicating that nutrient intake and nutrient utilization were the same. On the high-carbohydrate diet *RQ* was systematically lower than *FQ* ($p <$ 0.001), indicating that the oxidation of fat was higher than fat intake, assuming there was no net protein synthesis or oxidation. Hill *et al.* (1991) measured nutrient utilization in subjects after 3 and 7 days on three different diets, a mixed diet, a low-fat diet, and a high-fat diet.

Unfortunately, they did not report whether the subjects were in energy balance, and only presented data on diet composition and nutrient utilization. On all diets there was a tendency for *RQ* to be lower than *FQ*. The authors explained the discrepancy by pointing at the difficulty in using indirect calorimetry combined with food-table nutrient analysis.

4   NUTRIENT ADDITION AND NUTRIENT BALANCE

Dallosso and James (1984) increased the energy intake for one week by 50 per cent with fat, mainly in the form of double cream, after a one-week observation period on a maintenance diet with a c/p/f ratio of 57/13/30 energy%. The subjects were observed on days of low activity and of high activity respectively. The carbohydrate and protein balance appeared not to be affected by the addition of fat to the diet. The 5 MJ fat supplement was mainly stored as fat judging from the change in fat balance from +14±25 to +163±27 and from –14±19 to +151±33 g per day on the low-activity and high-activity days respectively. The mean net fat storage in terms of energy calculated from these figures is 5.5 MJ on low-activity and 5.1 MJ on high-activity days. Schutz *et al.* (1989) carried out a comparable experiment. They measured energy expenditure and substrate utilization over two-day intervals in subjects eating a maintenance diet on the first day (c/p/f ratio 50/15/35 energy%). On the second day they consumed a diet with the same amounts of carbohydrate and protein but with twice the amount of fat. On the first day the mean difference between intake and expenditure was 0.1±0.2 MJ, indicating subjects were in energy balance. The 24-hour *FQ* and mean *RQ* values were 0.87 and 0.85±0.01 respectively, indicating the subjects were in nutrient balance as well. On the second day the energy expenditure and substrate utilization were not changed by the fat supplement. The energy balance became +4.1±0.3 MJ as compared to the extra intake of 4.1±0.1 MJ, in other words, the fat supplement was stored. Acheson *et al.* (1988) overfed subjects for one week with carbohydrate after a three-day interval on a restricted diet to deplete the glycogen stores. The energy-restricted diet provided 6.7 MJ/day, with a c/p/f ratio of 10/15/75 energy%. On the third day the energy expenditure was 9.6 MJ. Subsequently the subjects received a diet with a c/p/f ratio of 86/11/3 energy%, increasing the energy intake over 7 days from 15.5 to 21.0 MJ/day. On the first day of overfeeding the energy surplus was fully stored as glycogen. From the 5th day of overfeeding onwards, *de novo* lipogenesis started to use up all the energy surplus. At the end of the overfeeding period there was a glycogen gain of 0.7 kg and a fat gain of 1.1 kg, together representing 75 per cent of the energy consumed in excess of maintenance requirements.

5   CONCLUDING REMARKS

Based on the literature referred to here, several conclusions can be drawn. Firstly, the evidence points in the direction of a change in diet to more fat causing an increase in body weight. Combining this with the fact that obese people tend to eat more fat, the conclusion seems to be justified that overweight can be prevented by reducing the fat content of the diet. Secondly, there is evidence of the body having a limited ability to oxidize fat compared to its capacity to oxidize carbohydrate and protein. It is often suggested that this limitation is more pronounced in individuals prone to obesity. Thus, some individuals are more likely to become obese on a high-fat diet than others. On the other hand, intervention studies do not unequivocally support the theory of the body's limitations to burn up fat.

Short-term studies, measuring substrate utilization during up to 14 hours after a meal, show how carbohydrate oxidation increases after a high-carbohydrate meal while addition of fat to a meal does not influence fat oxidation (Table 20.2). In the long term, measuring substrate utilization over at least 24 hours indicates the opposite. On diets higher in carbohydrate there is a larger discrepancy between *FQ* and *RQ* such that *RQ* is lower than *FQ* (Table 20.3). This should mean that substrate utilization is closer to substrate intake on diets higher in fat than on diets containing more carbohydrate. Such a phenomenon cannot be readily explained.

Under conditions of perfect energy and nutrient balance, *FQ* must equal *RQ*. In the case of energy or nutrient imbalance a normal-weight adult stores or mobilizes, in the long term, nearly all energy in the form of body fat. Then, the body does not use its protein and carbohydrate reserves for energy storage or energy mobilization. The carbohydrate store in the form of liver and muscle glycogen fluctuates between 250 and 500 g or 4 and 8 MJ. Reference figures are a muscle mass of 30 kg with 7.5 kg protein or an energy equivalent of 120 MJ. However, the changes in muscle mass are insignificant in terms of energy when compared with changes in fat mass, unless the body fat is nearly depleted. Thus, in the long term, if *RQ* exceeds *FQ*, this implies conversion of carbohydrate or protein to body fat. Conversely, if *RQ* is lower than *FQ*, energy is mobilized from body fat.

Combining the observation of the lower than expected *RQ* on a high-carbohydrate diet with the fact that *RQ* measured over a long-term interval can only be lower than the *FQ* by mobilizing body fat, leads to the conclusion that high-carbohydrate diets induce body fat loss. Apart from the studies referred to earlier, there is a recent study suggesting that the macronutrient composition of the diet plays a role in the energy requirement for weight maintenance. Prewitt *et al.* (1991) reported on 18 women consuming a standard diet (c/p/f ratio of 44/19/37 energy%) during 4 weeks, followed by 20 weeks on a low fat diet (c/p/f ratio of 60/19/21 energy%). The energy intake of the subjects was adjusted to maintain body weight throughout the experimental period, i.e. the intake was increased or decreased when the body weight decreased or increased by more than 1 kg respectively. Comparing the initial four-week interval on the standard diet with the last four weeks on the low-fat diet, the mean energy intake showed an increase of 19 per cent whereas the mean body weight decreased with 2 kg. Thus, a high-carbohydrate diet resulted in a significant reduction in body weight despite a substantial increase in energy intake aimed at weight maintenance.

6   SUMMARY

The possible maintenance of energy balance is related to the macronutrient composition of the diet.

A change in diet to relatively more fat causes increase in body weight, as obese people tend to eat relatively more fat, overweight can possibly be reduced by reducing the relative fat content of the diet.

The body has a limited ability to oxidize fat compared to its oxidation of carbohydrate and protein. Relatively high carbohydrate diets appear to be able to induce body fat loss.

21

Body weight regulation

Humans are discontinuous eaters and continuous metabolizers. Thus, part of the energy intake has to be stored before it is used. Energy intake usually balances energy expenditure over several days. Within a day intervals with a positive energy balance, usually some hours after a meal, alternate with intervals with a negative energy balance. Humans do not balance energy intake and energy expenditure on a daily basis either (Chapter 18.4). An example of long-term cycles in energy balance are seasonal changes in body weight, for instance as a consequence of food availability. Finally, some people restrict their food intake deliberately for some time to lose weight. This often results in weight cycling, i.e. periods with a negative balance alternating with periods with a positive energy balance.

This chapter discusses the consequences of a discrepancy between energy intake and energy expenditure for the maintenance of energy balance and the regulation of body weight.

1   FEEDING FREQUENCY AND NUTRIENT UTILIZATION

An animal that takes its food in meals, such as a human, periodically takes in food in excess of its physiological needs even when it is in (daily) energy balance. Figure 21.1 shows the state of energy balance in a person eating three times a day and using 10 MJ/day. If the reference point of balance is at midnight, there will be an energy deficit of about 2 MJ before breakfast and a surplus of about 2 MJ once the evening meal has been absorbed. Periodic hyperphagia alternates with energy deficiency. During hyperphagia after a meal, the metabolites are stored to be mobilized during the following intervals of energy deficiency, i.e. before the next meal starts. The pattern of intermittent feeding and fasting has consequences for energy expenditure as shown in Figure 21.1. During and after a meal the energy expenditure increases due to the processing of the ingested food (Chapter 16). The energy deficiency before a new meal is started can lead to a reduction of energy expenditure (Chapter 19). Normally, the latter probably does not occur if the energy deficiency is only brief. However, people tend to be less energetic when the inter-meal intervals are prolonged. This section first discusses the evidence for a daily pattern of nutrient storage and mobilization. This will be followed by a discussion on the evidence for a relationship between meal pattern or number of meals per day and the maintenance of energy balance.

Diurnal rhythm of nutrient storage and mobilization
Verboeket-van de Venne *et al.* (1991) investigated whether there is a diurnal rhythm of nutrient storage and mobilization in humans and if so, how this is affected. The subjects were 13 student volunteers whose age, height, weight and body mass index (*BMI*) are presented in Table 21.1. They were all non-smokers.

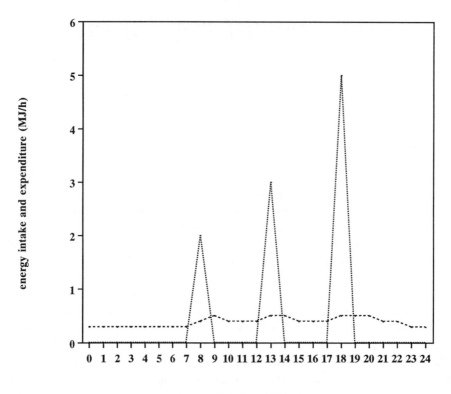

clocktime (hours)

FIGURE 21.1

**Energy intake (...) and energy expenditure (------) throughout the day in a person using 10
MJ and ingesting three meals providing 2, 3, and 5 MJ respectively**

TABLE 21.1

**Physical characteristics of subjects participating in a study on diurnal rhythms in nutrient
storage and mobilization**

Subject	Sex	Age (yr)	Height (m)	Weight (kg)	BMI (kg/m^2)
1	M	23	1.84	66.3	19.6
2	M	22	1.81	72.5	22.1
3	F	18	1.77	61.0	19.5
4	F	19	1.60	59.0	23.0
5	F	19	1.71	62.2	21.3
6	F	20	1.78	65.3	20.6
7	F	19	1.70	64.8	22.4
8	F	20	1.75	59.6	19.5
9	F	19	1.65	50.0	18.4
10	F	22	1.59	66.0	26.1
11	F	20	1.70	67.0	23.2
12	F	18	1.68	57.4	20.3
13	F	23	1.65	50.2	18.4

(From: Verboeket-van de Venne et al., 1991)

The experiment consisted of two periods of two consecutive days during one of which the total daily intake was consumed in two meals (the so-called gorging pattern) and during the other in seven meals (nibbling pattern). Seven subjects started on the gorging pattern followed by the nibbling pattern, the other six subjects did the experiment in reverse order. The interval between the two experimental periods was at least 7 days. The first day of each period consisted of a day of adaptation, during which the food provided was consumed at home according to the prescribed eating pattern (either gorging or nibbling). This was followed by a day in a respiration chamber. In this chamber oxygen consumption, carbon dioxide production and urinary nitrogen excretion were measured (see below). Before the subjects participated in the study, their sleeping metabolic rate (*SMR*) was measured during an overnight stay in the respiration chamber. The *SMR* values were used to estimate the total energy need of each subject over 24 hours assuming the 24-hour energy expenditure (24-h *EE*) to be $1.4 \times SMR$. The subjects were fed to energy balance during the two two-day periods based on this value. Figure 21.2 illustrates the experimental design schematically.

Two eating patterns

The two eating patterns were characterized by a minimal and a maximal spreading of the energy intake throughout the day. The gorging pattern consisted of two large meals: lunch at 12.00 h which contained 60 per cent of the total energy intake, and at 18.00 h a meal with bread, fruit and orange juice making up the remaining 40 per cent. The nibbling pattern consisted of seven small meals: at 7.30 h, 10.00 h and 20.30 h a meal with bread (15, 10 and 10 per cent of the energy intake respectively) at 12.00 h and 18.00 h a small dinner (25 per cent each); at 14.00 h a dessert (10 per cent) and at 16.00 h a piece of fruit (5 per cent). The total energy intake was the same for the two feeding patterns with the same menu and macro-nutrient composition. The food provided 16 per cent of the total energy content from protein, 38 per cent from fat and 46 per cent from carbohydrate. The value of the food quotient (*FQ*), was 0.84.

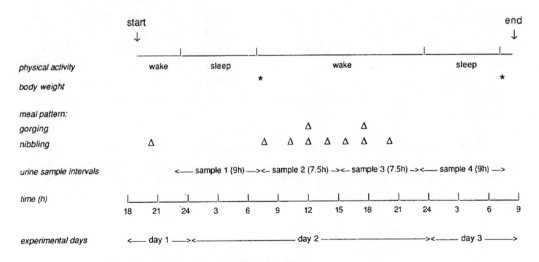

FIGURE 21.2
Experimental design of the stay in the respiration chamber

Oxygen consumption and carbon dioxide production were measured in a respiration chamber. During daytime the subjects were allowed to move freely, to sit, lie down, study, telephone, listen to the radio and watch television, only sleeping and strenuous exercise were not allowed. The subjects weighed themselves (without clothing) in the morning upon rising, after voiding and before any food/drink consumption, on a digital balance accurate to 0.1 kg. Thus, body mass values were obtained at the start and end of both 24-hour observation periods in the respiration chamber. The urine was collected in 4 portions after the following intervals: from 22.30 h day 1 to 7.30 h day 2; from 7.30 h day 2 to 15.00 h day 2; from 15.00 h day 2 to 22.30 h day 2; from 22.30 h day 2 to 7.30 h day 3. Oxygen consumption and carbon dioxide production (and hence *RQ* and *EE*) were calculated over 3-hour intervals during the observation period. Urinary nitrogen excretion was determined for the same intervals, taking into account the different collecting periods. The diurnal pattern of nutrient utilization was studied by comparing the *RQ*, reflecting the composition of the oxidized fuel mixture, with the food quotient (*FQ*) of the diet. Body weight stability is achieved when the mean 24-hour *RQ* equals the *FQ* of the diet. When *RQ* is higher than *FQ* the energy balance is positive; the reverse implies a negative energy balance. With regard to the *RQ/FQ* ratio, periods of lipogenesis (*RQ* > *FQ*) and lipolysis (*RQ* < *FQ*) were defined.

Energy intake and energy expenditure

For a correct interpretation of the effect of a change in meal frequency, it is important that the subjects are in energy balance during the observation period. An adult is supposed to be in energy balance when the difference between energy intake (*EI*) and energy expenditure (*EE*) is less than 600 kJ/day. The energy balance was determined by subtracting *EE* (measured from 7.00 h to 7.00 h) from *EI*. For the gorging pattern *EI* - *EE* was +348 ± 162 kJ/day, for the nibbling pattern +429 ± 217 kJ/day. Individual differences were higher but between the two regimens this difference was highly reproducible thus allowing further comparison. Figure 21.3 shows *EE* (in kJ/min) over 3-hour intervals from the night preceding the second experimental day to the early morning of the third day. There was no statistical difference in the 24-hour *EE* between the gorging and the nibbling pattern (5.57 ± 0.16 kJ/min for the gorging pattern vs 5.44 ± 0.18 kJ/min for the nibbling pattern). However, when *EI* was consumed in two meals per day, *EE* was significantly elevated during the postprandial interval ($EE_{12-15h} > EE_{9-12h}$, $p < 0.001$ and $EE_{18-21h} > EE_{15-18h}$, $p < 0.05$). From 15.00 h to 18.00 h *EE* had decreased significantly, compared to the previous interval ($p < 0.01$). When *EI* was spread throughout the day, *EE* did not show significant changes from 12.00 h to 21.00 h.

Summarizing the results, no statistical difference in 24-hour *EE* was observed between the gorging and nibbling pattern. Throughout the day, *EE* was significantly higher from 12.00 h to 15.00 h and from 18.00 h to 21.00 h when *EI* was consumed in two meals per day ($p < 0.01$ and $p < 0.05$).

Respiratory quotient and food quotient

Figure 21.4 illustrates the course of *RQ* from the night preceding the second experimental day to the early morning of the third day for both gorging and nibbling patterns. There was no statistical difference in the mean 24-hour *RQ* between the two feeding patterns (0.833 ± 0.009 for the gorging pattern; 0.854 ± 0.011 for the nibbling pattern).

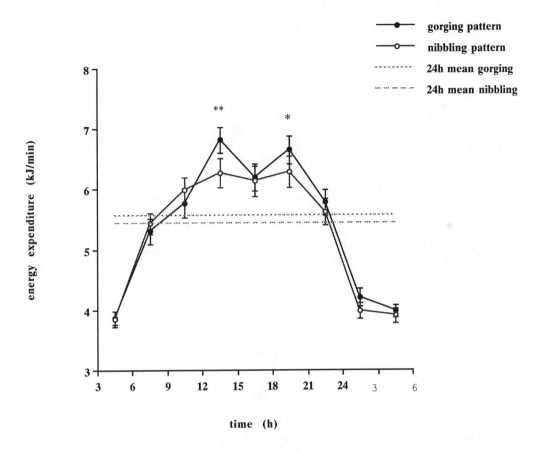

gorging pattern

nibbling pattern

24h mean gorging

24h mean nibbling

FIGURE 21.3

Energy expenditure (in kJ per min) over 3-hour intervals from 3.00 h on day 2 to 6.00 h on day 3, for both gorging and nibbling patterns (mean ± *SEM*, *n* = 13)

Statistical significance: * $p < 0.05$; ** $p < 0.01$.

These values for the mean 24-hour *RQ* matched the *FQ* of the diet (= 0.84). Although the feeding frequency had no effect on the mean 24-hour *RQ*, significant changes did occur in the diurnal pattern of *RQ*. When the *EI* was consumed in two large meals per day, a strong diurnal fluctuation in *RQ* was observed. From rising in the morning until the first meal at 12.00 h, *RQ* was below *FQ* indicating lipolysis. Between 12.00 h and 24.00 h *RQ* exceeded *FQ* (lipogenesis) and after 24.00 h *RQ* dropped quickly to a night level of 0.80. When *EI* was spread throughout the day (nibbling pattern) the fluctuation of *RQ* was considerably smaller. During the active period of the day *RQ* remained at a constant level which is indicative of lipogenesis. Comparing the results, the overall effect was a significantly lower *RQ* ($p < 0.001$) during the preprandial interval from 9.00 h to 12.00 h when the food was consumed in two meals per day.

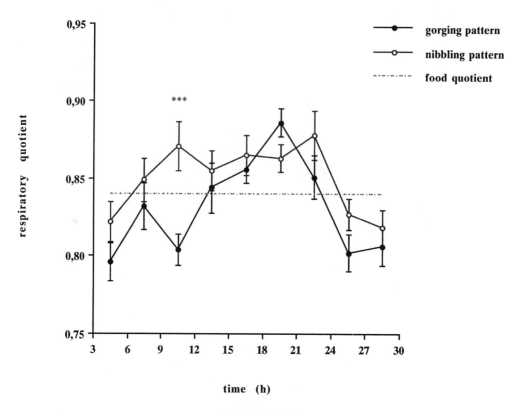

FIGURE 21.4

RQ fluctuation over 3-hour intervals from 3.00 h on day 2 to 6.00 h on day 3, for both gorging and nibbling patterns (mean ± SEM, n = 13)

Statistical significance: *** $p < 0.001$.

Carbohydrate and fat oxidation

By measuring oxygen consumption, carbon dioxide production and urinary nitrogen excretion, it is possible to calculate the overall nutrient utilization. Analysis of the non-protein *RQ* (*NPRQ* = *RQ* corrected for protein oxidation) was used to determine the proportion of oxygen consumption used for fat and carbohydrate oxidation respectively. A *NPRQ* of 0.707 is indicative of pure fat oxidation, a *NPRQ* of 1.0 represents pure carbohydrate oxidation. The effect of meal frequency then became even clearer: from 9.00 h to 12.00 h the *NPRQ* was significantly lower ($p < 0.01$) for the gorging pattern, whereas its value was significantly higher ($p < 0.05$) from 18.00 h to 21.00 h (Figure 21.5).

Another striking result was the quick drop in *NPRQ* after the second meal of the gorging pattern. During the night *NPRQ* was lower in the gorging pattern. To express the effects of feeding frequency on nutrient utilization in figures, the oxidation of protein, carbohydrate and fat was calculated (see Chapter 16.2 for calculation procedure) over 3-hour intervals from 3.00 h on day 2 to 6.00 h on day 3 (Table 21.2). There were no statistically significant changes in protein oxidation during the day between the two feeding patterns. In the gorging pattern an elevated protein oxidation was observed, especially from 15.00 h to 21.00 h, but this effect was not significant. Statistically significant changes in carbohydrate and fat oxidation between the two feeding patterns were observed.

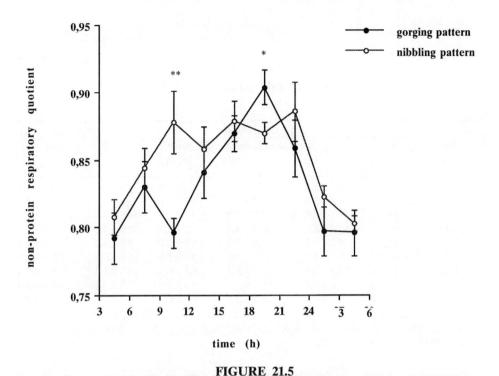

FIGURE 21.5

NPRQ **fluctuation over 3-hour intervals from 3.00 h on day 2 to 6.00 h on day 3, for both gorging and nibbling patterns (mean ± *SEM*, *n* = 11).**
Statistical significance: * $p < 0.05$, ** $p < 0.01$.

In the nibbling pattern carbohydrate and fat oxidation remained relatively constant during the active hours of the day. A feeding pattern of two meals per day resulted in a significantly lower rate of carbohydrate oxidation during the time interval from 9.00 h to 12.00 h, compared to the nibbling pattern ($p < 0.001$). Subsequently, carbohydrate oxidation increased significantly from 12.00 h to 15.00 h and from 18.00 h to 21.00 h, compared to the preceding hours ($p < 0.01$ and $p < 0.05$ respectively). In agreement with these results, fat oxidation was significantly elevated from 9.00 h to 12.00 h (15.9 ± 1.0 g for the gorging pattern vs 10.2 ± 2.2 g for the nibbling pattern; $p < 0.05$). The fasting period in the gorging pattern (from rising in the morning until the first meal at 12.00 h), resulting in a diminished rate of carbohydrate oxidation, thus seemed to be compensated by increased fat oxidation.

Lipogenesis and lipolysis
The results of this showed that changing feeding frequency had no significant effect on the mean 24-hour *RQ*, although there was a marked difference in diurnal *RQ* pattern between the two feeding regimens. With an *EI* of two meals per day the *RQ/FQ* ratio indicated lipogenesis over 4 intervals (i.e. from 12.00 h to 24.00 h) and lipolysis from 24.00 h to 12.00 h. In the nibbling pattern lipogenesis lasted longer (i.e. from 6.00 h to 24.00 h); lipolysis occurred from 24.00 h to 6.00 h. Although the duration of lipogenic activity was prolonged in the nibbling pattern, in the gorging pattern its intensity was increased, which is reflected in a 20 per cent higher carbohydrate oxidation rate between 18.00 h and 21.00 h ($p < 0.01$).

TABLE 21.2

Nutrient oxidation for the gorging and nibbling pattern

time (h)	Gorging pattern			Nibbling pattern		
	protein (g)	fat (g)	carbohydrate (g)	protein (g)	fat (g)	carbohydrate (g)
3-6	9.6 ± 1.8	9.7 ± 1.2	8.7 ± 1.6	10.2 ±1.6	8.4 ± 0.6	10.9 ± 1.6
6-9	8.7 ± 1.1	11.9 ± 1.5	19.7 ± 2.6	9.2 ± 0.6	11.0 ± 1.3	22.8 ± 2.2
9-12	7.7 ± 1.0	15.9 ± 1.0*	17.4 ± 2.6***	8.2 ± 0.7	10.2 ± 2.2	31.0 ± 3.0
12-15	7.7 ± 1.0	15.0 ± 2.1	30.0 ± 4.4	8.2 ± 0.7	12.6 ± 1.9	29.4 ± 1.9
15-18	13.3 ± 2.3	9.6 ± 1.1	30.0 ± 3.3	10.2 ± 1.1	9.8 ± 1.4	32.0 ± 2.5
18-21	13.3 ± 2.3	8.0 ± 1.2	38.6 ± 2.9**	10.2 ± 1.1	10.8 ± 0.9	32.3 ± 1.8
21-24	11.5 ± 1.0	10.3 ± 1.8	25.5 ± 3.3	9.4 ± 0.8	8.5 ± 1.6	30.6 ± 3.4
24-3	9.8 ± 1.8	9.6 ± 0.9	11.7 ± 2.0	8.6 ± 0.8	8.6 ± 0.6	13.6 ± 0.9
3-6	9.8 ± 1.8	9.3 ± 1.0	10.3 ± 2.1	8.6 ± 0.8	9.5 ± 0.5	11.1 ± 0.7

Figures as means ± *SEM*, over 3-hour intervals from 3.00 h on day 2 to 6.00 h on day 3

$* p < 0.05;** p < 0.01; *** p < 0.001$.

Summarizing the effects of feeding frequency on nutrient utilization and consequently on *EE*, there were no significant effects on either the 24-hour nutrient utilization or the 24-hour *EE*, although the diurnal fluctuations were markedly different in the two feeding patterns. A gorging pattern of *EI* resulted in a stronger diurnal periodicity of lipogenesis and lipolysis with possible consequences for the maintenance of long-term energy balance.

Effect of few but ample meals

There are indications that a dietary pattern of few but ample meals, ingested after longer time intervals, can manifest itself in overweight independently of the total energy intake. Fábry *et al.* (1964) investigated the relationship between the frequency of food intake and overweight. The subjects were 440 males aged 60-64 years. They were selected by random sampling to be representative of the male population of the particular age group in a typically urban district of Prague in the (now) Czech Republic. Sixty one subjects suffering from diseases or conditions calling for a special diet which had a bearing on meal frequency were eliminated from the original group.

The number of meals per day in this investigation was evaluated with the aid of a questionnaire, designed especially for the purpose. This was filled in by a dietician after the interview with the subjects. The subjects were divided into sub-groups according to the reported number of meals: three meals per day or less (group 1; 42 subjects), three to four meals (group 2; 206 subjects), three to four meals plus snacks between main meals (group 3; 67 subjects), three to four meals plus a small bedtime snack (group 4; 25 subjects) and finally subjects who reported that they took five or more meals per day (group 5; 39 subjects). Overweight was examined by measuring bodyweight, height and skinfold thickness. The skinfold on the arm and back was measured by means of calipers.

TABLE 21.3
Prevalence of overweight in relation to frequency of meals

Group	Frequency of meals	Body weight (kg)	Height (m)	Broca's index[1]	skinfold thickness (cm) arm	back
1	3 or less	77±10	1.67±0.05	114±16	1.0±0.5	1.7±0.7
2	3 - 4	73±11	1.68±0.04	108±14	0.9±0.4	1.4±0.6
3	3 - 4[2]	73±11	1.69±0.08	107±14	0.8±0.4	1.3±0.8
4	3 - 4[3]	75±10	1.70±0.05	108±12	0.9±0.4	1.4±0.5
5	5 or more	73±13	1.69±0.10	106±17	0.7±0.3	1.3±0.5
				1 vs 2 **	1 vs 2 *	1 vs 2 **
				1 vs 3 **	1 vs 3 **	1 vs 3 **
				1 vs 5 *	1 vs 5 **	1 vs 5 **
					2 vs 5 *	
					4 vs 5 *	

Figures are average ±SD
[1] weight (kg)/height (m) less 1
[2] additional snacks between meals
[3] additional snacks at bedtime
* $p < 0.05$; ** $p < 0.01$

It is apparent from Table 21.3 that the proportion of overweight subjects rises markedly with declining meal frequency. The difference between the extreme groups (1 and 5) is statistically significant in all parameters. The results suggest that there is a relationship between the time distribution of food intake and the tendency to develop obesity.

Additionally, a similar investigation was performed including measurements of total daily energy intake. A random sample of 100 clinically healthy men, aged 35-50 years, was selected. All were, and had been for many years, employees of the state railway company, working on long-distance lines as engine drivers or as assistants. In the course of the study 11 subjects had to be excluded, mostly on account of inadequate co-operation. Food consumption was investigated during a 2-week period. A dietician visited the subjects daily and assessed the type and weight of the consumed foods, and the size of the helpings and left-overs. The same procedure was applied to foods the subjects took to work. The subjects - and, where possible, the dietician also - recorded the exact time when the meal was consumed. After the survey had been completed, the entire group was divided into subgroups according to their average meal frequency. The subjects with an average meal frequency of less than 4 were included in group 1 (27 subjects), subjects with 4-5 meals in group 2 (46 subjects) and subjects taking more than 5 meals per day in group 3 (16 subjects).

The daily energy intake was lowest in group 1 and increased with the number of meals (Table 21.4). The results of the somatometric examinations are of interest in relation to the food consumption data. The height in the sub-groups did not differ statistically (in the two extreme groups it is practically identical) but marked differences were found in body weight, Broca's index (a measure for overweight comparable with the earlier mentioned body mass index, i.e. weight corrected for height) and skinfold thickness.

TABLE 21.4

Energy intake and prevalence of overweight in relation to frequency of meals

Group (n)	*Mean number of meals* (24h)	*Body weight* (kg)	*Height* (m)	*Energy intake* (MJ/d)	*Broca's index*[1]	*Skinfold thickness* arm (cm)	*Skinfold thickness* back (cm)
1 (27)	3.6	83±8	1.73±0.06	15.0±3.3	114±11	1.4±0.5	1.9±0.6
2 (46)	4.5	80±9	1.71±0.06	16.6±2.5	114±13	1.3±0.4	1.8±0.6
3 (16)	5.2	77±10	1.73±0.06	18.3±1.4	106±13	1.1±0.3	1.6±0.4
				1 vs 2 **			
				1 vs 3 ***	1 vs 3 *	1 vs 3 *	1 vs 3 *
				2 vs 3 **		2 vs 3 *	

Figures are average ± *SD*

[1] weight (kg)/height (m) less 1

* $p < 0.05$; ** $p < 0.01$; *** $p < 0.001$

Broca's index and skinfold thickness, indicating a certain degree of overweight, were highest in the group with the smallest number of meals (as described above for another population) and also with the lowest energy intake.

In conclusion, the results of these studies suggest that a dietary pattern with a smaller number of ample meals, ingested after longer time intervals, can manifest itself in overweight. The mechanism is probably an enhanced lipogenesis and a tendency to a higher deposition of fat reserves. This may, to a certain extent, occur independently of the total energy intake.

2   CYCLES IN ENERGY BALANCE

The energy balance in humans shows cycles within a period of one week and one year. In women, there is an additional change in energy balance related to the menstrual cycle.

2.1   THE MENSTRUAL CYCLE

Taggart (1962) reported data on changes in weight and food intake related to the menstrual cycle in one subject. No change related to the menstrual cycle was found but the data collected by Taggart are illustrative of the extent of daily fluctuations in weight and food intake. Daily measurements of food and water intake, body weight and urine output were taken in one subject during a period of 80 days. The usual food habits and activities were maintained throughout the observation interval. The subject, a healthy female aged 31 years, was doing full-time laboratory work. Her weight was, and had been since adolescence, about 64 kg with little noticeable change, and her height was 168 cm, i.e. she had a normal body mass index of 22.7 kg/m². Menstruation was regular, with bleeding for 3-4 days in a 25-day cycle. Her daily activity was light; during working hours from 9 to 17.30 h and recreation the time was divided about equally between sitting activities and activities which involved moving around, mostly indoors. Hours of work were regular, with a half-day on Saturday and no work on Sunday.

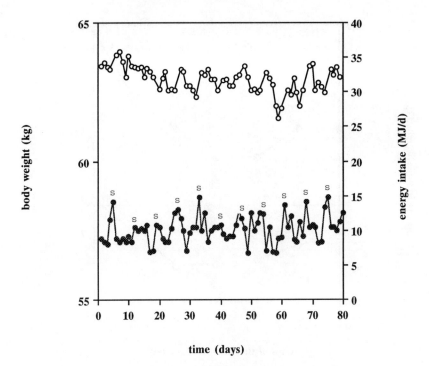

time (days)

FIGURE 21.6

Daily measurements in one woman of body weight (o) and energy intake (•) over a period of 80 days

Sundays are marked s.

Food intake was measured by weighing all food and drink on a balance (±2 g). The pattern of eating was dictated by choice and custom alone. The energy value of the food was calculated from food tables. Body weight was measured on weekdays in the laboratory between 9 and 9.30 in the morning. The indoor clothes were weighed on a domestic food balance at home each day before dressing and their weight was deducted from the weight of the clothed body subsequently measured in the laboratory. To ensure standard conditions the subject voided at 8 h and no food or drink was taken between midnight and 9.30 h. This was the only alteration in the usual eating habits.

Body weight was about the same at the beginning and end of the 80-day period indicating that energy balance was roughly maintained (Figure 21.6). However, there was quite a large day-to-day variation in weight, from 61.5 to 63.9 kg. Even changes of 0.8 kg between two consecutive weighings were recorded. Unfortunately, weight was not measured on Sundays. The daily energy intake ranged from 6.6 to 14.6 MJ and, as shown in Figure 21.6, there were peaks at intervals of about 1 week. The mean energy intakes were usually much greater at weekends than in mid-week, an average of about 3.5 MJ more being taken on Sundays than on Wednesdays and Thursdays (Figure 21.7). The subject did not realize that she ate more at weekends until the values had been calculated at the end of the 80-day observation period. As might be expected, the day-to-day body-weight changes ran parallel to the food-intake changes, resulting in the same weekly cycle. The mean difference between minimum and maximum weight was nearly 1 kg.

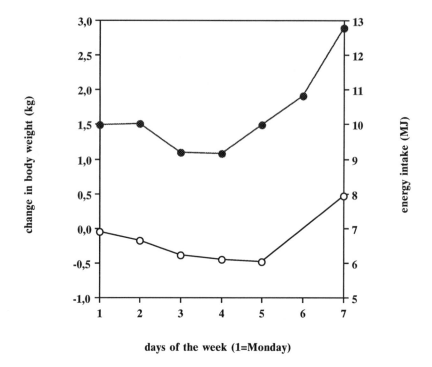

FIGURE 21.7

Mean values for energy intake (•) and change in body weight (o) for each day of the week as measured the following morning in one female subject

2.2   THE ANNUAL CYCLE

Annual cycles in energy balance were studied by Van Staveren *et al.* (1985). A random sample of 114 women aged 29-32 years was selected from a small industrial town in the Netherlands. Food consumption was estimated once a month with a 24-hour recall method. The subjects weighed themselves every day, after the visit of the dietician for the 24-hour recall before breakfast and after the bladder had been emptied, on a supplied high-accuracy balance (±0.5 kg) without clothes.

The study did not demonstrate systematic changes in energy intake throughout the year although there was a relatively lower fat consumption in the summer months (37.2 energy% versus 39.4 energy% in winter). The body weight showed small seasonal fluctuations (Figure 21.8) with an increase before winter and a decrease before summer. This result was suggested to be a consequence of changes in physical activity, since the subjects were more physically active in spring and summer than in winter and autumn. This might also explain the higher efficiency of slimming programmes conducted in spring and summer.

Prentice *et al.* (1981) showed the more pronounced annual cycle of body weight in developing countries where food availability changes throughout the year. In the Gambia, energy intake in the wet season was clearly inadequate while in the dry season intake was the approximate norm. Thus, most adults undergo seasonal body weight changes with weight loss during the the wet season and weight regain during the dry season when food supplies are relatively abundant.

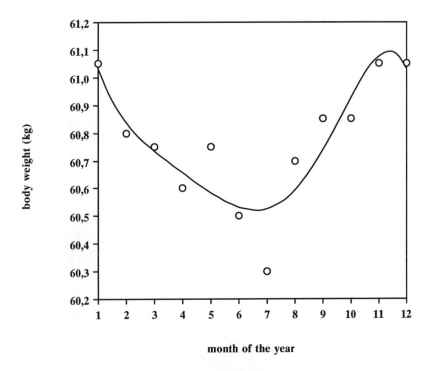

FIGURE 21.8

Mean fluctuation of body weight in young adult women in the Netherlands as assessed monthly throughout one year

($n = 114$)

Data obtained by weighing all non-pregnant, non-lactating women of child-bearing age in a rural subsistence farming community, at 2-months intervals, illustrate this annual pattern of weight change (Figure 21.9). Pregnant women showed an inadequate gain of 0.4 kg per month during the wet season compared to a 1.4 kg per month weight gain during the dry season, which is in accordance with conventional standards. All women, including women at different stages of pregnancy, utilized body fat during the wet season, as measured by the sum of skinfold thicknesses.

When the annual cycle of body weight in a European country is compared with that in a developing country where food availabilty is suboptimal in part of the year, the difference is pronounced. The women in the Gambia show a mean annual weight change of 1.5 kg while for a comparable group of Dutch women it is only 0.5 kg. The situation in the Gambia illustrates the importance of body fat reserves in women to allow for reproduction, including pregnancy and a lactation period of more than 1 year.

The function of body fat reserves in women is used mostly to explain the gender differences in body composition. The function of fat is more than to make a woman look attractive. In evolution it has been important as a determinant of reproductive success, which explains why women have more fat than men from puberty onwards (see Figure 17.5).

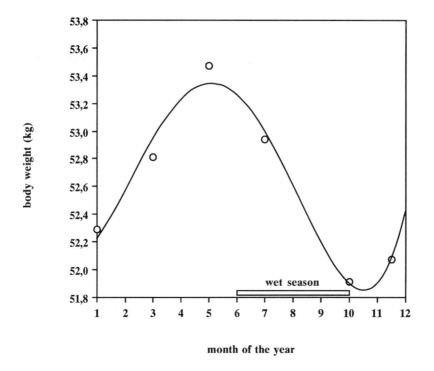

FIGURE 21.9

Mean fluctuation of body weight in young adult Gambian women as assessed bi-monthly throughout one year

(*n* = 53)
(adapted from Prentice, 1981)

3   WEIGHT CYCLING

Repetitive weight loss and weight regain is thought to induce adaptive changes in the body to resist a negative energy balance. In a population survey in 1985, one third of American women aged 19-39 reported to diet at least once a month and one sixth considered themselves perpetual dieters. Another survey reported that one quarter of men and nearly twice as many women are trying to lose weight. In animal research it has often been shown that weight loss is slower during a second restriction period than during the first. Even while the energy intake is the same in both cycles, the second restriction period typically takes more than twice as long as the first period to reach the same weight loss. Weight regain per gram food consumed is also higher after a second restriction period. These observations suggest that frequent dieting makes weight loss more difficult. The mechanism by which this happens is yet unknown.

Blackburn *et al.* (1989) compared weight loss in obese patients who participated more than once in a weight-loss program. Patients had to regain more than 20 per cent of their first-cycle weight loss during the interdiet interval to be included in the analysis. It was necessary to match the number of dieting days in each period. This was achieved by comparing weight loss during the number of days of the shortest of the two cycles. The subjects were placed on a very-low-energy diet (< 3.3 MJ/day), and treated either as inpatient or outpatient. Comparing the weight loss of members in the outpatient group (*n* = 43), the mean weight loss was 0.19±0.03 kg/day in cycle 1 and 0.15±0.03 kg/day in cycle 2 (difference $p < 0.01$). The mean length of the dieting cycle was 72 days and the patients regained on average 120 per cent of their weight loss during the interdiet interval which lasted on average 810 days. The results in the strictly controlled inpatient group ($n = 14$) supported this observation, i.e. a higher weight loss in the first cycle (0.47 kg/day) compared to that during the second cycle (0.37 kg/day). Here too, the subjects regained more than 100 per cent of their weight loss in the interdiet period.

The results of this study support findings in animal studies. The speed of weight loss is slower during a later diet cycle with weight regain in between. This phenomenon has already been mentioned in Chapter 18 with respect to athletes who frequently limit their energy intake to lose weight before a competition in weight-class sports. One explanation is a further lowering of the resting metabolic rate in a later dieting period as mentioned in Chapter 19. Another explanation could be a (further) lowering of *DEE* and *AEE* in a later dieting period. Whatever the reason may be, on a day-to-day basis the differences are very small but they have important long-term consequences. This makes it a very difficult phenomenon for research with the techniques available now, while many people are confronted with the consequences.

4   SUMMARY

Humans are discontinuous eaters and continuous metabolizers.

Part of the energy intake has to be stored before it is used.

Within a day intervals with a positive energy balance, usually some hours after a meal, alternate with intervals with a negative energy balance.

Periods with a negative balance alternating with periods with a positive energy balance can have consequences for the maintenance of energy balance and the regulation of body weight.

A dietary pattern with a smaller number of ample meals, ingested after longer time intervals, can manifest itself in overweight. The mechanism is probably an enhanced lipogenesis and a tendency to a higher deposition of fat reserves. This may, to a certain extent, occur independently of the total energy intake.

Humans do not balance energy intake and energy expenditure on a daily basis. Weekly cycles of body weight changes of 1 kg with a high food intake at weekends is a common phenomenon.

Annual cycles are pronounced in developing countries as a consequence of food availability.

Repetitive weight loss and weight regain is thought to induce adaptive changes in the body to resist a negative energy balance.

IV

Food intake, energy expenditure and evolution

22

Effects of nutrition on reproductive capacity

This chapter reviews the role of nutrition on reproductive capacity. The presence of normal menstrual cyclicity serves as an endogenous and easily measurable marker of reproductive capacity. Two commonly measured aspects of menstruation are the menarche, and the menopause. Normally, women have menarche at 12.8 ± 1.2 years of age and subsequently have regular monthly cycles until the time of menopause which typically occurs at 45-50 years of age. Fertility is possible only during periods of normal cyclicity.

Nutritional inadequacy or restriction of food intake is associated with the loss of both reproductive development and capacity to reproduce in females and males. Conditions in which weight loss predominates, whether it be accidental, self-imposed as in anorexia nervosa, or related to illness, are particularly marked by the loss of menstrual function. Weight loss associated with malnutrition during lactation, and intense exercise are also examples of nutrition-related diminished reproductive capacity.

This chapter reviews data which support the concept that reproductive economy resulting from undernutrition is (a) mediated from within the central nervous system at or above the level of the hypothalamus; (b) reversible with improved nutrition; and (c) probably adaptive to the survival of the species. The first section gives a review of the normal reproductive axis since an understanding of the normal physiology will enable the reader to understand the pathophysiology incurred on the reproductive axis with undernutrition.

1   ENDOCRINE COMPONENTS OF REPRODUCTION

The central structure of the brain, the hypothalamus, integrates the inputs obtained from the internal and external environment. Certain aspects of the physiology of the central nervous system have already been mentioned in Chapter 6. The following is especially concerned with nutrition and reproduction. The hypothalamus is responsible for the control of body temperature, food and water intake, and reproduction. The hypothalamus receives and relays information to many areas within the brain including neurons that send their projections to autonomic centres within the brainstem where heart rate, blood pressure, and gastric function are controlled; to the limbic system, a site important for learning involving emotional experiences; and to the pituitary, which is involved with endocrine responses. The relationship of the hypothalamus to the pituitary is critical both for the onset of menarche and for the maintenance of menstruation. In essence, the hypothalamus is in a strategic place for the mediation of homeostasis, which is the capacity to maintain the internal environment stable, without significant variability, despite a constantly changing external environment.

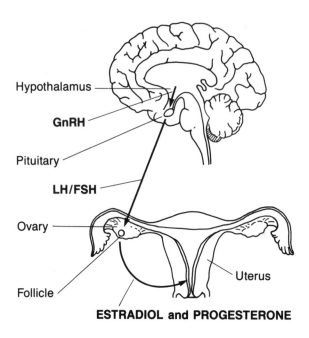

Hypothalamus

GnRH

Pituitary

LH/FSH

Ovary

Follicle

Uterus

ESTRADIOL and PROGESTERONE

FIGURE 22.1

Anatomical structures and hormones essential for reproduction in a well-nourished state

The anatomical structures and hormones discussed in this section are shown in Figure 22.1. The hypothalamus contains the neurosecretory neurons responsible for sending signals, via releasing hormones, to the pituitary. The pituitary gland, also centrally located (between the ears and behind the eyes), is anatomically connected to the hypothalamus by a small stalk of neural tissue via which hormones from the hypothalamus are able to reach the pituitary. Under the stimulus of hypothalamic hormones, the pituitary in turn releases numerous hormones which serve to stimulate the activity of various target glands distributed throughout the entire body. For example, the hypothalamus releases gonadotropin releasing hormone (GRH) in a pulsatile fashion, which reaches the pituitary and releases two other hormones: luteinizing hormone (LH) and follicle stimulating hormone (FSH). Pulses of LH and FSH are released into the peripheral circulation and reach a target gland, the ovary, where follicles, each of which contains one egg cell, are stimulated to grow and produce estrogen. At mid-cycle, LH release by the pituitary increases acutely, the so-called LH surge, and stimulates an egg to be released by the dominant follicle. This event is known as ovulation. The empty follicle, now called the corpus luteum, produces another hormone called progesterone for 14 days thereafter. The corpus luteum, just as the follicle, requires frequent LH pulses to function adequately. Estrogen and progesterone from the ovary are necessary to prepare the lining of the uterus for the implantation of the embryo if the egg is fertilized with a male sperm cell. When the egg is not fertilized, the uterine lining is shed as soon as the corpus luteum stops secreting estrogen and progesterone on day 14. The shedding of the lining of the uterus is called menstruation. Estrogen, among other hormones, is also necessary for preparing the breast to respond to prolactin, another hormone released from the pituitary which in turn stimulates the mammary glands of the breast to produce milk.

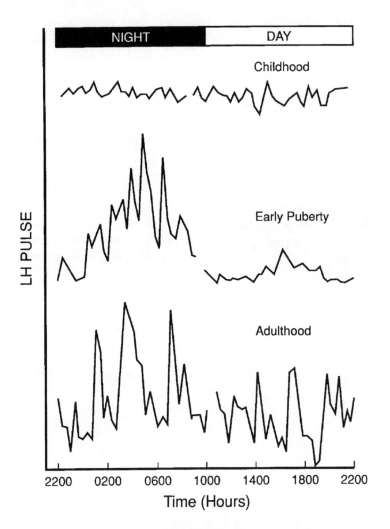

FIGURE 22.2

Pattern of pituitary LH release during the night and day in 3 stages of development

Upper panel displays the relatively quiescent pulse frequency and amplitude of LH pulses during childhood. Middle panel shows pattern of early puberty with LH pulses occurring only during the night. Lower panel shows adult pattern of LH release with consistent pulsatility throughout the entire day. The LH-pulse pattern mirrors hypothalamic GRH-secreted pulses.

(Modified with permission from Boyar *et al.*, 1974)

It is possible to study the activity of the GRH neurons in the hypothalamus even though GRH levels are undetectable in the blood. Because each GRH pulse received by the pituitary is followed by a correspondung LH pulse which is secreted into the peripheral circulation, one is able to detect these GRH-derived LH pulses with frequent blood-sampling and thus assess the function of the reproductive axis. Under normal circumstances the GRH pulse, and thus the LH pulse frequency, is 1 pulse every 2 hours in the first half of the menstrual cycle, and 1-2 pulses every 12 hours in the second half of the cycle. Prior to the onset of puberty, i.e. in childhood, the hypothalamic GRH neurons are quiescent, resulting in low levels of LH, FSH, estrogen, and progesterone. Just prior to puberty episodic pulses of LH and FSH are secreted but only at night. Finally the axis matures sufficiently to allow LH and FSH secretion to occur throughout the day (Figure 22.2). The central and/or peripheral signals which trigger the increased activity of the reproductive axis resulting in menarche, remain obscure.

2   NUTRITIONAL DEFICIENCY STATES

Having reviewed the basic components of the reproductive axis, it is now possible to return to the original premise that nutrition affects reproductive function. There are various situations in life where nutritional deficits are prevalent: simple weight loss due to illness, or lack of food; poor nutrition during lactation, in which stored body fat is diverted to the infant; imposed food deprivation as is found in anorexia nervosa, and the relative nutritional deficiency found in athletes. All these situations are associated with low reproductive capacity because the reproductive axis stops working properly. This dysfunction is manifested in young girls by a delay of menarche, and in adult women by the loss of menstruation or amenorrhea.

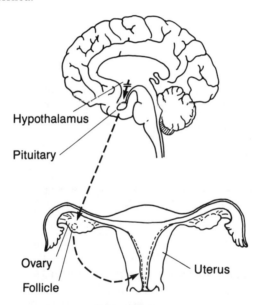

FIGURE 22.3
Anatomical structures of the reproductive axis without their hormones in nutritional deficiency states

3   PATHOPHYSIOLOGY OF REPRODUCTIVE CAPACITY LOSS

The basis of nutrition-related menstrual abnormalities appears to be hypothalamic dysfunction, although higher brain centers may also be involved (Figure 22.3). This assumption is supported by the finding of low levels of the pituitary hormones LH and FSH, and the ovarian steroids estradiol and progesterone, and a normal to hyper response of these hormones to exogenously administered GRH (Baker *et al.*, 1981; Veldhuis *et al.*, 1985; Vigersky *et al.*, 1977; Warren & Van de Wiele, 1973). In most situations of nutrition-induced reproductive dysfunction, LH pulses are infrequent and slow. Depending on the severity of the nutritional deficit there is a corresponding regression in the pulsatile patterns of LH and FSH, resembling those of pre-pubertal and childhood stages (Boyar *et al.*, 1974). These findings suggest that there is a central, inhibitory stimulus to the GRH neurons in nutritional deficiency states. However, each particular situation described may affect these neurons to variable extents.

4   DELAY OF MENARCHE: FRISCH HYPOTHESIS

As mentioned previously, the signals that activate the reproductive axis for the first time, or maintain it are not totally understood. Nonetheless many hypotheses have been proposed (Conte *et al.*, 1981; Frisch, 1976; Frisch & Revelle, 1970; Grave & Mayer, 1974; Grumbach, 1980; Grumbach *et al.*, 1974; Hohlweg & Dorhn, 1932; Reiter & Grumbach, 1982). Frisch has been the leading proponent of the theory that the onset of menarche depends on a critical body weight (Frisch & Revelle, 1971). There are two observations which support her thesis: firstly, poor nutritional conditions delay, while good nutritional conditions enhance, the onset of menstruation (Frisch, 1972); and secondly, marked increases in body fat accompany the onset of the growth spurt and menarche in girls (Frisch *et al.*, 1973). In collaboration with McArthur, Frisch extended these studies to examine the relationship between menarche and body fat and found that menarche does not occur unless body fat represents at least 17 per cent of body weight (Figure 22.4).

There is a certain percentage of body fat that each individual must achieve before they can maintain regular menstrual cycles or recover from secondary amenorrhea caused by nutritional deficiency. This percentage of body fat is not predictable, but that set point is consistent for each individual. For example, a woman who is 165 centimeters tall would have to weigh at least 49 kilograms before menstruation would resume, if she had cessation of her periods after menarche. Similarly, there is a certain percentage of body fat that each individual must have before they can experience menarche, and that percentage body fat is lower than the percentage of body fat necessary to maintain normal menstrual cycles. This difference is shown by the lowest dotted diagonal line for the same group for women when they went through menarche (Figure 22.4). In cases where menstruation begins but terminates due to nutritional insufficiency (known as secondary amenorrhea), body fat must reach 22 per cent body weight before menses resume (Frisch & McArthur, 1974). The causal relationship between body fat and the onset of puberty proposed by Frisch has been difficult to ascertain by direct measurements (Billewicz *et al.*, 1976; Cameron, 1976; Johnston *et al.*, 1975). Nonetheless, there is evidence that moderately obese girls have onset of menarche at an earlier age (Zacharias *et al.*, 1976). In anorexia nervosa (Boyar *et al.*, 1974) and in simple weight loss (Kapen *et al.*, 1981) there is regression to an immature or childhood pattern of LH and FSH release, which is reversible after returning to normal weight (Figure 22.5). Warren found that women with anorexia nervosa have a progressive improvement in the response of LH and FSH to GRH during recovery toward ideal body weight (Warren *et al.*, 1975). These studies have supported Frisch's hypothesis that a certain amount of fat is necessary for the onset and subsequent normal function of the hypothalamic-reproductive axis.

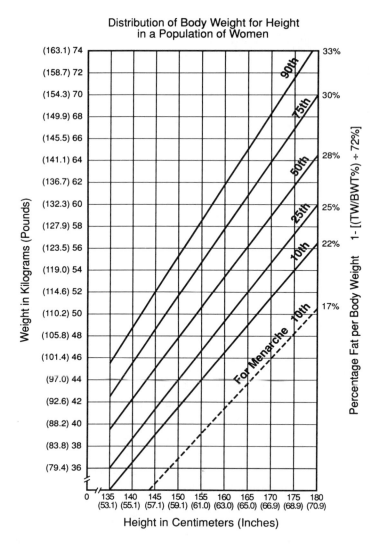

Distribution of Body Weight for Height
in a Population of Women

FIGURE 22.4

Distribution of body weight for height in a population of women

The diagonals are percentiles of total water (*TW*) as a percentage of body weight (*BW*) which is an index of body fat. From the amount of body water one can determine the weight of lean mass (72 percent of *TW*) and therefore calculate percentage body fat. For simplicity sake the percentiles of total water as a percent of body weight in the original figure have been converted to per cent body fat using the equation 1-[(*TW/BW*%)/72%]. This is shown in the right side of the nomogram. The upper 5 diagonal lines show the height-weight distribution for an 18-25 year old female population. The 50th percentile diagonal represents the average body weight per height of this group of women. (Modified with permission from Frisch, 1988)

Warren also found that young athletes exhibit delay in the onset of menarche, but proposes that factors other than percentage body fat modulate menstrual function. In her studies on young ballerinas, these dancers were older and slightly heavier, 22-24 per cent body fat, at the onset of menarche (Warren, 1980).

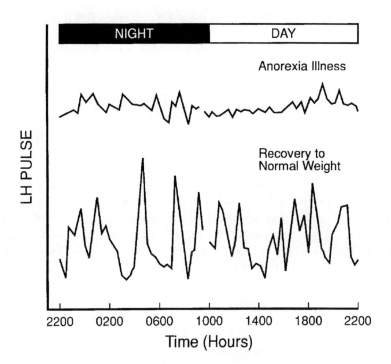

FIGURE 22.5

Pattern of LH pulses obtained from a patient with anorexia nervosa

Upper panel shows the LH pulse pattern during acute illness, reminiscent of the childhood-prepubertal patterns shown in Figure 22.2. Lower panel shows the same woman after recovery from anorexia nervosa. LH pulses return to the adult pattern shown in Figure 22.2.
(Modified with permission from Boyar *et al.*, 1974)

Furthermore, many ballerinas had onset of their menses or recover from secondary amenorrhea during periods of rest when they were not as active, while they showed no apparent changes in weight (Warren, 1980). Warren has attributed these findings to the high activity level of these athletes. She postulates that the hypothalamus is sensing intensity of activity, rather than percentage body fat (Warren, 1980).

4.1   FAT AS A METABOLIC FUEL

The pressure for a decreased amount of body fat by Western society is directly opposed by the evolutionary pressure at the cellular and physiological level to maintain a certain amount of body fat . Fat is needed as substrate for the synthesis of various components of our bodies including cell membranes and steroids such as estrogen and progesterone (Strauss *et al.*, 1981). Secondly, fat is a very compact way to store energy, for example there are 37.6 kJ of energy in a gram of fat compared to 16.7 kJ in a gram of carbohydrate (Schmidt-Nielsen, 1979). Thirdly, fat storage is particularly important for pregnancy and lactation. In addition to serving as a source of nutrition for the infant, fat cells, such as those found in the breast have the capacity to form estrogen (Nimrod & Ryan, 1975), which may have a functional role in promoting lactation. Frisch argues that adequate body fat ensures the energy required to yield a viable pregnancy, and to secure the subsequent survival of its product through lactation (Frisch, 1990). Fourth, fat storage also reflects the availability of nutrients in the environment. Generally, food intake is matched by the energy requirements of the animal, such that excess energy intake is stored as fat. In times of dearth, in order to meet metabolic needs, the fat storage is mobilized. Because fat is the most labile component of body weight, some postulate that it may serve as a endogenous marker gauging the safety of the environment for reproduction (Frisch, 1984).

4.2   GLUCOSE AND AMINO ACIDS AS METABOLIC FUELS

Fat may not be the only metabolic fuel that gauges the safety of the environment for reproduction. Cahill (1970) conducted studies to assess fuel utilization during starvation in man, and determined that despite the large fat stores present, amino acids obtained from muscle protein were preferentially utilized over fat. These amino acids were converted to glucose for brain utilization. This is because in mammals glucose cannot be made from fat, and early in a fast the brain can only utilize glucose!

Evidence supporting the exquisite sensitivity of the reproductive axis to the ambient concentration of metabolic fuels, probably not related to fat content, comes from studies on both male rhesus *(Cynomolgus)* monkeys and humans. Both slow their LH-pulse frequency and decrease their testosterone blood levels after a 24-48 hour fast (Cameron & Nosbisch, 1991; Cameron, 1991). In agonadal lambs, limited nutrition decreases LH pulsatility, but pulses return to normal after 14 days of nutritional repletion (31). Furthermore, on limited nutrition these animals retain normal LH pulsatility if infused with either amino acids, glucose or both. Cameron found similar results in female rhesus monkeys in which the absolute quantity of calories, rather than quality, determined the presence of normal hypothalamic function, and thus LH pulsatility (Parfitt *et al.*, 1990). These data strongly suggest that hypothalamic GRH neurons depend on the availability of metabolic fuels, in absolute quantity and on a reliable basis, for normal functioning.

5   WEIGHT LOSS AND ANOREXIA NERVOSA

Anorexia nervosa, covered in depth in Chapters 5 and 13, is a disorder of altered body image marked by voluntary, stringent, food restriction due to a relentless desire for thinness. This disorder results in individuals who have body weight approximately 25 per cent below ideal weight. Anorexia nervosa, relative to the other nutritional deficient states, results in the most severe form of hypothalamic dysfunction. In this disorder all of the hypothalamic-pituitary interactions are altered including other non-pituitary related functions such as temperature control (Vigersky *et al.*, 1977). Here, the emphasis will be on the endocrine factors. The regression of LH secretion to prepubertal patterns has been shown in particular for anorexia nervosa (Boyar *et al.*, 1974; Marshall & Kelch, 1979). In these patients their LH pulses returned to normal after recovery to normal weight (see Figure 22.5). Similar to anorexia nervosa, investigators have shown that LH and FSH release is inhibited in simple weight loss without anorexia nervosa (Boyar *et al.*, 1974; Warren *et al.*, 1975), but the hypothalamic dysfunction in simple weight loss is not as severe (Vigersky *et al.*, 1977).

6   EXERCISE: A STATE OF RELATIVE UNDERNUTRITION

Warren and others include exercise-induced menstrual dysfunction as one of the many categories of nutritionion-derived loss of menstrual cyclicity (amenorrhea) (Warren, 1983). Exercise as a state of relative undernutrition also shows its effects on the endocrine functions. Female athletes who stop menstruating, maintain a slow and irregular LH-pulse frequency throughout the month (Fisher *et al.*, 1986; Marshall & Kelch, 1986; Veldhuis *et al.*, 1985; Figure 22.6). Even runners who maintain normal periods and have normal LH-pulse frequency prior to a run, slow the frequency of pulses of LH after a run (Cumming *et al.*, 1985a,b). Moreover, Frisch showed that athletes who began their athletic training before the onset of menarche had menarche at a mean age of 15, as compared to post-menarcheal trained athletes who experienced menarche at a mean age of 12.8 years. Each year of pre-menarcheal training delayed menarche by five months (Frisch *et al.*, 1981).

Bullen and McArthur (1985) serially recorded the hormonal changes that occur with increased athletic training. Previously untrained women with normal menstrual cycles began an 8-week athletic training period. The running distance was increased so that in the last 4 weeks the women were running 13-16 km/day (Figure 22.7). In the first 4 weeks of training the duration of the menstrual cycle was shortened, due to a decreased duration of progesterone release by the corpus luteum. This was probably the result of a decreased frequency of LH pulses which are required for normal corpus luteum functioning.

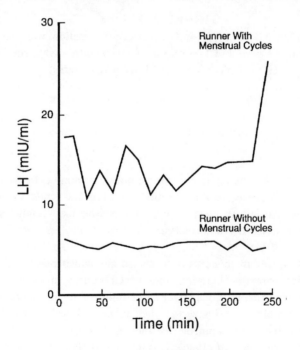

FIGURE 22.6
LH-pulse frequency in 2 women runners monitored over 4 hours
Upper panels shows runner who continues to have menstrual cycles, lower panel shows woman who does not. (Modified with permission from Fisher *et al.*, 1986)

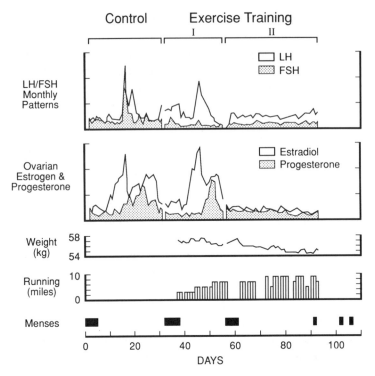

FIGURE 22.7

**Hormonal patterns of LH, FSH, estrogen and progesterone secretion, weight and running
mileage monitored over a period of 3 months in one woman trained to run increasing
mileage over a period of 8 weeks following the control cycle**

See text for details. Note that in the first 4 weeks of exercise training cycle I there is an abbreviated
LH secretion after the LH surge (the maximum rise of LH) and decrease in FSH secreted. In cycle II
there is no LH surge and no rise in ovarian estrogen or progesterone. Note this is accompanied by a
decrease in her body weight and a delayed and short menstrual period (menses).
(From: Bullen *et al.*, 1985)

Later, in the last 4 weeks, the big rise in the pituitary hormone LH which is required
for ovulation disappeared. Therefore no ovulation occurred and levels of estrogen and
progesterone were low. Among the test group, women who lost weight had a higher
incidence of more severe menstrual abnormalities.

Warren (1980) introduced the concept of 'energy drain' which suggests that there is a
relative metabolic deficit or relative undernutrition which occurs as a result of high energy
expenditure during intense exercise. On a larger and absolute scale numerous investigators
have noted that women athletes who have lost their menstrual cycles have lower energy
intake than their non-amenorrheic controls for their degree of activity (Duester *et al.*,
1986; Marcus *et al.*, 1985; Nelson *et al.*, 1986). This may be partly due to pressure to maintain a lower weight to ensure better performance.

In support of the role of undernutrition in exercise-related loss of reproductive capacity are the findings by Cameron and colleagues which showed that exercise-related loss of
menstrual function could be reversed by increasing food intake by 24-54 per cent in vigor-
ously trained female *(Cynomolgus)* rhesus monkeys, using a similar protocol to that which

Bullen used in humans (1985). These data suggest that there is indeed a certain degree of undernutrition occurring during intense exercise which is sensed by the hypothalamus. Based on these data one can postulate that intense activity, possibly through a relative decrease of metabolic substrate available to the GRH neurons at the time of exercise, affects the function of these neurons acutely in some, and chronically in other individuals.

7   OTHER FACTORS RELATED TO LOSS OF REPRODUCTION WITH UNDERNUTRITION

Various factors which are difficult to dissociate from the direct effect of weight loss, undernutrition, athletic activity or the stress of these have been implicated in the etiology of nutrition-related amenorrhea. Of these, two seem relevant to the biological existence of reproductive economy during undernutrition: opiates, and the hypothalamic-pituitary-adrenal (HPA) axis. The HPA axis, similar to the reproductive axis, is composed of a hypothalamic releasing hormone, corticotropin releasing hormone (CRH), which stimulates the release of adrenocorticotropin releasing hormone (ACTH) by the pituitary into the systemic circulation. ACTH reaches its target gland, the adrenal, and induces the secretion of cortisol, a hormone needed in low concentration under normal physiological circumstances, but released in large quantity during stress. Both opiate and HPA axis are activated during stress and are thought to help the animal or human deal with stress.

Opiates appear to play a role in the normal menstrual cycle (Ferin *et al.*, 1984; Quigley & Yenn, 1980; Ropert *et al.*, 1981; Rossmanith & Yen, 1987; Veldhuis *et al.*, 1984). Under normal circumstances opiates are believed to slow the activity of the hypothalamic neurons which secrete GRH at different times during the cycle. Because the loss of normal menstrual cyclicity with undernutrition is associated with a very slow LH frequency, some investigators have suggested that a similar opiate-mediated mechanism may be involved (Marshall & Kelch, 1986). In fact, a number of studies support this idea because it is possible to increase the frequency of LH pulses with an opiate-receptor antagonist in some, but not all, women who suffer from nutrition-related menstrual dysfunction (Baranowska *et al.*, 1984; Khoury *et al.*, 1985; McArthur *et al.*, 1980; Sauder *et al.*, 1984).

Despite the compelling evidence suggesting an opiate-mediated inhibition of the hypothalamic GRH neuron in these amenorrheas, it is impossible to conclude from these studies that the increased opiates cause amenorrhea. Firstly, because opiate-receptor antagonist do not consistently result in normal LH-pulse frequencies in humans. Also, they have not been shown to be effective in activating the GRH neurons in nutritionally deprived animals, suggesting that other mechanisms are effecting reproductive dysfunction.

The hypothalamic-pituitary-adrenal axis is activated in stress situations (Brown *et al.*, 1982; Lenz *et al.*, 1987; Rivier & Vale, 1983). Undernutrition represents a stressful state to the animal. In addition, recent studies show that stress interferes with the reproductive axis (Rivier *et al.*, 1986). Because of these reasons it becomes relevant to assess the role of stress in nutrition-related loss of reproductive function.

Athletes without menses were more likely to associate higher psychological stress with their running (Schwartz *et al.*,1981). Malina *et al.* (1978) noted that the higher the competitive level of training, the later the menarche occurred. Warren and associates (1980) found that ballerinas regained their menses during periods when they could not exercise due to secondary injuries, but she attributed this effect to energy drain rather than to stress alone, since young musicians with equally intense, goal oriented, lives did not experience delay in the onset of menarche.

Elevated levels of cortisol, which are found normally during stress, have been noted in athletes (Cumming & Rebar, 1983; Ding *et al.*, 1988; Villaneuva *et al.*, 1986) as well as in patients with anorexia nervosa (Gold *et al.*, 1986; Kaye *et al.*, 1987). Women with anorexia nervosa also have elevated levels of corticotropin releasing hormone (CRH) in their cerebrospinal fluid (Kaye *et al.*, 1987). This peptide hormone has also been implicated in the modulation of appetite in humans (Morley, 1987) and in animals (Britton *et al.*, 1982; Krahn *et al.*, 1986, 1988; Levine *et al.*, 1983). In addition, central CRH can inhibit gonadotropin (LH and FSH) secretion from the pituitary (Olster & Ferrin, 1987). Villaneuva *et al.* (1986) found plasma cortisol levels consistently elevated in runners with and without menstrual periods compared to normal non-exercising controls. This finding casts doubt on the relevance of excess cortisol in these nutrition-related reproductive disorders, at least in those which are exercise-related.

Although a large body of evidence suggests that stress causes menstrual dysfunction, it is inherently difficult to study stress as an isolated component because weight loss, caloric privation, anorexia nervosa and intense exercise all have, integrally related, a stress component. Cameron (1991) has attempted to dissect stress from caloric privation and have shown resumption of normal LH pulses in fasted monkeys when nutrients were replaced by intra-gastric infusion, despite the persistent display of stressed behavior. These hungry monkeys were stressed because they could see, but did not have access to, a plate of monkey chow during the duration of the experiment. However, cortisol levels were not measured in these studies (Schreihofer *et al.*, 1990).

Other central factors are also being evaluated for causality in nutrition-related loss of reproduction, namely changes in neurotransmitters such as glutamate and its agonists (Ebbing *et al.*, 1990). It remains plausible that nutrition-related GRH-neuron dysfunction is due to either too much inhibition, or lack of excitation by neurotransmitters produced by neurons that may be sensitive to changes in the nutritional and metabolic milieu. It is clear that this problem is not easy to decipher because many factors are participating in concert to mediate the inhibition of the reproductive axis with undernutrition.

8   SUMMARY

Nutritional inadequacy in a relative or absolute way results in a decline in the activity of hypothalamic GRH neurons which are exquisitely sensitive to the ambient metabolic milieu. As a result, the pulsatile release of LH and FSH by the pituitary decreases such that adequate ovarian follicular maturation, and estrogen and progesterone synthesis are prevented. Therefore, reproductive capacity is aborted during times of undernutrition.

The loss of normal reproductive development and established menstrual cyclicity with undernutrition is reversible with weight gain (Frisch and McArthur, 1974; Warren, 1987), better nutrition (Van der Walt *et al.*, 1978), or recovery from anorexia nervosa (Boyar *et al.*, 1974).

The transient and reversible nature of the menstrual dysfunction associated with undernutrition is indirect evidence for the sensitivity of the reproductive axis to the nutritional environment. Others have suggested that these alterations in fertility in association with undernutrition are simply nature's way of birth control in times of scarcity (Frisch, 1990).

23
Feeding and evolution

From the theory of evolution, we know that species become adapted to their environments by a process of natural selection. This adaptation goes on within species as local populations adjust to local conditions. Ecology studies the interactions that determine the distribution and abundance of organisms, in other words how organisms are adapted to their environment. This definition implies that ecologists are interested in questions such as where organisms are found and in what quantities and why. 'Organisms' here include all organisms i.e. viruses, bacteria, plants and animals.

Since this book focuses on feeding, we shall restrict ourselves to ecological examples of feeding strategies in animals, i.e. the analysis of mechanisms by which animals adapt to their biotic and abiotic environment as far as feeding is concerned. Adaptations can be expressed in vital functions, such as body shape and behavior.

The study of behavioral adaptation, encompasses two types of questions. Firstly, what is the function of the adaptation in the light of evolution, in other words: What is the survival value of a behavior (ultimate factors)? The second question is: How are animals able to express the appropriate behavior at the right moment (proximate factors)? The examples described in this chapter will show that the question of survival value can be studied by measuring the costs and benefits of certain behaviors, assuming that by way of natural selection animals have reached the optimal solution. The outcome of these measurements can then be used to understand the proximate factors, our second question. In general, this approach can be used for each environmental factor, but we shall restrict ourselves to the relationships between herbivorous animals and their food source. Food is one of the four main (biotic) factors determining animal numbers, the others being predation, disease and competition. Feeding and adaption will be studied in three case studies involving three different species, in fact three different classes of animals. The first two examples show how reproductive cycles are geared to seasonality in plant growth and deal with migrating birds and sedentary reptiles respectively. The third case study involves mammals and shows how the reproductive cycles of different species are integrated.

1   WILD GEESE

During autumn, winter and early spring wild Brent Geese stay in the areas around the Dutch Wadden Sea. Brent Geese are herbivorous and feed on the salt marshes of the islands in the Wadden Sea and the coastal zone of the mainland.

Towards the end of May they leave for the breeding grounds on Taymyr Peninsula in the Arctic (Figure 23.1). The summer there is very short and food availability is restricted to the short period of time when the snow is melted. In September the geese return to their wintering areas in the Wadden Sea. One of the advantages of breeding in the Arctic is that, in the growing season the quality of the food there is high. The young take advantage of this high quality which allows them to grow rapidly.

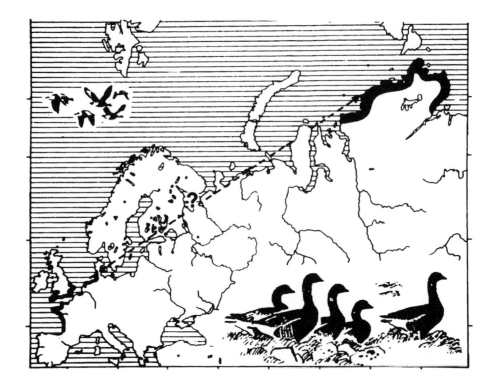

FIGURE 23.1
Spring migration of dark-bellied Brent Geese from the Dutch Wadden Sea to the breeding grounds on the Taymir Peninsula, Northern Siberia (arrow)
Wintering and spring staging areas along the coasts of western Europe (lower left) indicated in black. Breeding areas in coastal Northern Siberia (upper-right corner) also given in black.

The question arises how in such a short time the adult animals manage to lay their eggs, breed and raise their young so that they can fly all the way back to the Netherlands, a distance of some 5000 km.

The solution can be found in the possibility to build up considerable body reserves in the form of proteins and fat before leaving the Wadden Sea. In the course of spring, in a period of only five weeks, the body mass of the geese increases rapidly by about 400 g (Figure 23.2). This amounts to 30 to 40 per cent of their initial body mass which is the maximum load they can carry in the form of body weight. When heavier, the birds simply cannot take off any more. In order to understand how this rapid body mass increase is achieved, three important aspects should be studied: the amount of food available during spring time, the (grazing) behavior, and the animal's body mass.

1.1 BODY MASS

In order to determine body mass birds are caught by cannon nets during spring. The captured animals are weighed and a colored numbered ring is put around one leg. They are then released. The ring allows the birds to be followed and recognized from a great distance with a telescope.

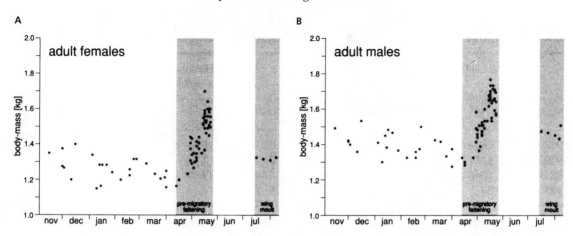

FIGURE 23.2

Changes in body mass in the course of winter and spring of adult female (A) and male (B) Brent geese

Since geese have their fat deposits near the tail, the amount of deposit can be estimated visually with a telescope (Figure 23.3).

There is a clear relationship between the amount of fat deposit and body mass. From the animals caught a reference collection has been made of the abdomal index/body weight ratio (the so-called fat index), so that the fat deposits of individual geese can be evaluated without disturbing the animals (Figure 23.4). In this way changes in body mass of 16 g or more can be estimated.

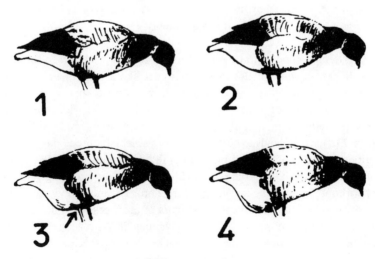

FIGURE 23.3

Four stages of the abdomal profile

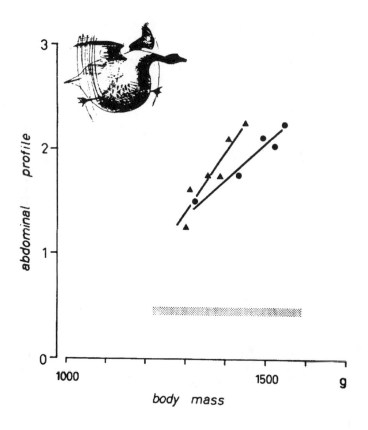

FIGURE 23.4
Abdomal profile as scored by observers prior to capture, related to body mass
of captured animals

The rate of fattening appears to be related to the geese's behavior. In this species the geese forage in flocks, in which males and females act like a pair. In the flock the male protects his female against other geese to provide her with a supply of the preferred patches of food. The female of a male that wins the competition with other males most times, increases her body weight at a higher rate than a female of a male with a lower success rate (Figure 23.5), which leads to earlier and more successful breeding.

This relationship can be illustrated by plotting the body mass of the females in spring at the time of departure to the breeding areas (May) against the breeding success, i.e. the return with or without young in the autumn (Figure 23.6). The mean body mass of successful females is much higher than the mean body mass of unsuccessful females. The deposition of enough fat reserves thus presents a great advantage in the raising of offspring.

The question now arises whether the variations in breeding success between years and individuals can be explained by diet. Therefore, more accurate measurements on the food availability and quality are necessary. On the marshes the geese mainly feed on two grass species (*Festuca rubra* and *Puccinellia maritima*) and two other species (*Triglochin maritima* and *Plantago maritima*). Determination of food composition can be carried out by determining which leaves have been eaten after a flock of geese has passed an area of the salt marshes, with the aid of stereo photography.

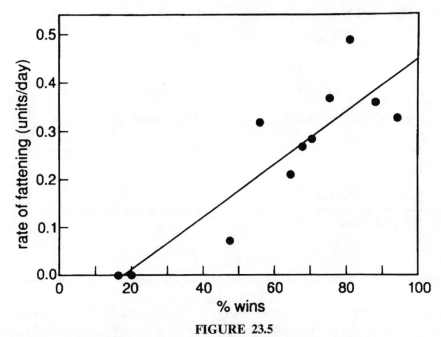

FIGURE 23.5

Rate of fattening of females during spring in the Dutch Wadden Sea area in relation to the percentage of the male's total number of successful fights

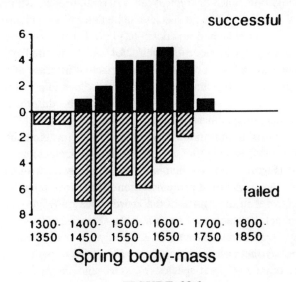

FIGURE 23.6

Distribution of spring body mass of successful (above X-axis), and failed (below X-axis) breeding female Brent Geese

These leaves are sampled at regular intervals and dried and analyzed for dry matter content and chemical composition. During grazing a movie camera records the number of bites taken. Combining these data (amount of leaves removed, dry mass and nutritional content per g leaf, and number of bites per minute) allows us to calculate the intake rate. Direct observation reveals the total time the birds spend grazing.

PARAMETER	METHOD
% time spent foraging	observation
intake rate/feeding minute	film + stereo photos
diet	fecal analyses
digestibility; ME (kJ/day)	feeding trials with captive geese

RESULT

$$\frac{(ME - maintenance)}{synthesis\ costs} \Rightarrow \textbf{g/d}\ body\ mass\ increment$$

FIGURE 23.7

Methods used to determine energy intake and body-mass increment of wild geese

In the laboratory digestibility tests on geese give information on the amount of nutrients and energy the animals extract from their food (see below), how much food the bird needs to convert it to body mass increment can then be calculated. Finally, during grazing it is very hard to tell which species are grazed. This problem is solved by analysing the droppings by microscope. Since the typical cell wall structures can still be recognized in the dropping material, the ingested plant species can be determined (Figure 23.7).

On the basis of the data collected it is possible to predict the increase in body weight during spring. The body-mass increase of a bird for which detailed observations were available, was predicted to be 9.4 g per day, while the actual increase after recapture was 10 g per day, which is remarkably similar. It must now be possible to explain the differences in fattening and breeding success between years on the differences in vegetation (growth rate of the food plants, total availability, etc.) during spring.

The quality of the birds' food is mainly determined by its protein content. The metabolizable energy they can extract from the food is positively correlated to the concentration of protein in the food (Figure 23.8). Young plant shoots usually have a higher protein content than older leaves. As a result the protein content of the four main food plant species mentioned above decreases in the course of the growing season (Figure 23.9), resulting in a gradual decrease in metabolizable energy (*ME*, kJ/g) (Figure 23.10). In April the protein content of the grasses (*Festuca* and *Puccinelia*) is already less than 25 per cent. Below that value the geese cannot extract enough nutrition from the food to build up energy reserves. During that time the other two plant species (*Plantago* and *Triglochin*) still have higher protein contents.

Combining food quality and bite size shows that the most common species, the grasses, have a very low rate of return (Figure 23.11), while the plants from which the geese can extract enough energy and nutrients, are far less abundant and their distribution is patchy. These species are very heavily grazed. In fact, they become totally depleted when the geese are feeding on the marsh.

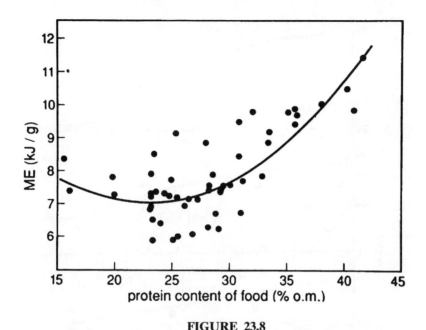

FIGURE 23.8
Relationship between metabolizable energy (*ME*) and protein in the food
Metabolizable energy is the amount of energy the animals can extract from a given amount of food.

To show the importance of the relatively rare food plants *Plantago* and *Triglochin*, a model has been constructed (Figure 23.12). This model shows the difference in body-mass increase between geese feeding on a diet containing only their staple food *Puccinelia* and geese on a diet of the actually observed intake. Through May the weight increase declined when the diet consisted of *Puccinellia* only. In geese on an average diet, including *Plantago* and *Triglochin,* the body-mass gain peaked at a much higher level and was maintained until the end of the month. Although the percentage of *Plantago* and *Triglochin* in the diet was less than 20 per cent, half of the body-mass increase appeared to be due to these high-quality species. By the end of May the grazing pressure on these plants becomes so heavy that they are virtually depleted. These plants thus play an important role in the fattening of the geese prior to their flight to the Arctic, which in turn is crucial for reproduction.

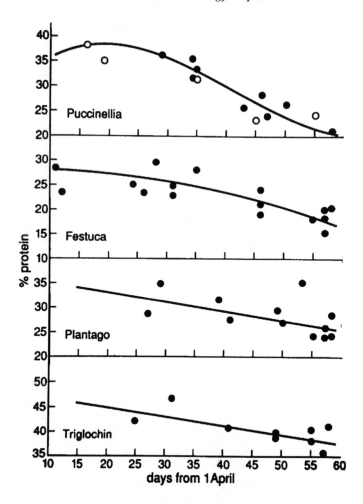

FIGURE 23.9

Protein contents of the four main food species during spring

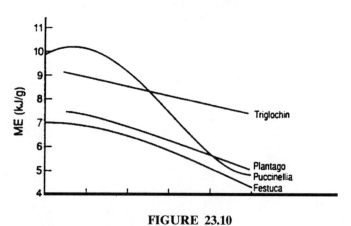

FIGURE 23.10

Metabolizable energy (*ME*) of the four food species during spring

ME largely follows the trend in protein content.

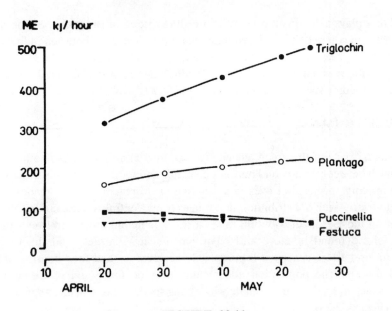

FIGURE 23.11
Energetic return in kJ/hour of the main food plants

Clearly *Triglochin* and *Plantago* are much more profitable. These species, however, are much less abundant than *Puccinellia* and *Festuca*.

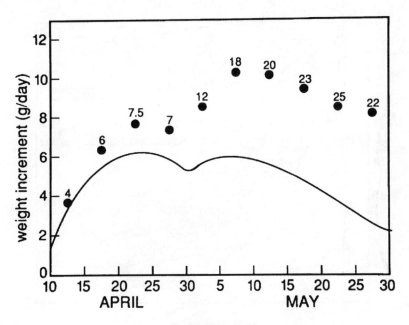

FIGURE 23.12
Model of mass gain of Brent Geese feeding on *Puccinellia* only (solid line), or supplemented with *Plantago* or *Triglochin* (less than 20 per cent) as actually occurs in the field (dots)

Note that, despite the relatively small amount of supplementation, these plants contribute up to 50 per cent of the body mass increase.

Food thus plays a dominant role in the breeding success of Brent Geese. The shortage of food results in competition between individual birds. The rank between individuals determines to a large extent the amount of food they can obtain and is related to breeding success. The (food) situation in the Dutch Wadden Sea area during April and May determines the breeding success one month later and 5000 km away.

2   GREEN IGUANA

The annual cycle of the green iguana *(Iguana iguana)* may serve as another example of a mismatch between energy and nutrient demands and food supply. This tropical reptile is not a migratory species, but lives in a seasonal environment. Green iguanas are large herbivorous lizards with a distribution that ranges from Central America to Brazil and the Caribbean. In most of its distribution range it inhabits humid areas and lives along river banks. It is also found in more arid environments in Venezuela and the Caribbean. Contrary to most lizard species it is completely herbivorous and food consists of leaves, fruits and flowers. The population under study lives on the island of Curaçao in the Caribbean which has a semi-arid climate with strong spatial, seasonal, and annual variation in rainfall (Figure 23.13).

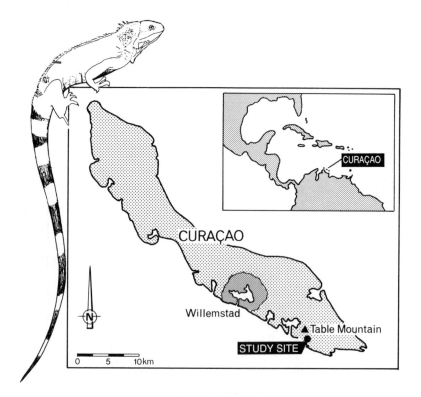

FIGURE 23.13
Location of the green iguana studies

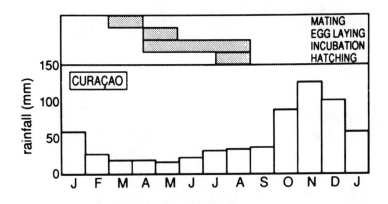

FIGURE 23.14
Reproductive cycle of green iguanas and the average monthly rainfall
on the island of Curaçao

During the long dry period from February to May hardly any rain falls. From May onwards the probability of rainfall increases, with the main rains in October to January. Rainfall has a strong impact on the vegetation and plant production and thus the food availability and food choice of the iguanas. Unlike the Brent Geese the iguanas on Curaçao do not migrate over large distances but, like the geese, they face great fluctuations in food availability.

The iguanas show a distinct reproductive cycle where courtship and mating take place in the first half of the dry season (March, April), followed by oviposition at the end of April and in May (Figure 23.14). The males are territorial throughout the year. During the mating period, they defend their territory rigorously against male intruders. Moreover, they keep their females company during foraging and attempt to mate with her. As a result they hardly feed themselves and loose much weight during this time (Figure 23.15). In contrast, the females behave more like they do in the rest of the year and spent most of their time basking and feeding. However, during the egg-laying stage in April-May the females leave their rocky habitat in search of an area with soft soil. Having found a suitable spot, they dig a burrow and lay their eggs. They cover the burrow carefully and return to their living site. The females reach their minimum body weight in this period, independent of the weight loss due to the clutch (Figure 23.15). Both males and females regain weight after the first rains and the accompanying leaf flush in June-August. The eggs are incubated by the heat of the sun. After 3 months of incubation the eggs hatch in July-August after which the young dig their way up. The hatchlings are totally independent of their parents. They are herbivorous and must rely on the food-plant availability.

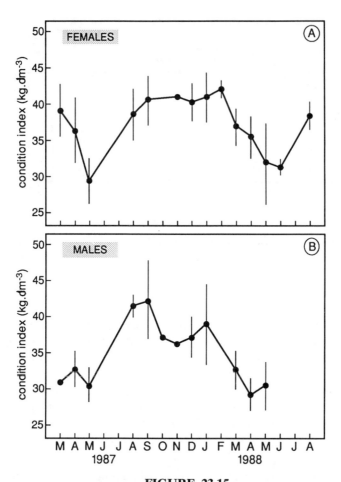

FIGURE 23.15
Changes in the condition index (body mass/snout-vent length³) of (*a*) females and (*b*) males in the course of the year

The reproduction cycle seems to be geared to the seasonality in rainfall and the corresponding plant production. The striking fact of this reproductive cycle is that the females must reproduce and lay their eggs in a time of the year when food availability is minimal. Two questions arise: Why do they lay their eggs in the dry season and not in the wet season when food is much more abundant? And: What are the implications of this annual cycle for the iguanas, or: How do they manage to survive? In order to get more insight into the timing and the repercussions of the reproductive cycle we need to know how energy and nutrient needs and intake are related to each other in the different seasons.

The question of how the body-mass changes, as shown in Figure 23.15, can be explained by a variation in energy expenditure or by an change in food (energy) intake.

Daily energy expenditure free-living iguanas is determined by the doubly-labeled water method (see Chapter 16). First the animal must be caught in order to take a background blood sample and give the dose of labeled water. The animals are weighed and color-marked for individual recognition with beads sewn on their crest. Some animals are equipped with temperature-sensitive transmitters for obtaining field body temperatures (see below).

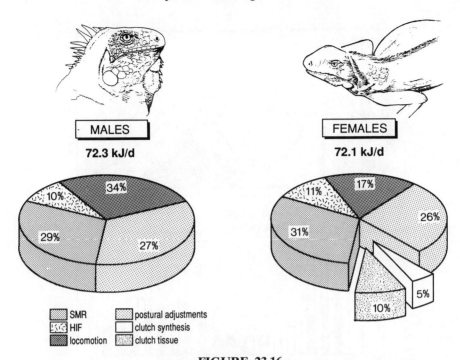

FIGURE 23.16
Average daily energy allocation in male and female green iguanas calculated
on a yearly basis
Both sexes spend approximately the same amount of energy, but in females egg production requires
14 per cent, while males spend more energy on social activities and locomotion.

After taking an initial blood sample the animal is released, but must be recaptured
after 2-3 weeks to take a final sample. During the time of energy measurement the animals
are observed by telescope in order to register their behavior, and study the allocation of
energy.

The results from the doubly-labeled water studies revealed that the daily energy expenditure was *not* different between the seasons and the sexes, but there was a marked difference in the allocation of energy between males and females (Figure 23.16). In females
approximately 13 per cent of their annual energy expenditure is accounted for by egg
clutch formation, while the males spent much more time and energy on social and locomo-
tor activities. On average, they covered twice the distance of the females.

2.1   FOOD CHOICE

The fluctuations in body weight throughout the year are thus likely to be caused by
differences in energy intake. Food intake was determined by direct observations on grazing animals combined with monthly sampling of fresh plant material for chemical analysis
and biomass (standing crop) determination.

Food availability is strongly related to rainfall (Figure 23.17). The first rains in May
cause an outburst of young leaves. The iguanas' diet mainly consists of these leaves.
Throughout the wet season the availabilty of young leaves decreases so that mature leaves
and flowers must be eaten. During the end of the wet season and the beginning of the dry
season the diet consists mainly of old leaves and, in relatively wet years, of berries.

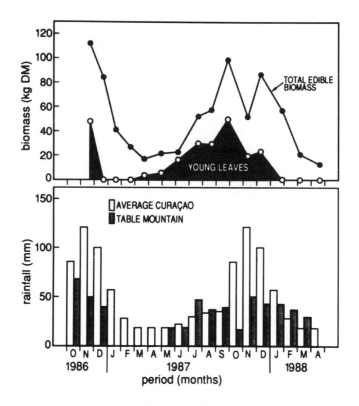

FIGURE 23.17
Total edible biomass and young leaf availability and monthly rainfall at the study site during the study period, and mean monthly rainfall on Curaçao

In the dry season there are hardly any young or mature leaves available, but some trees produce flowers during this period, like *Acacia tortuosa* (Wabi), which then form the main constituent of the iguanas' diet.

Contrary to animal food, plant material is generally hard to digest and varies much in composition during the course of a year. Most plant parts have a much lower protein content than animal food and, depending on the season, the water content can also be low. Plant material consists to a large extent of cell wall components, like celluloses and hemi-celluloses which form the components that are diffucult to digest or undigestible at all (Figure 23.18).

Green iguanas have an enlarged colon which contains microorganisms of the same species and in comparable concentrations as those in herbivorous mammals, for example cows. Digestion of the food thus depends to a large extent on the activity of the microorganisms. In addition, in ectotherms the body temperature plays an important role, as bacterial activity is temperature-dependent and body temperature in ectotherms is not as fixed as in mammal herbivores.

In a series of cage experiments (Figure 23.19) in controlled laboratory circumstances the influence of food composition and body temperature on digestive efficiency was investigated (Van Marken-Lichtenbelt, 1992). The digestibility of the food was determined by registering precisely the amount of food consumed and the amount of feces and uric acid produced.

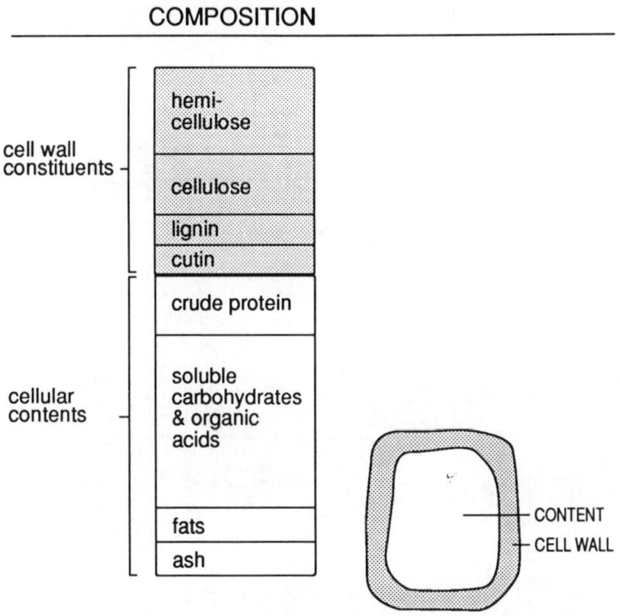

COMPOSITION

FIGURE 23.18

Approximate plant food composition

Cell contents are generally readily digestible by enzymes of the herbivore itself. Cell walls are in part digested by enzymes of the microorganisms in the intestinal tract.

Chemical analysis of samples of food and excreta revealed the digestibility of the different nutrients. During the experiments the body temperature was measured continuously. The body termperature could be manipulated by varying the temperature of the environment. The results show that iguanas digest their food as efficiently as mammals (55-80 per cent, depending on the type of food). A remarkable result was that the percentage digested food was independent of the body temperature. However, the passage time was inversely related to body temperature (Figure 23.20). Thus, at lower body temperatures the food is digested as efficiently but the amount of food that can be processed (ingested) per unit of time is less as a result of slower emptying of the intestines. The reduction in transit time at higher temperatures is a direct thermal effect, and goes hand-in-hand with a difference in degree of fill of the digestive tract.

The intake capacity thus depends on body temperature: at higher body temperature the animals can process more food material (Figure 23.21). Measurements of oxygen consumption and CO_2 production of iguanas in a respiration chamber (see Chapter 16) show that energy expenditure increases with body temperature. From Figure 23.21 it is obvious that at a temperature of 36.5 °C the potential net energy intake is maximal. From the animals in the field of which both intake and body temperature are known, it appears that the iguana with the highest food intake has a body temperature of 36.6 °C, remarkably close to the model's prediction. Iguanas with a lower food intake appear to have a lower body temperature. Because these animals do not necessarily need to empty their intestines, they can save energy by lowering their body temperature!

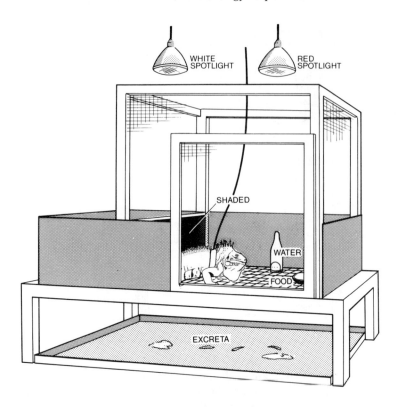

FIGURE 23.19
Cage used for digestibility tests

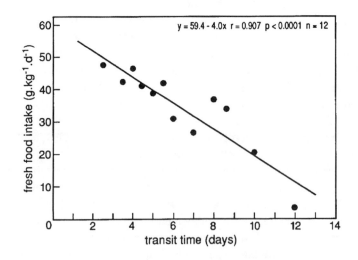

FIGURE 23.20
Relationship between the amount of maximum daily fresh food intake and the transit time of the food in the intestinal tract

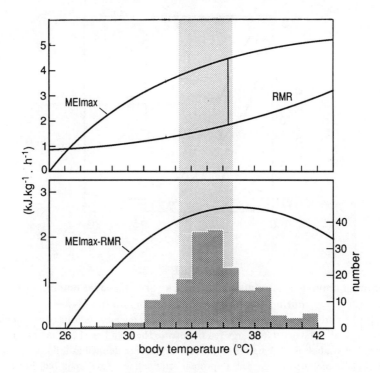

FIGURE 23.21

Resting metabolic rate (*RMR*), maximal energy intake (*MEI*$_{max}$), and the difference between *MEI*$_{max}$ and *RMR* as functions of body temperature in the green iguana

Note that the net energy yield is highest at approximately 36.6 °C.

2.2   ANNUAL BALANCE

In most cases food choice is not determined by one parameter but by a combination of characteristics of the food and the momentary needs of the animal. In search for an explanation for the observed food choice, we need to include the apparently important parameters in our analyses. In the study on iguanas linear programming was used as an optimization model. From the model the optimal diet under the given circumstances can be derived and compared to the actual, observed intake. As an example, we shall discuss the food choice of iguanas during the dry period with a choice of two food types: leaves and flowers (Figure 23.22).

Food intake is constrained by the capacity of the intestinal tract, which in turn depends on body temperature, as shown above. Line C in Figure 23.22 represents this capacity. It is known that 14 g dry matter of flowers is the maximum daily quantity the animals can consume. The daily limit for leaves is 15 g. The intestinal capacity for a combination of leaves and flowers can be read along line C. Line M represents the food intake that covers the daily energy requirement of the iguanas. To calculate this line, the digestive efficiency and the costs of foraging were included. In order to obtain enough energy, the iguanas need to eat 19 g leaves per day which is more than the intestines can hold (line C).

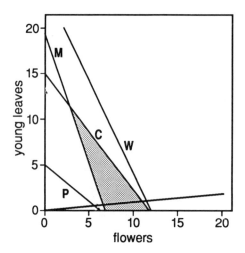

FIGURE 23.22
Linear programming constraints solved for 1 kg green iguanas maintaining body weight
during part of the dry period for two food types

The x- and y-axis are expressed in g dry food matter. The gray area indicates diet combination which satisfy protein and metabolizable energy requirements. The bold line shows the observed proportion in the diet. The symbols describing the various constraints are defined as follows: C, digestive capacity; M, metabolizable energy required; P, protein required; W, water required. For explanation see text.

Due to the low density of the leaves during the dry season the costs of foraging are high and the iguanas can not obtain enough energy by consuming leaves only. For flowers, which are abundant during the dry season, 7 g per day is nutritionally sufficient. The grey area shows the combinations of flowers and leaves that can be consumed to cover the daily protein and energy requirements. The bold line shows the actual proportion of leaves and flowers in the diet during the dry period. At that point the diet consists of almost 100 per cent flowers. The model indicates that food choice in this period is determined by water maximization. It can be extended by adding other nutrients and food types, in which case the number of dimensions increases. The solution of the various combinations in the model indicate that during the dry season food choice is mainly determined by water requirement, while in the period after the first rains and during the wet season protein is the main determinant. During the dry season there is no drinking water available, and the iguanas do indeed obtain their water from their food. On Curaçao some flowers contain just enough water to cover the iguanas' needs. After the first rain in spring the young leaf flush provides food with a high protein content. The animals then choose these leaves in order to replenish their protein and energy stores.

The importance of the period after the first rains becomes evident when the results obtained from the detailed observations in the linear programming model, including the results on behavior, field energetics, digestion and body temperature, are used to reconstruct the iguanas' energy and protein balance throughout an average year.

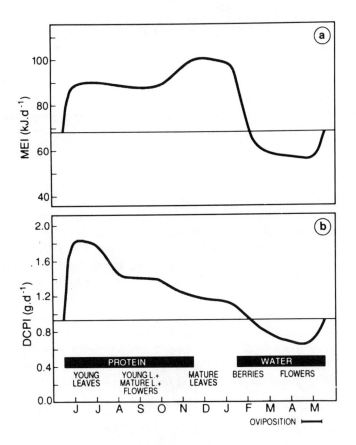

FIGURE 23.23

Reconstruction of (a) the daily metabolizable energy intake (*MEI*) and (b) the daily digestible protein intake (*DCPI*) in the course of an average year

Note that *MEI* and *DCPI* during the dry period are too low to meet the requirements, while during the wet period energy and protein reserves can be replenished. Below are given the nutrients that play a dominant role in food selection, being proteins and water depending on the season, and the main food components in the iguana's diet.

The reconstruction shows that the metabolizable energy index (*MEI*) exceeds the daily energy expenditure (*DEE*) from July until January, but during the dry season *MEI* is too low to cover daily energy expenditure (Figure 23.23). The pattern of digestible crude protein intake (*DCPI*) is comparable, but protein intake is relatively high during June/July, when young leaves are consumed. Thus, in an average year energy will be stored from July to the end of the year. From January to May the animals generally are in negative energy balance and depend on the previously stored reserves. Protein storage mostly coincides with the period of energy storage, but reaches peak levels in summer. As already mentioned, the reproduction cycle is geared to the seasonal rainfall. This is true for adults but even more so for the young, since the latter cannot rely on large body reserves. They can only survive if there is enough high-quality food available during the first days and weeks of their life, i.e. in the period of young leaf flush. Female green iguanas on the island of Curaçao lay their eggs from the end of April through May. Since egg development requires about 90 days the time of laying ensures that the hatchlings can profit from the late summer peak in availability of young leaf material. Oviposition thus coincides with the second half of a period in which intake is below maintenance, which is reflected in the condition indices (Figure 23.15). This shows that reproducing females must store energy and proteins long before the time of oviposition. The period of young leaf flush may be crucial for replenishing the protein stores. This indicates a 8-10 month interval between the acquiring the main portion of the protein required for egg synthesis and laying the eggs.

2.3 CONCLUSION

Temporary shortage of the food component of primary production limiting the herbivores could be a feature common to all terrestrial vegetation/herbivore systems (Sinclair, 1975). The seasonality in plant growth often results in a discrepency between time of food supply and energy demands for reproduction (Drent & Prins, 1987). Maternal budgets are strained in reptiles during egg laying, in birds during egg laying and incubation, and in mammals during lactation. The case studies of the geese and iguanas show that the timing of reproduction is set to allow the young to take advantage of the flush of vegetative growth. This necessitates storage of nutrients and energy to allow optimal timing of hatching. Similarly, in other herbivores reproductive success has been shown to depend heavily on resource acquisition at a much earlier phase of the cycle (for mammals see for example Morton Boyd & Jewell, 1974; Clutton-Brock *et al.*, 1982; Drent & Daan,1980; Ebbinge, 1989). It seems to be a general principle that reproductive success in herbivores depends on previously stored body reserves.

3 INTERACTING REPRODUCTIVE STRATEGIES

Each of the many species of animals on earth (either herbivorous, omnivorous or carnivorous) has its own unique evolutionary history. To some extent the same applies to variations between different populations within a species. The dietary and climatic characteristics of environments are diverse, resulting in a large variation in seasonal reproductive strategies. As was shown above, even the precise seasonal patterns between the sexes may differ.

This last case study shows how reproductive cycles of various animal species in one location differ and are attuned to one another (Bronson, 1989). In this case study too, the timing of the reproductive cycles is to a large extent set by food availability. The study area is located in Kansas where annual patterns of reproduction of three species of mammals have been studied: the coyote *(Canis latrans)*, the black-tailed jackrabbit (*Lepus californicus*, and the prairie vole (*Microtus ochrogaster*). In Kansas the winters are cold with little rainfall and reduced plant growth. The rise in temperature and precipitation in spring is clearly reflected in the production of plants (Figure 23.24).

Vole and jackrabbit are strictly herbivorous. Coyotes prey on them. The vole has a life expectancy of a few weeks to 3 months. It matures quickly, and produces many litters in quick succession and continuously. Depending on the temperature, voles may even reproduce in winter. In general, voles are opportunistic and reproduce as soon as food is available. As a result, the numbers of voles can vary greatly from season to season and from year to year (Gaines & Rose, 1976), and the population size pattern shown in Figure 23.24 may not occur every year.

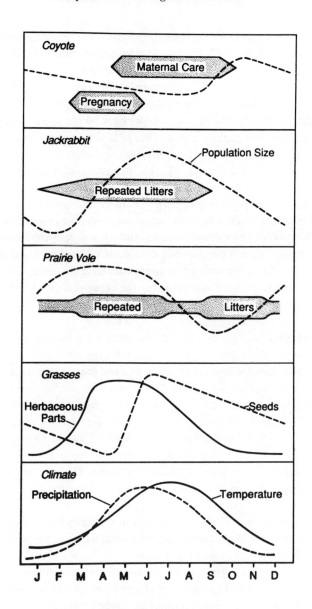

FIGURE 23.24
Annual patterns of reproduction in mammals living in central Kansas, modeled by
Bronson (1989)

The jackrabbit is larger, has a longer lifespan than voles (3-6 months), and reproduces only during a limited part of the year. They can also produce several offsprings but it takes longer to produce a weaned litter. The jackrabbits start mating well before the beginning of the period of maximum food availability. As shown in Figure 23.24 the jackrabbit's repeated periods of lactation coincide with the maximum availability of the vegetative part of grasses and seeds. To start breeding early in the year carries certain risks for the jackrabbit. Prolonged blizzards in March and April will block all access to food, and massive prenatal and neonatal mortality will be the result (Bronson & Tiemeyer, 1958). The jackrabbit thus shows some degree of opportunism, but less than the shorter-lived vole.

The coyote has a lifespan of 2-3 years. This longer lifespan is balanced by the production of a single litter each year, an effort requiring 7 or 8 months. The coyote also starts mating well before the onset of spring, resulting in a period of maternal care that coincides with the maximum availability of jackrabbits, but not with that of voles, because the numbers of voles are less predictable. The coyote with its longer lifespan has an annual cycle that is more rigidly programmed than that of the jackrabbit and the vole.

So, in one type of environment three different reproductive strategies each with their own timing have evolved. The strategy of each species includes a balance between the females' life expectancy and her reproductive potential, and is geared to food availability.

From an ultimate perspective, most cases of seasonal reproduction reflect variation in a complex of interacting biotic (dietary) and abiotic (climatic) factors (food, rainfall, and temperature). Reproductive processes are, among other processes, limited by the amount of available food. Rainfall and temperature determine plant growth and hence the amount of food available to herbivores, which in turn provide food for carnivores. In environments characterized by seasonal variation in climate, natural selection will tend to favour reproductive cycles that maximize the potential for success.

4   SUMMARY

Food, predation, disease and competition are the four factors that determine animal numbers.

Feeding and adaptation studies in herbivores show how reproductive cycles are geared to seasonality of food availability, i.e., plant growth. Food plays a dominant role in the breeding success of Brent Geese. Shortage of food results in competition between individual birds. The rank between individuals determines to a large extent the amount of food they can obtain and is related to breeding success. The food situation in the Dutch Wadden Sea area during April and May determines the breeding success one month later, in the Arctic.

Seasonality in plant growth often results in a discrepancy between time of food supply and demands for reproduction. The timing of reproduction is set to allow the young to take advantage of the flush of vegetative growth.

Reproductive success in herbivores depends on previously stored body reserves. From an ultimate perspective most cases of seasonal reproduction reflect variation in a complex of interacting biotic (e.g. dietary) and abiotic (e.g. climatic) factors.

Glossary

Abdominal - concerning the portion of the body between the diaphragm and the pelvis

Abiotic - non-living

Abnormal behavior - behavior which is unusual or infrequent in a particular culture

Acetylcholine - chemical transmitter substance released by nerve endings

Acidosis - condition in which the blood has a higher hydrogen ion concentration than normal with a decreased pH

Acromegaly - abnormal pattern of bone and connective-tissue growth characterized by enlarged hands, face, and feet, and associated with excessive pituitary growth hormone that is secreted after the epiphyseal cartilages have been replaced

Activity induced energy expenditure (AEE) - energy expenditure due to physical activity

Adaptation - decrease in sensitivity during sustained presentation of a stimulus

Addiction - physical or psychological dependence upon one or more drugs. Once addicted to a particular drug, individuals cannot function normally in its absence

Adiposity (obesity) - excessive accumulation of fat in the body

Adrenalectomy - the removal of an adrenal gland

Adrenalin - trade name for epinephrine.

Adrenal cortex - the outer covering of the adrenal gland

Adrenal glands - hormone-producing glands located superior to the kidneys; each consists of medulla and cortex areas

Adrenal medulla - the central part of the adrenal gland

Adrenergic - norepinephrine-sensitive

α-Adrenergic receptor - one of several types of norepinephrine-sensitive binding sites

β-Adrenergic receptors - one of several types of norepinephrine-sensitive binding sites

Adrenoceptor - adrenergic receptor

Adrenocorticotropic hormone - a hormone that influences the activity of the adrenal cortex and is released by the anterior portion of the pituitary

Adrenodemedullation - surgical removal of the adrenal medulla

***Ad libitum* food** - free access to food

Afferent - tranporting or conducting toward a central region

Agonadal - without gonads

Agonist - substance reacting with a receptor leading to physiological and biochemical changes and ultimately resulting in a biological effect

Alcoholism - type of conduct disorder involving heavy drinking that interferes with the individual's ability to function at home and at work

Algorithm - set of rules that will eventually guarantee the successful solution of a problem

Alloxan - chemical substance that destroys the insulin-producing cells in the islets of Langerhans

Amenorrhea (primary/secondary) - absence of menstruation; when menstruation has not been established at the time when it should have been, it is primary amenorrhea; absence of menstruation after it has once commenced is referred to as secondary amenorrhea

Aminostatic theory - eating behavior regulates a deficit or surplus of amino acids in the blood plasma

Amnesia - failure of memory often due to some traumatic injury

Amyline - hormone co-localized with insulin in the same secretory granules

Anabolism - the energy-requiring building-up phase of metabolism in which simpler substances are synthesized to more complex substances

Anorexia nervosa - eating disorder characterized by intentional starvation, distorted body image, excessive amounts of energy, and an intense fear of gaining weight

Antagonist - substance reacting with a receptor resulting in a blockade of the functioning of the agonist

Antidiuretic hormone (vasopressin) - pituitary gland hormone that controls the reabsorption of water by the kidney

Aphagia - inability to swallow

Appetite - disposition or orientation in behavior towards some object or objective such as for instance eating some food

Atropine - alkaloid of belladonna; parasympatholyticum; drug capable of neutralizing the effect of parasympathetic stimulation

Atwater factors - the available energy in the main nutrients: 17 kJ/g for protein, 38 kJ/g for fat, 17 kJ/g for carbohydrate, 29 kJ/g for alcohol

Autonomic nervous system - the part of the nervous system that functions involuntarily and is responsible for innervating cardiac muscle, smooth muscle, and glands

Autotrophs - organisms able to synthesize food from inorganic substances by utilizing the energy of the sun or of inorganic compounds

Average daily metabolic rate - total energy expenditure measured under daily living conditions

Aversive conditioning - type of counterconditioning in which punishment is used in order to associate negative feeling with an undesirable response

Avoidance conditioning - process through which organisms learn to avoid unpleasant consequences by engaging in prevantative actions. Often, external stimuli signal the necessity for such responses, but in some cases such warning signals are internally generated by the passage of time

Axon - the process of a nerve cell by which impulses are carried away from the cell

Basal metabolic rate - rate of energy expenditure necessary for maintaining the basal functions of the body

Behavior modification - altering of overt behavior in accordance with the principles of learning and by means of manipulating the conditions under which learning takes place (e.g., by having an individual learn unpleasant associations - say, an electric shock - with undesirable acts - smoking, perhaps)

Behaviorism - view, first expressed by John B. Watson, that psychology should focus upon the study of behavior, overt actions capable of direct observation and measurement

Biofeedback - technique in which minute changes occurring in the body or brain are amplified and displayed to individuals experiencing them. By means of such feedback, many persons can learn to exert voluntary control over their internal bodily processes

Biotic - living

Blood-brain barrier - mechanism that inhibits passage of substances from the blood into brain tissues and cerebrospinal fluid

Body impedance - resistance to an electrical current through the body

Body mass index (Quetelet Index; BMI) - defined as body weight in kilograms divided by the square of the height in meters (kg/m^2); an index for fatness

Bomb calorimetry - determination of the energy content of food; a foodstuff is placed in a small chamber or bomb and exposed to a high pressure of oxygen, all the organic material is burnt and the heat liberated can be measured

Bradycardia - reduction in heart rate from the normal level

Brainstem - the portion of the brain consisting of the medulla, pons, and midbrain

Broca's index - defined as body weight (kg) divided by Broca weight (kg); Broca weight is body height (cm) - 100 in men and body height (cm) - 110 in women; it is an index for fattness

Brown adipose tissue - tissue of which the fat cells are more vascularized and contain more mitochondria than in white adipose tissue; adipose tissue with a high metabolic rate; relevant in animals, maybe relevant in neonates, not relevant in human adults

Bulimia - eating disorder characterized by periodic binging and purging, the latter usually taking the form of self-induced vomiting or laxative abuse

Calorie - amount of heat energy required to raise the temperature of 1 g of water by 1 °C from 14.5 °C to 15.5 °C

Carbohydrate cravers - (obese) people with a great appetite for sweet food

Catabolism - process in which living cells break down more complex substances into simpler substancers

Catecholamines - any of a group of amines which are secreted in the human body to act as neurotransmitters; examples are epinephrine, norepinephrine, and dopamine

Central nervous system (CNS) - the brain and the spinal cord

Cephalic response - response that has its origin in a stimulation of sensory receptors whose neuronal output is processed by the central nervous system to produce physiological changes

Cephalic phase - rapid, physiological change (e.g., an increase in insulin response) as a result of mainly the sensory perception of food

Cerebellum - part of the hindbrain; controls movement coordination

Cerebral cortex - outer surface layer of the cerebrum

Cerebrospinal fluid - fluid produced in the cerebral ventricles; fills the ventricles and surrounds the CNS

Cerebrum - largest part of the brain

Chemotrophs - organisms producing organic matter from mineral substances without the use of solar energy; however, most organisms draw energy from other organic compounds

Chemoreceptors - receptors sensitive to various chemical stimulations and changes

Cholecystokinin - hormone which contracts the gallbladder; secreted by the upper intestinal mucosa

Cholinergic - acetylcholine-sensitive

Circadian rhythm - rhythm with a periodicity of 24 h

Cocaine - stimulant drug that also has anesthetic properties

Computed tomography - computer-constructed imaging technique of a thin slice through the body, derived from X-ray absorption data

Conditioned response - response evoked by a conditioned stimulus

Conditioned stimulus - stimulus which acquires the capacity to evoke particular responses through repeated pairing with another stimulus capable of eliciting such reactions

Control group - group in an experiment which is not exposed to the independent variable under investigation. The behavior of subjects in this condition is used as a base-line against which to evaluate the effects of experimental treatments

Corpus luteum - yellow ovarian glandular body that arises from a major follicle that has released its ovum; it secretes progesterone; if the ovum released has been fertilized, the corpus luteum grows and secretes during gestation; if not, it atrophies and disappears

Correlation coefficient - statistic which indicates the degree of relationship between two or more variables. The larger the correlation (the more it departs from 0.00), the stronger the observed relationship

Correlational method of research - method of research in which variables of interest are observed in a careful and systematic manner in order to determine whether changes in one are associated with changes in the other

Correlation studies - studies designed to yield information concerning the degree of relationship between two variables

Corticosteroids - steroid hormones released by the adrenal cortex

Corticosterone - steroid hormone, released by the adrenal cortex

Corticotrophin-releasing factor - hormone released by the hypothalamus that activates the anterior portion of the pituitary to release adrenocorticotropic hormone

Cortisol - glucocorticoid produced by the adrenal cortex

Cotransmitters - factors that are colocalized and coreleased with the classical neurotransmitters in the autonomic nerve endings; they are released during stimulation of the autonomic nerves; vaso-inhibitory peptide and galanin are parasympathetic cotransmitters; neuropeptide-Y, galanin, adenosine-tri-phosphate and adrenalin are sympathetic cotransmitters

Counterconditioning - behavior modification technique in which a desired response is substituted for an undesirable one by means of conditioning procedures

Cranial - pertaining to the skull

Cranial nerves - 12 pairs of nerves arising from the brain and the brain stem, connecting the outlying parts of the body and their receptors with the CNS

Craving - extremely strong selective appetite

Criterion validity - type of validity in which the test is compared to some independent criterion

Cross-sectional studies - type of research design in which subjects of different ages are studied at one point in time

Dawn - end of the night

Decision criterion - determinant of how likely an observer is to report detecting a signal. The higher the criterion, the less likely a signal will be reported, regardless of the intensity of the stimulus

2-Deoxy-D-glucose - analog of glucose

Dependence - condition that occurs when a drug becomes incorporated into the functioning of the body's cells so that is is needed for 'normal' functioning

Dependent variable - variable (usually some measure of behavior) which is expected to change in a psychological experiment as one of more additional factors (the independent variables) are changed or varied

Dexamethasone - drug acting on glucocorticoid receptors

Diabetes mellitus - disease caused by deficient insulin release, leading to failure of the body tissue to oxidize carbohydrates at a normal rate

Diet-induced energy expenditure - energy cost of processing food

Diet-induced thermogenesis (diet-induced energy expenditure) - energy cost of processing food

Dietary recall - method to measure food intake at individual level; a subject's report of food intake over the previous 24h-period (24h-recall) or the report of customary food intake over the previous week up to the past year(s) (diet history)

Dietary record - method to measure food intake at individual level; a subject's report of types and amounts of all foods consumed over a given time interval

Direct calorimetry - determination of energy expenditure by measuring heat production of the body

Discriminative cue - any stimulus which serves as a signal for the performance of a specific form of behavior. Responses in the presence of the discriminative stimulus yield reinforcement, while responses in its absence fail to produce such consequences

Discriminative stimulus - stimulus which indicates that some particular form of behavior will or will not be reinforced

Displacement - defense mechanism in which responsibility is shifted from oneself to another person. Also, the substitution of one fear for another

Diuresis - increased secretion of urine

Dopamine - intermediate in norepinephrine synthesis; possibly a CNS transmitter

Dorsomedial hypothalamic areas - certain areas in the hypothalamus

Double blind - experimental design in which neither the subjects nor those who dispense the treatment condition know who receives the treatment and who receives the placebo

Dummy meal - meal without nutritional value; for example a mixture of paraffin oil, vaseline and cellulose, all of them indigestible

Dusk - beginning of the night

Eating disorder - any serious and habitual disturbance in eating behavior that produces unhealthy consequences

Efferent - carrying away from a center

Endocrine glands - ductless glands that empty their secretions directly into the blood

Endogenous - arising within the body

Endorphins - naturally occurring neurochemicals whose effects resemble the opiates

β-Endorphin - one of several types of neuropeptides that exhibit morphine-like actions

Endotherms - organisms keeping their body temperature high by means of their metabolic rate

Energostatic theory - energy produced by the liver from oxidative metabolism of all the absorbed and mobilized nutrients is monitored to control food intake

Energy balance - energy intake minus energy expenditure

Energy cost of arousal - difference in energy expenditure between resting metabolic rate and basal metabolic rate

Energy expenditure - heat production of a subject measured by indirect calorimetry

Energy expenditure over 24h - total energy expended by a subject during 24 hours

Energy intake - food energy available for energy metabolism

Enteric nervous system - part of the autonomic nervous system involved in gastro-intestinal coordination

Enteroglucagon - glucagon produced by specialized cells in the intestinal mucosa

Enterohepatic - going from liver to liver, passing by the intestines and the blood stream

Entero-insular axis - all interactions between intestines and islets of Langerhans

Epinephrine - generic name for the catecholamine released from the adrenal cortex; also known by the trade name Adrenalin

Estradiol - synthetic estrogen

Estriol - estrogen metabolite present in the urine of pregnant women

Estrogen (estrogen)- ovarian hormones: estrone, estriol, estradiol; it is responsible for the female secondary sex characteristics; it also prepares the reproductive system for fertilization and implantation of the ovum

Estrone - hormone similar to estradiol

Estrous cycle - periodic episodes of estrus, marked by sexual receptivity in mature females of most mammalian species

Eumenorrhea - normal menstruation pattern

Exocrine glands - glands that have ducts through which their secretions are carried to a particular site

Exocytosis - fusion of the vesicle membrane to the surface membrane and subsequent expulsion of the vesicle content to the cell exterior

Extirpation - complete removal or destruction of a part

Extrinsic - originating from outside an organ or part

Fat-free mass - mass of the body minus the total mass of fat

Fat mass - total of ether-extractable substances in the body; the mass of the body minus the fat-free mass

Feedback - return of output to the input part of a system. In negative feedback, the sign of the output is inverted before being fed back to the input so as to stabilize the output. In positive feedback, the output is unstable because it is returned to the input without a sign inversion, and thus becomes self-reinforcing, or regenerative

Fenfluramine - prescription diet drug : an appetite suppressant

Fenoterol - β_2-selective adrenoceptor agonist

Fermentation - enzymatic decomposition; anerobic transformation of nutrients without net oxidation or electron transfer

Fistula - abnormal passage between organs or between a body cavity and the outside

Flow chart - schematic rendering of a series of events or operations occurring sequentially

Follicle stimulating hormone (FSH) - hormone produced by the anterior pituitary that stimulates ovarian follicle production in females and sperm production in males

Food-specific satiety - short-term satiety specific to a food that just has been consumed

Food quotient (FQ) - the ratio of the calculated carbon dioxide production and oxygen consumption when the consumed food is fully oxidized

Galanin - sympathetic and parasympathetic cotransmitter

Ganglion - group of nerve cell bodies, usually located in the peripheral nervous system

Ganglion coeliacum - an important ganglion for the innervation of the intestinal organs

Ganglion nodosum - an important ganglion by which feedback messages from the gastro-intestinal system are transferred

Gap junction - intercellular specialization with the cell membranes of adjacent cells only 20 Å apart

Gastrin - a protein hormone that is liberated by the gastrin cells of the pyloric gland and induces Gastric secretion and motility

Gastro-inhibitory polypeptide - a gastro-intestinal hormone released into the blood stream from the duodenal mucosa, inhibiting gastric secretion and motility

Gastro-intestinal - pertaining to the stomach and intestine

Gigantism - excessive growth due to hypersecretion of pituitary growth hormone from birth

Glands of Brunner - exocrine glands that are located in the intestinal mucosa and secrete an alkaline mucoid fluid

Glands of Lieberkühn - exocrine glands located in the intestinal mucosa

Glossopharyngeal nerve (IX) - cranial nerve supplying the tongue and the pharynx

Glucagon - protein hormone formed by the α-cells of the pancreatic islets; raises the glucose level of blood

Glucocorticoids - adrenal cortex hormones that affect metabolism of fats and carbohydrates

Gluconeogenesis - synthesis of carbohydrates from noncarbohydrate sources, such as fatty acids or amino acids

Glucopenia - lack of glucose for the tissues

Glucoprivation - decreased glucose availability

Glucostatic theory - regulation of the use of glucose by certain privileged cells

Glucosurea - excretion of excessive amounts of glucose in the urine

Glycogenesis - synthesis of glycogen

Glycogenolysis - breakdown of glycogen to glucose-6-phosphate

Gold thioglucose - toxic agent

Gonadotrophin-releasing factor (GRF) - hormone released by the hypothalamus that activates the anterior portion of the pituitary to release gonadotropic hormones

Gonadotropic hormones - the gonad-stimulating hormones produced by the anterior pituitary

Gonads - glands or organs producing gametes; an ovary or testis

Gorging - pattern of food intake characterized by a few large meals daily

Gross energy - ingested energy, including what is lost in feces and urine

Growth hormone (somatotropic hormone) - hormone that is secreted by the anterior pituitary and stimulates growth; directly influences protein, fat, and carbohydrate metabolism and regulates growth rate

Growth hormone releasing factor - hormone released by the hypothalamus that activates the anterior portion of the pituitary to release growth hormone

Habituation - progressive loss of behavioral response probability with repetition of a stimulus

Hedonic rating - expression of emotional experience with a causal role in choice.

Hedonic theory - eating behavior does not necessarily compensate for a deficiency in a homeostatic manner, but may originate from the motivation of the organism to find pleasure, reward and satisfaction

Heterotrophs - organisms with nutritional dependency on other organisms

High-density lipoproteins - plasma protein relatively high in protein, low in cholesterol; it is involved in transporting cholesterol and other lipids from plasma to the tissues

Hirsutism - excessive growth of hair at sites in which body hair is normally found

Homeostatis - state of internal stability or balance

Homeotherms - animals (mammal, bird) that regulate their own internal temperature within a narrow range, regardless of the ambient temperature

Hormone - chemical compound synthesized and secreted by an endocrine tissue into the blood stream; it influences the activity of a target tissue

Hunger - desire for food; motivation to acquire and ingest food

5-Hydroxytryptamine (serotonin, 5-HT) - neurotransmitter

Hyperglycemia - excessive blood glucose levels

Hyperinsulinemia - excessive plasma insulin levels

Hyperphagia - conditioning of gross overeating produced by damage to the ventromedial hypothamalus

Hyperthyroidism - increased thyroid activity

Hypoglossal nerve (VII) - cranial nerve located beneath the tongue and supplying the intrinsic and extrinsic muscles of the tongue

Hypoglycemia - low blood glucose levels

Hypophagia - undereating

Hypophysis (pituitary gland) - complex endocrine organ situated at the base of the brain and connected to the hypothalamus by a stalk; it has a variety of functions including regulation of the gonads, thyroid, adrenal cortex, and other endocrine glands

Hypothalamic-pituitary-adrenal axis - the sequence of the following glands and hormones: the hypothalamus releasing corticotrophin-releasing factor, the pituitary releasing adrenocorticotropic hormone and the adrenal cortex releasing corticosterone

Hypothalamus - part of the diencephalon that forms the floor of the third ventricle of the brain; it is the highest centre of the autonomic nervous system and contains centres controlling various physiological functions such as emotion, hunger, thurst and circadian rhythms; it also has an important endocrine function, producing releasing and some inhibiting hormones that act on the anterior pituitary and regulate the release of its hormones

Hypothermia - state of abnormally low body temperature

Hypothesis - proposition that seeks to place certain facts (or variables) within a construct that will explain or predict relationships between these facts. A prediction regarding the relationship between two variables is tested by conducting research if the findings offer support for the hypothesis, confidence in its accuracy may by increased, while if findings fail to offer such support, confidence in its accuracy may be reduced

Hypothyroidism - reduced thyroid activity

ICI 118.551 - β_2-selective adrenoceptor antagonist

Incentive - external stimulus which activates a motive

Independent variable - factor in an experiment which is varied by the researcher in a systematic manner in order to determine its effect(s) on the dependent variable

Indirect calorimetry - determination of energy expenditure by measuring oxygen consumption and possibly carbon dioxide production and converting it to heat production

Insulin - protein hormone formed by the β-cells of the pancreatic islets; lowers the glucose level of blood; it also influences lipid and amino acid metabolism

Insulin-like growth factor (IGF-1; somatomedin-C) - growth stimulating substance produced in the liver

Interaction - refers to instances in which the effects of one independent variable are determined or influenced by another

Intracardiac - within the heart

Intracerebral - within the cerebrum

Intragastric - within the stomach

Intraperitoneal - within the peritoneal cavity

Intraportal - within the portal vein

Intravenous - within or into a vein

Islets of Langerhans - microscopic endocrine structures dispersed throughout the pancreas; they consist of three cell types: the α-cells, which secrete glucagon; the β-cells, which secrete insulin; and the δ-cells, which secrete gastrin

Isoenergetic - containing the same amount of energy

Joule - SI unit of energy, 1 joule = 0.2388 calorie

Just noticeable difference, or difference threshold - sensitivity of the output to a particular input over the tested range. Threshold for detection of a difference

Ketonemia - presence of ketone bodies in the blood

Lactose intolerance - due to loss of enzyme lactase in the intestinal wall the lactose in milk cannot be digested and absorbed and it passes to the lower intestine where it is fermented by bacteria which creates discomfort and illness

Lateral hypothalamic areas - certain areas in the hypothalamus

Lean body mass - all tissues of the body, except for fat tissue

Limbic system - brain area playing a major role in emotional responses

Lipectomy - removal of fat

Lipogenesis - formation of fat

Lipolysis - breakdown of fat

Lipoprivation - decreased availability of free fatty acids

Lipostatic theory - amount of fat in the body, particularly the triglyceride stored in adipose tissue, is regulated by food intake and by energy use by the body

Luteinizing hormone (LH) - anterior pituitary hormone that stimulates maturation of cells in the ovary and acts on interstitial cells of the male testis

Luxuskonsumption - elevated energy expenditure in overfeeding; adaptation in energy expenditure following an increased food intake

Mechanoreceptors - receptors sensitive to mechanical pressures such as touch, sound, or contractions

Median - midpoint of a set of scores; 50 per cent of the scores fall above the median, 50 per cent below

Medulla oblongata - cone-shaped neural mass connecting the pons and the spinal cord

Melatonin - catecholamine hormone produced by the pineal gland

Menarche - onset of menstruation

Menopause - physiological termination of menstrual cycles

Metabolizable energy - gross energy minus energy in faeces and urine

Monoamine - amine with one amino group; the neurotransmitters serotonin, norepinephrine and dopamine are monoamines

Morphine - active ingredient of opium; a drug used for stopping pain and making people calmer

Motilin - hormone released by the gastro-intestinal tract

Motorneurons - nerve cells that innervate muscle cells

Motor cortex - part of the cerbral cortex that controls motor function

Mucoid (mucous) - pertaining to or containing mucus

Mucosa - mucous membrane facing a cavity or the exterior of the body

Mucous (mucoid) - pertaining to or containing mucus

Mucus - sticky, thick fluid secreted by mucous glands

Muscarinic receptors - one of several types of acetylcholine-sensitive binding sites; sensitive to muscarine but not to nicotine

Myogenic - having the potential to contract automatically without nervous stimulation

Naloxone - analog of morphine that acts as an opioid antagonist

Neuropeptide - peptide molecule identified as a neurotransmitter substance

Neurotensin - hormone released by the gastro-intestinal tract

Neurotransmitter - chemical mediator released by a presynaptic nerve ending that interacts with receptor molecules in the postsynaptic membrane; this process generally induces a permeability increase to an ion or ions and thereby influences the electrical activity of the postsynaptic cell

Nibbling - pattern of food intake characterized by a lot of small meals daily

Nicotine - stimulant drug contained in tobacco; addictive ingredient in tobacco smoke.

Nicotinic receptors - one of several types of acetylcholine-sensitive binding sites; sensitive to nicotine but not to muscarine

Non-protein respiratory quotient - respiratory quotient that is corrected for protein oxidation

Norepinephrine (noradrenalin) - neurohumor secreted by the peripheral sympathetic nerve terminals, some cells of the CNS, and the adrenal medulla; neurotransmitter released by the peripheral nerve endings of the sympathetic nervous system

Nucleus - group of nerve cell bodies, located in the central nervous system

Nucleus tractus solitarius - important nucleus belonging to the vagus nerve

Obesity (adiposity) - excessive accumulation of fat in the body

Olfactory - pertaining to the sense of smell

Opiate - opium-derived narcotic substance

Opioid - substance that exerts opiate-like effects

Orexia - appetite

Oropharyngeal - pertaining to the mouth and pharynx

Osmosis - passage (diffusion) of a solvent through a membrane from a dilute solution into a more concentrated one

Overnight metabolic rate - average metabolic rate from 23.30 - 8.00 h

Palatability - sensed food qualitities

Pancreas - gland located behind the stomach, between the spleen and the duodenum

Paracrinia - release of hormonal substances by endocrine cells into the surrounding extracellular fluid with diffusion to neighbouring 'target' cells (and not into the circulation)

Paralysis - loss of muscle function or sensation

Parasympathetic nervous system - craniosacral part of the autonomic system

Paraventricular nucleus - area in the hypothalamus

Pericaryon - cell body of a neuron

Perinatal - currently used to describe the weeks before a birth, the birth and the succeeding few weeks

Pharynx - cavity at the back of the mouth

Phentolamine - α-adrenergic antagonist

Physical activity index (PAI) - ratio of average daily metabolic rate and basal metabolic rate

Pica - eating earth

Pinealectomy - removal of the pineal body

Pineal body - small structure on the dorsal surface of the midbrain; its functions are not fully understood, but there is some evidence that it secretes melatonin which appears to inhibit secretion of the luteinizing hormone

Pituitary gland see hypophysis

Motive - acquired motivational system

Placebo effect - changes in behavior stemming from conditions or procedures which accompany, but are not directly related to, independent variables in an experiment. For example, changes in behavior following injections of a specific drug may result from the act of being injected, rather than from the drug itself

Plexus - network of vessels or nerves

Plexus of Auerbach - important plexus in the coordination of the peristaltic contractions of the stomach

Plexus of Meissner - important plexus in the coordination of the peristaltic contractions of the stomach

Poikilotherms - animals whose body temperature more or less follows the ambient temperature

Polydipsia - excessive thirst

Polyuria - excretion of an excessive amount of urine

Postganglionic - situated after a ganglion

Postpartum - after birth

Postprandial - following a meal

Postsynaptic - located distal to the synaptic cleft

Preabsorptive insulin response - rapid increase of insulin after oral food intake

Preference - selection of specific foods from alternatives

Preganglionic - preceding or in front of a ganglion

Prenatal - pertaining to the period between the last menstrual period and birth of the child

Preoptic area - important hypothalamic area in the monitoring of body temperature

Preprandial - before a meal

Pressor receptors - nerve endings sensitive to vessel stretching

Presynaptic - located proximal to the synaptic cleft

Prevalence - proportion of a population that has a disease or disorder at a specific point in time

Producing both endocrine and exocrine secretions

Progesterone - hormone of the corpus luteum responsible for preparing the uterus for the fertilized ovum

Prolactin - hormone secreted by the anterior pituitary that stimulates milk production and lactation

Prostaglandins - family of recently discoverd natural fatty acids that arise in a variety of tissues and are able to induce contraction in uterine and other smooth muscle, lower blood pressure, and modify the actions of some hormones

Pylorus - distal stomach opening, ringed by a sphincter, that releases the stomach contents into the duodenum

Raphe nuclei - nuclei in the lower brainstem involved in serotonergic projections on the hypothalamus

Regurgitation - backward flow, for example of stomach contents into, or through, the mouth

Reliable - extent to which a test or other measuring instrument yields consistent results

Respiratory quotient (RQ) - ratio between inspired oxygen and expired carbon dioxide during a specified time

Resting metabolic rate - rate of energy expenditure at rest measured under less standardized conditions as basal metabolic rate

Sacral cord - lower portion of the spinal cord

Satiation - process of the inhibition of food intake as a consequence of eating and the tendency to refuse food as a result of that process

Satiety - state or degree of inhibition of food intake as a result of ingestion;the situation after the process of satiation that follows food intake and before the desire for food occurs again

Secretin - polypeptide hormone secreted by the duodenal and jejunal mucosa in response to the presence of acid chyme in the intestine; it induces pancreatic secretion into the intestine

Serotonin hydroxytryptamine, 5-HT - neurotransmitter

Set point - in a negative feedback system, the state to which feedback tends to bring the system

Sham feeding - type of feeding in which ingested food leaves the esophagus of stomach via a fistula

Shivering thermogenesis - production of heat by shivering

Significant difference - statistical concept of probability, specifically, that the likelihood of a large difference occurring in the behavior of subjects in various groups of an experiment by chance alone is quite low. When such a difference does occur, it is assumed to reflect some aspect of the experimental manipulation of conditions and therefore (since chance is ruled out) can be used as a reliable basis for further work

Sleeping metabolic rate - rate of energy expenditure during sleeping usually measured from 3.00 h to 6.00 h

Somatic nervous system - part of the peripheral nervous system, also called the voluntary nervous system

Somatomedin - growth-stimulating substance produced in the liver

Somatostatin - growth hormone release-inhibiting factor

Somatotropic hormone see growth hormone

Specific hunger - nutrient selection

Specific dynamic action see diet-induced energy expenditure

Sphincter - circular muscle surrounding and enclosing an orifice

Sphincter of Oddi - sphincter localized where the common duct of pancreas and liver enter the doudenum

Splanchnic - pertaining to or supplying the viscera

Statistics - form of mathematic by which scientists evaluate the findings of their research. When employed appropriately, statistics permit investigators to determine whether the findings of their research are reliable or trustworthy (i.e. whether they would be likely to occur again if the research were repeated)

Stimulants - drugs such as caffeine, nicotine, and amphetamines which increase the functioning of the nervous system and thereby facilitate physical and mental activity

Stimulus - anything that elicits a physiological or psychological activity (i.e., a response).

Stimulus discrimination - ability to tell the difference between two or more stimuli

Stimulus generalization - tendency for a response conditioned to a particular stimulus to be elicited by similar stimuli as well

Subcutaneous - beneath the skin

Submucosa - layer of connective tissue beneath a mucous membrane

Suprachiasmatic nucleus - small area above the chiasma opticum involved in generating the circadian rhythm

Sympathetic nervous system - thoracolumbar part of the autonomic nervous system

Sympathoadrenal system - combined hormonal and neural outflow of adrenalin and noradrenalin

Thermic effect of food see diet-induced thermogenesis

Thermodynamics - study of the laws that govern the conversion of energy from one form to another, the direction in which heat will flow, and the availability of energy to do work

Thermogenesis - production of body heat by metabolic means such as muscle contraction during shivering

Thermoreceptor - receptor sensitive to temperature changes

Thermostatic theory - metabolic heat production is the variable regulated by the control of food intake

Thirst - dryness of the mouth; desire for fluid, motivation to acquire a drink

Thyroid gland - one of the largest of the body's endocrine glands; it secretes thyroxine. Thyroid glands raises the cellular metabolic rate, as does thyroxine

Thyroid stimulating hormone - adenohypophysial hormone that stimulates the secretory activity of the thyroid gland

Thyroxine - iodine-bearing, tyrosine-derived hormone that is synthesized and secreted by the thyroid gland; it raises cellular metabolic rate

Timolol - β-adrenoceptor antagonist

Torpor - state of inactivity, often with lowered body temperature and reduced metabolism, that some homeotherms enter into so as to conserve their energy stores

Total body electrical conductivity - the intrinsic property of the body to conduct electrical current

Trigeminus nerves (V) - cranial nerves; the major sensory nerves of the face

Triiodothyronine - iodine-bearing tyrosine derivative synthesized in and secreted by the

Tryptophan - one of the essential amino acids necessary for growth; it is a precursor of serotonin

Unconditioned response - response elicited by an unconditioned stimulus

Unconditioned stimulus - any stimulus possessing the capacity to elicit reactions from organisms in the absence of prior conditioning

Vagus nerve (X) - major cranial nerve that sends sensory fibers to the tongue, pharynx, larynx, and ear; motor fibers to the esophagus, larynx, and pharynx; and parasympathetic and afferent fibers to the viscera of the thoracic and abdominal regions

Valid - accurate; extent to which a test or other measuring instrument measures what it is supposed to measure

Variance - average squared distance of the scores of a distribution from the mean of that distribution

Vaso-inhibitory peptide - parasympathetic cotransmitter

Vasoconstriction - the narrowing of blood vessels

Vasodilatation - the relaxation of the smooth muscles of the vascular system producing dilated vessels

Vasopressin see antidiuretic hormone

Ventromedial hypothalamus - areas in the hypothalamus

Very low calorie diet - commercially prepared liquid formulations providing less than about 600 kcal/day and taken for 7 days or longer

Very low density lipoproteins - plasma lipoproteins mainly involved in triglyceride transport

Visceral - pertaining to the internal organs

Waist to hip girth ratio - anthropometric measurement giving an indication of fat distribution: in the abdominal or in the gluteofemoral region

Weber-Fechner Law - sensation increases arithmetically as stimulus increased geometrically; the least perceptible change in stimulus intensity above any background is proportional to the intensity of the background

Yohimbine - α_2-selective adrenergic antagonist

References

Abbott, W.G.H., B.V. Howard, G. Ruotolo and E. Ravussin, Energy expenditure in humans: effects of dietary fat and carbohydrate. *Am. J. Physiol.* 258 (1990), E347-51.

Acheson, K.J., I.T. Campbell, O.G. Edholm, D.S. Miller and M.J. Stock, The measurement of food and energy intake in man - an evaluation of some techniques. *Am. J. Clin. Nutr.* 33 (1980), 1147-54.

Acheson, K.J., J.P. Flatt and E. Jéquier, Glycogen synthesis versus lipogenesis after a 500 gram carbohydrate meal in man. *Metab.* 31 (1982), 1234-40.

Acheson, K.J., Y. Schutz, T. Bessard, E. Ravussin, E. Jéquier and J.P. Flatt, Nutritional influences on lipogenesis and thermogenesis after a carbohydrate meal. *Am. J. Physiol.* 246 (1984), E62-70.

Acheson, K.J., Y. Schutz, T. Bessard, K. Anantharaman, J.P. Flatt and E. Jéquier, Glycogen storage capacity and de novo lipogenesis during massive carbohydrate overfeeding in man. *Am. J. Clin. Nutr.* 48 (1988), 240-47.

Anand, B.K. and J.R. Brobeck, Hypothalamic control of food intake in rats and cats. *Yale J. Biol. Med.* 24 (1951), 123-46.

Anderson, G.H., Control of protein and energy intake: Role of plasma aminoacids and brain neurotransmitters. *Can. J. Physiol. Pharmacol.* 57 (1979), 1043-57.

Andersson, B. and B. Larson, Influence of local temperature changes in the preoptic area and rostral hypothalamus on the regulation of food and water intake. *Acta Physiol. Scand.* 52 (1961), 75-89.

Armstrong, S., A chronometric approach to the study of feeding behavior. *Neurosci. Biobehav. Rev.* 4 (1980), 27-53.

Astrup, A., S. Toubro, N.J. Christensen and F. Quaade, Pharmacology of thermogenic drugs. *Am. J. Clin. Nutr.* 55 (1992), 246S-248S.

Baker, B.J., J.P. Duggan, D.J. Barber and D.A. Booth, Effect of *dl*-fenfluramine and xylamidine on gastric emptying of maintenance diet in freely feeding rats. *Eur. J. Pharmacol.* 150 (1988), 137-42.

Balkan, B., A.B. Steffens, J.H. Strubbe and J.E. Bruggin, Biphasic insulin secretion after intravenous but not after intraportal CCK-8 infusion in rats. *Diabetes* 39 (1990), 702-06.

Baranowska, B., G. Rozbicka, W. Jeske and M.H. Abdel-Fattah, The role of endogenous opiates in the mechanism of inhibited luteinizing hormone (LH) secretion in women with anorexia nervosa: The effect of naloxone on LH, follicle-stimulating hormone, prolactin and beta-endorphin secretion. *J. Clin. Endocrinol. Metab.* 59 (1984), 412.

Basiotis, P.P., S.O. Welsh, F.J. Cronin, J.L. Kelsay and W. Mertz, Number of days of food intake records to estimate individual and group nutrient intakes with defined confidence. *J. Nutr.* 117 (1987), 1638-41.

Beaumont, P.J.V., M.S. Alami and S.W. Touys, The evolution of the concept of anorexia nervosa. *Handbook of Eating Disorders, Part 1. Anorexia and Bulimia Nervosa*, Eds. P.J.V. Beumont, G.D. Burrows and R.C. Casper. New York, Elsevier (1987), 105-38.

Bereiter, D.A., W.C. Engeland and D.S. Gann, Peripheral venous catecholamines versus adrenal secretory rates after brain stem stimulation in cats. *Am. J. Physiol.* 251 (1986) (*Endocrinol. Metab. 14*): E14-E20.

Billewicz, W.S., H.M. Fellowes and C.A. Hytten, Comments on the critical metabolic mass and the age of menarche. *Ann. Human Biol.* 3 (1976), 51.

Birch, L.L., Effects of peer models' food choices and eating behaviors on preschoolers' food preferences. *Child Dev.* 51 (1980), 489-96.

Birch, L.L., The acquisition of food acceptance patterns in children. *Eating Habits*. Eds. Boakes, R.A., M.J. Burton, D.A. Popplewell, Chichester: Wiley (1987), 107-30.

Birch, L.L. and M. Deysher, Caloric compensation and sensory specific satiety: evidence for the self regulation of food intake by young children. *Appetite* 7 (1986), 323-31.

Birch, L.L. and D.W. Marlin, I don't like it: never tried it: Effects of exposure on two-year-old children's food preferences. *Appetite* 3 (1982), 353-60.

Birch, L.L., S. Zimmerman and H. Hind, The influence of social affective context on preschool children's food preference. *Child Dev.* 51 (1980), 856-61.

Birch, L.L., J. Billman and S. Richards, Time of day influences food acceptability. *Appetite* 5 (1984), 109-16.

Blackburn, G.L., G.T. Wilson, B.S. Kanders, L.J. Stein, P.T. Lavin, J. Adler and K.D. Brownell, Weight cycling: the experience of human dieters. *Am. J. Clin. Nutr.* 49 (1989), 1105-09.

Blair A.J., V.J. Lewis and D.A. Booth, The relative success of offical and informal weight reduction techniques: retrospective correlation evidence. *Psychol. Health* 3 (1989), 195-206.

Blundell, J.E., Serotonin manipulations and the structure of feeding behavior. *Appetite* 7 (1986), 39-56.

Blundell, J.E., Appetite disturbances and the problems of the overweight. *Drugs* 39 (1990), Suppl. 3, 1-19.

Blundell, J.E. and V.J. Burley, Evaluation of the satiating power of dietary fat in man. *Progress in Obesity Research*, Eds Y. Oomura, S. Tarui, S. Inoue and T. Shimazu. London, John Libbey (1990), 453-58.

Blundell, J.E., P.J. Rogers and A.J. Hill, Artificial sweeteners and appetite in man. *Low Calorie Products*, Eds. Birch, G.G. and M.G. Lindley. London, Elsevier (1988).

Booth, D.A., Food preferences and nutritional control in rats and people. Annual Meeting of the British Association for the Advancement of Science (Exeter, UK), 1969, 90.

Booth, D.A, Conditioned satiety in the rat. *J. Comp. Physiol. Psychol.* 81 (1972a), 457-71.

Booth, D.A., Caloric compensation in rats with continuous or intermittent access to food. *Physiol. Behav.* 8 (1972b), 891-99.

Booth, D.A., Conditioned reactions in motivations. *Analysis of Motivational Processes*. Eds. F.M. Toates and T.R. Halliday. London, Academic Press (1980), 77-102.

Booth, D.A., Holding weight down: Physiological and psychological considerations. *Medicographia* 7(3), 1985, 22-25 & 52.

Booth, D.A., Central dietary "feedback onto nutrient selection": Not even a scientific hypothesis. *Appetite* 8 (1987a), 195-201.

Booth, D.A., Cognitive experimental psychology of appetite. *Eating Habits*. Eds. Boakes, R.A., M.J. Burton, D.A. Popplewell. Chichester, Wiley (1987b), 175-209.

Booth, D.A., Objective measurement of determinants of food preference: Sensory, physiological and psychosocial. *Food Acceptance and Nutrition*. Eds. J. Solms, D.A. Booth, R.M. Pagnborn and O. Raunhardt. London, Academic Press (1987c), 1-27.

Booth, D.A., Culturally corralled into food abuse: The eating disorders as physiologically reinforced excessive appetites. *The Psychobiology of Bulimia Nervosa*. Eds. K.M. Pirke, W. Vandereycken & D. Ploog. Berlin, Springer-Verlag (1988a), 18-32.

Booth, D.A. and J.D. Davis, Gastro-intestinal factors in the acquisition of oral sensory control of satiation. *Physiol. Behav.* 11 (1973), 23-29.

Booth, D.A. and A.M. Toase, Conditioning of hunger/satiety signals as well as flavor cues in dieters. *Appetite* 4 (1983), 235-36.

Booth, D.A., F.M. Toates and S.V. Platt, Control system for hunger and its implications in animals and man. *Hunger*. Eds. D. Novin, W. Wyrwicka and G.A. Bray. New York, Raven Press (1976), 127-142.

Booth, D.A, A.L. Thompson and B. Shahedian, A robust, brief measure of an individual's most prefered level of salt in an ordinary foodstuff. *Appetite* 4 (1983), 301-12.

Boulos, Z., A.M. Rosenwasser and M. Terman, Feeding schedules and the circadian organization of behavior in the rat. *Behav.Brain Res.* 1 (1980), 39-65.

Boyar, R.M., J. Katz, J.W. Finkelstein, S. Kapen, H. Weiner, E. Weitzman and L. Hellman, Anorexia nervosa: Immaturity of the 24-hour leutinizing hormone secretory pattern. *N. Eng. J. Med.* 291 (1974), 861.

Brala, P.M.and R.L. Hagen, Effects of sweetness perception and caloric value of a preload on short-term intake. *Physiol. Behav.* 30 (1983), 1-9.

References

Brazeau, P., W. Vale, R. Burgus, N. Ling, M. Butcher, J. Rivier and R. Guillemin, Hypothalamic polypeptide that inhibits the secretion of immunoreactive growth hormone. *Science* 179 (1973), 77-79.

Britton, D.R., G.F. Koob, J. Rivier and W. Vale, Intraventricular corticotropin-releasing factor enhances behavioral effects of novelty. *Life Sci.* 31 (1982), 363.

Brobeck, J.R., J. Tepperman and C.N.H. Long, Experimental hypothalamic hyperphagia in the albino rat. *Yale J. Biol. Med.* 15 (1943), 831-853.

Bronson, F.H., *Mammalian Reproductive Biology*. Chicago, Univ Chicago Press, 1989.

Brouns, F., W.H.M. Saris, J. Stroecken, E. Beckers, R. Thijssen, N.J. Rehrer and F. ten Hoor, Eating, drinking, and cycling. A controlled Tour de France simulation study. Part II. Effect of diet manipulation. *Int J Sports Med.* 10 (1989), S41-S48.

Brouwer, E., On simple formulae for calculating the heat expenditure and the quantities of carbohydrate and fat oxidized in metabolism of men and animals from gaseous exchange (oxygen intake and carbonic acid output) and urine-N. *Acta. Physiol. Neerl.* 6 (1957), 795-802.

Brown, M.R., L.A. Fisher, J. Spiess, C. Rivier, J. Rivier and W. Vale, Corticotropin-releasing factor: actions on the sympathetic nervous system and metabolism. *Endocrinology* 111 (1982), 928.

Bruce, D.G., L.H. Storlien, S.M. Furler and D.J. Chisholm, Cephalic phase metabolic responses in normal weight adults. *Metabolism* 36 (1987), 721-8.

Bullen, B.A., G.S. Skrinar, I.A. Beitins, G. von Mering, B.A. Turnbull and J.W. McArthur, Induction of menstrual disorders by strenuous exercise in untrained women. *N. Eng. J. Med.* 312 (1985), 1349.

Burley, V.J., A.R. Leeds and J.E. Blundell, The effect of high and low-fibre breakfasts on hunger, satiety and food intake in a subsequent meal. *Int. J. Obes.* 11 (1987), Suppl. 1, 87-93.

Buwalda, B., C. Nyakas, J.M. Koolhaas and B. Bohus, Neuroendocrine and behavioral effects of vasopressin in resting and mild stress conditions. *Physiol and Behav.*, 1993.

Cabanac, M., The phsyiological role of pleasure. *Science* 173 (1971), 1103-07.

Cahill, G.F., Jr., Starvation in man. *N. Eng. J. Med.* 282 (1970), 668.

Cameron, N., Weight and skinfold variation at menarche and the critical body weight hypothesis. *Ann. Human Biol.* 3 (1976), 279.

Cameron, J.L. and C. Nosbisch, Suppression of pulsatile LH & test secretion during short-term food restriction in the adult male rhesus monkey, *J. Clin. Endocrinol. Metab.*, 1990.

Cameron, J.L., T. Weltz, C. McConaha, D.L. Heilmreich and W.H. Kaye, Slowing of pulsatile LH secretion in man after 48 hrs fasting, *J. Clin. Endocrinol. Metab.*, 1990.

Campfield, L.A. and F.J. Smith, Systemic factors in the control of food intake. Evidence for patterns as signals. *Handbook of Behavioral Neurobiology, vol 10. Neurobiology of Food and Fluid Intake*, Ed. E.M. Stricker. New York, Plenum Press (1990), 183-203.

Clutton-Brock, T.H., F.E. Guinness and S.D. Albon, *Red deer: Behavior and Ecology of Two Sexes*. Edinburgh, Edinburgh Univ. Press, 1982.

Colditz, G.A., E. Giovannucci, E.B. Rimm, M.J. Stampfer, B. Rosner, F.E. Speizer, E. Gordis and W.C. Willet, Alcohol intake in relation to diet and obesity in women and men. *Am. J. Clin. Nutr.* 54 (1991), 49-55.

Conner, M.T., A.V. Haddon, E.S. Pickering and D.A. Booth, Sweet tooth demonstrated: individual differences in preference for both sweet foods and foods highly sweetened. *J. Appl. Psychol.* 7 (1988), 275-80.

Conte, F.A., M.M. Grumbach, S.L. Kaplan and E.O. Reiter, Correlation of LRF. Induced LH and FSH release from infancy to 19 years with the changing pattern of gonadotropin secretion in agonadal patients: Relation to the restraint of puberty. *J. Clin. Endocrin. Metab.* 50 (1981), 1163.

Crisp, A.H., The psychopathology of Anorexia nervosa: Getting the 'heat' out of the system. *Eating and its Disorders*, Eds. A.J. Stunkard and E. Stellar. New York, Raven Press (1984), 209-234.

Cumming, D.C. and R.W. Rebar, Exercise and reproductive function in women. *Am. J. Industr. Med.* 4 (1983), 113.

Cumming, D.C., M.M. Vickovic, S.R. Wall and M.R. Fluker, Defects in pulsatile LH release in normally menstruating runners. *J. Clin. Endocrinol. Metab.* 60 (1985a), 810.

Cumming, D.C., M.M. Vickovic, S.R. Wall, M.R. Fluker and A.N. Belcastro, The effects of acute exercise on pulsatile release of luteinizing hormone in women runners. *Am. J. Obs. Gyn.* 153 (1985b), 482.

Dallosso, H.M. and W.P.T. James, Whole-body calorimetry studies in adult men: 1. The effect of fat over-feeding on 24 h energy expenditure. *Br. J. Nutr.* 52 (1984), 49-64.

Davies, R.F., Long and short-term regulation of feeding patterns in the rat. *J. Comp. Physiol. Psychol.* 91 (1977), 574-485.

Debons, A.F., I. Krimsky, H.J. Likusky, A. From and R.J. Cloutier, Goldthioglucose damage to the satiety center: inhibition in diabetes. *Am. J. Physiol.* 214 (1968), 652-58.

Deutsch, J.A., Bombesin - satiety or malaise? *Nature* 285 (1980), 592.

Deutsch, J.A., Dietary control and the stomach. *Prog. Neurobiol.* 20 (1983), 313-32.

Deutsch, J.A., Food intake: Gastric factors. *Handbook of Behavioral Neurobiology, vol 10. Neurobiology of Food and Fluid Intake*, Ed. E.M. Stricker. New York, Plenum Press (1990), 151-178.

De Jong, A., J.H. Strubbe and A.B. Steffens, Hypothalamic influence on insulin and glucagon release in the rat. *Am. J. Physiol.* 233 (1977), 380-88.

Ding, J.R., C.B. Schekter, B.L. Drinkwater, M.R. Soules and W.J. Bremmer, High serum cortisol levels in exercise-associated amenorrhea. *Ann. Intern. Med.* 108 (1988), 530.

Drent, R.H. and S. Daan, The prudent parent: energetic adjustments in avian breeding. *Ardea* 68 (1980), 225-252.

Drent, R.H. and H.H.T. Prins, The herbivore as prisoner of its food supply. *Disturbance in Grasslands.* Eds. Andel, J. van et al. Dordrecht, Dr. Junk (1987), 133-49.

Dreon, D.M., B. Frey-Hewitt, N. Ellsworth, P.T. Williams, R.B. Terry and P.D. Wood, Dietary fat:carbohydrate ratio and obesity in middle-aged men. *Am. J. Clin. Nutr.* 47 (1988), 995-1000.

Drewnoski, A., Food palatability: The contribution of sugar and fat. *Progress in Obesity Research*, Eds. Y. Oomura, S. Tarui, S. Inoue and T. Shimazu. London, John Libbey (1990), 459-66.

Drewnoski, A., E.E. Shrager, C. Lipsky, E. Stellar and M.R.C. Greenwood, Sugar and fat: sensory and hedonic evaluations of liquid and solid foods. *Physiol. Behav.* 45 (1989), 177-83.

Driver, C.J.I., The effect of meal composition on the degree of satiation following a test meal and possible mechanisms involved. *Brit. J. Nutr.* 60 (1988), 441-9.

Duester, P.A., S.B. Kyle, P.B. Moser, R.A. Vigersky, A. Singh and E.B. Schoomaker, Nutritional survey of highly trained women runners. *Am. J. Clin. Nutr.* 45 (1986), 954.

Duncan, K.H., J.A. Bacon, and R.L. Weinsier, The effects of high and low energy density diets on satiety, energy intake, and eating time of obese and nonobese subjects. *Am. J. Clin. Nutr.* 37 (1983), 763-7.

Dunning B.E. and G.J. Taborsky Jr., Neural control of islet function by norepinephrine and sympathetic neuropeptides. *Fuel Homeostasis and the Nervous System.* Ed. M. Vranic et al. New York, Plenum Press, 1991.

Durnin, J.V.G.A. and J. Womersley, Body fat assessed from total body density and its estimation from skinfold thickness measurement on 481 men and women aged 16 to 72 years. *Br. J. Nutr.* 32 (1974), 77-97.

Ebbing, F.J., R.I. Wood, F.J. Karsch, L.A. Vannerson, J.M. Suttie, D.C. Bucholtz, R.E. Schall and D.L. Foster, Metabolic interfaces between growth and reproduction. III. Central mechanisms controlling pulsatile luteinizing hormone secretion in the nutritionally growth-limited female lamb. *Endocrinology* 126 (1990), 2719.

Ebbinge, B.S., A multifactoral explanation for variation in breeding performance of Brent Geese *(Branta bernicla)*. *Ibis* 131 (1992a), 196-204.

Ebbinge, B.S., *Population Limitation in Artic-breeding Geese*. PhD thesis State Univ. Groningen, 1992b.

Eckert, R., Neural processing and behavior. *Animal Physiology. Mechanisms and adaptations. 3rd ed.* Eds. Eckert, R. and D. Randall. New York, W.H. Freeman (1988), 219-66.

Eckert, R., Feeding, Digestion and Absorption. *Animal Physiology. Mechanisms and Adaptations. 3rd ed.* Eds. Eckert, R. and D. Randall. New York, W.H. Freeman (1988), 520-55.

Eckert, R., Animal energetics and temperature relations. *Animal Physiology Mechanisms and Adaptations. 3rd ed.* Eds. Eckert, R. and D. Randall. New York, W.H. Freeman (1988), 555-601.

Edholm, O.G., J.G. Fletcher, E.M. Widdowson and R.A. McCance, The energy expenditure and food intake of individual men. *Br. J. Nutr.* 9 (1955), 286-300.

Engelhardt, D. von (ed.), *Diabetes. Its Medical and Cultural History*. Berlin, Springer-Verlag, 1989.

Epstein, A.N., Oropharyngeal factors in feeding and drinking. *Handbook of Physiology section 6*

References

Alimentary canal Vol 1 Am. Physiol Soc., Washington D.C., 1967, 197-218.

Esler, M., G. Jennings, Leonard, N. Sacharias, F. Burke, J. Johns and Blomberry, Contribution of individual organs to total noradrenalin release in humans. *Acta Physiol. Scand.* 527, Suppl. 11-16, 1984a.

Esler, M., I. Willet, Leonard, G. Hasking, J. Johns, P. Little and G. Jennings, Plasma noradrenalin kinetics in humans. *J. Auton. Nerv. Syst.* 11 (1984), 125-44.

Fábry, P., Metabolic consequences of the pattern of food intake. *Handbook of Physiology*, vol. 1, *Control of Food and Water Intake; Section 6, Alimentary canal*, Eds. C.F. Code and W. Heidel. American Physiological Society, 1967, 31-50.

Fábry, P., *Feeding Pattern and Nutritional Adaptations*. Prague, Academia, 1969.

Falk, P., The effect of elevated temperature on the vasculature of mouse jejunum. *Br. J.Radiol.* 56 (1983), 41-49.

Ferin, M., D. Van Vugt and S. Wardlaw, The hypothalamic control of the menstrual cycle and the role of endogenous opioid peptides. *Recent. Prog. Hormone Res.* 39 (1984), 46.

Fernstrom, J.D., Carbohydrate ingestion and brain serotonin synthesis: Relevance to a putative control loop for regulating carbohydrate ingestion, and effects of aspartame consumption. *Appetite* 11 (1988), 35-41.

Fisher, E.C., M.E. Nelson, W.R. Frontera, R.N. Turksoy and W.J. Evans, Bone mineral content and levels of gonadotropins and estrogens in amenorrheic running women. *J. Clin. Endocrinol. Metab.* 62 (1986), 1232.

Fishman, J., Appetite and sex hormones. *Appetite and food intake*. Ed. T. Silverstone. Berlin, Dahlem Konfrerenzen. Life science research report no 2 (1976), 207-218.

Fishman, J., R.M. Boyar and L.Hellman, Effect of body weight on estradiol metabolism in young women. *J.Clin. Endocrinology* 41 (1975), 989-991.

Flatt, J.P., The difference in the storage capacities for carbohydrate and for fat, and its implications in the regulation of body weight. *Ann. N.Y. Acad. Sci.* 499 (1987), 104-23.

Flatt, J.P., E. Ravussin, K.J. Acheson and E. Jéquier, Effects of dietary fat on postprandial substrate oxidation and on carbohydrate and fat balance. *J. Clin. Invest.* 76 (1985), 1019-24.

Foltin, R.W., M.W. Fischman, T.H. Moran, B.J. Rolls and T.H. Kelly, Caloric compensation for lunches varying in fat and carbohydrate content by humans in a residential laboratory. *Am. J. Clin. Nutr.* 52 (1990), 969-80.

Fredrix, E.W.H.M., P.B. Soeters, I.M. Deerenberg, A.D.M. Kester, M.F. von Meyenfeldt and W.H.M. Saris, Resting and sleeping energy expenditure in the elderly. *Eur. J. Clin. Nutr.* 44 (1990), 741-47.

Friedman, M.I. and M. Stricker, The physiological psychology of hunger: A physiological perspective. *Psychol. Rev.* 83 (1976), 409-31.

Frisch, R.E., Weight at menarche: similarity for well-nourished and undernourished girls at differing ages, and evidence for historical constance. *Pediatrics* 50 (1972), 445.

Frisch, R.E., Fatness of girls from menarche to age 18 with a nomogram. *Hum. Biol.* 48 (1976), 353.

Frisch, R.E., Body fat, puberty and fertility. *Biol. Rev. (Lond.)* 59 (1984), 164.

Frisch, R.E., Fatness and fertility. *Sci. Am.*, March 1988, 88.

Frisch, R.E., Body fat, menarche, fitness and fertility. *Adipose Tissue and Reproduction Prog. Reprod. Biol. Med.* Ed. Frisch, R.E., Basel, Karger (1990), 14:1.

Frisch, R.E. and J.W. McArthur, Menstrual cycles: Fatness as a determinant of minimum weight for height necessary for their maintenance or onset. *Science* 185 (1974), 949.

Frisch, R.E. and R. Revelle, Height and weight at menarche and a hypothesis of critical body weights and adolescent events. *Science* 169 (1970), 397.

Frisch, R.E. and R. Revelle, Height and weight at menarche and a hypothesis of menarche. *Arch. Dis. Child* 46 (1971), 695.

Frisch, R.E., R. Revelle and S. Cook, Components of weight at menarche and the initiation of the adolescent growth spurt in girls: estimated total water, lean body weight and fat. *Hum. Biol.* 45 (1973), 469.

Frisch, R.E., G. Wyshak and L. Vincent, Delayed menarche and amenorrhea in ballet dancers. *N. Eng. J. Med.* 303 (1980), 17.

Frisch, R.E., A.V. Gotz-Welbergen, J.W. McArthur, Delayed menarche and amenorrhea of college athletes in relation to age of onset of training. *JAMA* 246 (1981), 1559.

Gaines, M.S. and R.K. Rose, Population dynamics of *Microtus ochrogaster* in eastern Kansas. *Ecology* 56 (1976), 1145-61.

Galbo, H., *Hormonal and Metabolic Adaptation to Exercise*. Stuttgart, Thieme, 1983.

Gardemann, A., G.P. Püschel and K. Jungermann, Nervous control of liver metabolism and hemodynamics. *Eur. J. Biochem.*, 1993.

Garrow, J.S., *Treat Obesity Seriously, a Clinical Manual*. Edinburgh, Churchill Livingstone, 1981.

Geliebter, A.A., Effects of equicaloric loads of protein, fat and carbohydrate on food intake in the rat and man. *Physiol. Behav.* 22 (1979), 267-73.

Geracioti, T.D. and R.A. Liddle: Impaired cholecystokinin secretion in bulimia nervosa. *N. Engl. J. Med.* 319 (1988), 683-88.

Gibbs, J. and R.C. Young and G.P. Smith, Cholecystokinin decreases food intake in rats. *J. Comp. Physiol. Psychol.* 84 (1973), 488-95.

Gibson, E.L., D.A. Booth, Acquired protein appetite in rats: dependence on a protein-specific need state. *Experientia* 42 (1986), 1003-4.

Gibson, E.L. and D.A. Booth, Dependence of carbohydrate conditioned flavor preference on internal state in rats. *Learn. Motiv.* 20 (1989), 36-47.

Gold, P.W., H. Swirtsman and P.C. Avgerinos, et al., Abnormal hypothalamic-pituitary-adrenal function in anorexia nervosa: pathophysiologic mechanisms in underweight and weight-corrected patients. *N. Eng. J. Med.* 314 (1986), 1335.

Goldberg, G.R., A.M. Prentice, H.L. Davies and P.R. Murgatroyd, Overnight and basal metabolic rates in men and women. *Eur. J. Clin. Nutr.* 42 (1988), 137-44.

Goldstein, D.S., Catecholamines in plasma and cerebrospinal fluid: sources and meanings. *Brain Peptides and Catecholamines in Cardiovascular Regulation*. Eds. J.P. Buckley and C.M. Ferrario. New York, Raven (1987), 15-25.

Grave, G.D. and F.E. Mayer (eds.), *Control of Onset of Puberty*. New York, John Wiley, 1974, 115.

Grumbach, M.M., The neuroendocrinology of puberty. *Neuroendocrinology*, Eds. Krieger, D.T. and J.C. Hughes. Sinauer Associates, Sunderland, Mass., 1980, 249.

Grumbach, M.M., J.C. Roth, S.L. Kaplan and R.P. Kelch, Hypothalamic-pituitary regulation of puberty in man: evidence and concepts derived from clinical research. *Control of Onset of Puberty*, Eds. Grumbach, M.M., G.D. Grave and F.E. Mayer. New York, John Wiley, 1974, 115.

Guy-Grand, B., Usefullness and limits of dexfenfluramine in the treatment of obesity. *Progress in Obesity Research*, Eds. Y. Oomura, S. Tarui, S. Inoue and T. Shimazu. London, John Libbey, 1990, 575- 580.

Guyton A.C., *Textbook of Medical Physiology*. Seventh edition. Philadelphia, W.B. Saunders, 1986.

Harris, J.A. and F.G. Benedict, *A Biometric Study of Basal Metabolism in Man*. Washington, Carnegie Institution, 1919.

Hetherington, A.W. and S.W. Ranson, Hypothalamic lesions and adiposity in the rat. *Anat.Rec.* 78 (1940), 149-72.

Hill, A.J., J.E. Blundell, Macronutrients and satiety: the effect of a high-protein or high-carbohydrate meal on subjective motiviation to eat and food preferences. *Nutr. Behav.* 3 (1986), 133-44.

Hill, A.J., L.D. Magson, J.E. Blundell, Hunger and palatability: tracking ratings of subjective experience before, during and after the consumption of preferred and less preferred food. *Appetite* 5 (1984), 361-71.

Hill, A.J., P.D. Leathwood and J.E. Blundell, Some evidence for short-term caloric compensation in normal weight human subjects: the effect of high- and low-energy meals on hunger, food preference and food intake. *Hum. Nutr. Clin. Nutr.* 41A (1987), 244-57.

Hill, J.O., J.C. Peters, G.W. Reed, D.G. Schlundt, T. Sharp and H.L. Greene, Nutrient balance in humans: effects of diet composition. *Am J Clin Nutr* 54 (1991), 10-7.

Hohlweg, W. and M. Dorhn, Uber die Beziehungen zwischen Hypophysen Vorderlappen und Keimdrusen. *Klin. Wochenschr.* 11 (1932), 233.

Hollenga, C. and J. Zaagsma, Direct evidence for the atypical nature of functional β-adrenoreceptors in rat adipocytes. *Br. J. Pharmacol.* 87 (1989), 1420-24.

Hsu, L.K.G., *Eating Disorders*. New York, Guilford Press, 1990.

Hunri, M., B. Burnand, Ph. Pittet and E. Jequier, Metabolic effects of a mixed and a high-carbohydrate low-fat diet in man, measured over 24h in a respiration chamber. *Br. J. Nutr.* 47 (1982), 33-43.

James, W.P.T., Appetite control and other mechanisms of weight homeostasis. *Nutritional Adaptation in Man.* Eds. Blaxter, K. and J.C. Waterlow. Londen, John Libbey, 1985, 141-54.

Janssen, G.M.E., C.J.J. de Graef and W.H.M. Saris, Food intake and body composition in novice athletes during a training period to run a marathon. *Int. J. Sports Med.* 10 (1989), S17-S21.

Jiang, C-L. and J.N. Hunt, The relation between freely chosen meals and body habitus. *Am. J. Clin. Nutr.* 38 (1983), 32-40.

Jimerson, D.C., M.D. Lesem, W.H. Kaye W.H., A.P.Hegg and T.D. Brewerton, Eating disorders and depression: Is there a serotonin connection? *Biol. Psychiatry* 28 (1990), 443-54.

Johnston, F.E., A.F. Roche, L.M. Schell and N.B. Wettenhall, Critical weight at menarche. *Am. J. Dis. Child* 129 (1975), 19.

Judd, F.K., T.R. Norman and G.D. Burrows, Pharmacotherapy in the treatment of anorexia and bulimia nervosa. *Handbook of Eating Disorders Part 1. Anorexia and Bulimia* Eds P.J.V. Beumont, G.D. Burrows and R.C. Casper, New York, Elsevier, 1987, 361-80.

Kandell, E.R., J.H. Schwartz, and T.M. Jessell, *Principles of Neural Sciences.* Third Edition. New York, Elsevier, 1991.

Kaneto, A, K. Kosaka and K. Nakao, Effects of stimulation of the vagus nerve on insulin secretion. *Endocrinology* 80 (1967), 530-36.

Kaneto, A., E. Miki and K. Kosaka, Effects of vagal stimulation on glucagon and insulin secretion. *Endocrinology* 795 (1974), 1005-10.

Kapen, S., E. Sternthal and L. Braverman, Case Report: A pubertal 24-hour luteinizing hormone (LH) secretory pattern following weight loss in the absence of anorexia nervosa. *Psychosom. Med.* 43 (1981), 177.

Kaye, W.H., H.E. Gwirtsman, D.T. George, *et al.,* Elevated cerebrospinal fluid levels of immunoreactive corticotropin-releasing hormone in anorexia nervosa relation to state of nutrition, adrenal function, and intensity of depression. *J. Clin. Endocrinol. Metab.* 64 (1987), 203.

Keesey, R.E., P.C. Boyle, J.W. Kemnitz and J.S. Mitchel, The role of the lateral hypothalamus in determining the body weight set-points. *Hunger: Basic Mechanisms and Clinical Applications,* Eds. D. Novin, W. Wyrwicka and G. Bray. New York, Raven press, 1976, 243-53.

Kersten A., J.H. Strubbe and N. Spiteri, Meal patterning of rats with changes in day length and food availability. *Physiol.Behav.* 25 (1980), 953-58.

Keys, A., J. Brozek, A. Henschel, O. Mickelsen and H.L. Taylor, *The Biology of Human Starvation.* Minneapolis, University of Minnesota Press, 1950.

Khoury, S., N. Reame, R.P. Kelch and J.C. Marshall, Reduced GnRH pulsatility in hypothalamic amenorrhea: Consistency of diurnal patterns and effects of opiate blockade and alpha adrenergic stimulation. *Clin. Res.,* 33 (1985), 875A.

Kirkwood, J.K., A limit to metabolisable energy intake in mammals and birds. *Comp. Biochem. Physiol.* 75A (1983), 1-3.

Kissileff, H.R. and T.B. Van Itallie, Physiology of the control of food intake. *Ann. Rev. Nutr.* 2 (1982), 371-419.

Kooy, K. van der, R. Leenen, Deurenberg, J. Seidell, K. Westerterp and J.G.A.J. Hautvast, Critical evaluation of 5 methods to assess changes in body composition in moderately obese subjects. *Int. J. Obes.* 1993.

Korte, S.M., S. van Duin, G.A.H. Bouws, J.M. Koolhaas, B. Bohus, Involvement of hypothalamic serotonin in activation of the sympathoadrenomedullary system and hypothalamo-pituitary-adrenocortical axis in male Wistar rats. *Eur. J. Pharmacol.* 197 (1991), 225-228.

Krahn, D.D., B.A. Gosnell, M. Grace, A.S. Levine. CRF antagonist partially reverses CRF- and stress-induced effects on feeding. *Brain Res. Bull.* 17 (1986), 285.

Krahn, D.D., B.A. Gosnell, A.S. Levine and J.E. Morley, Behavioral effects of corticotropin-releasing factor: localization and characterization of central effects. *Brain Res.* 443 (1988), 63.

Langer, S.Z., Presynaptic regulation of the release of catecholamines. *Pharmacol. Rev.* 32(4), 1981, 337-62.

Langer, S.Z. and Arbilla, Presynaptic receptors on peripheral noradrenergic neurons. *Ann. N.Y. Acad. Sci.* 604 (1990), 7-16.

Lean, M.E.J. and W.P.T. James, Metabolic effects of isoenergetic nutrient exchange over 24 hours in relation to obesity in women. *Int. J. Obes.* 12 (1988), 15-27.

Leblanc, J., L. Brondel, Role of palatability on meal-induced thermogenesis in human subjects. *Am. J. Physiol.* 248 (1985), E333-36.

Levine, A.S., B. Rogers, J. Kneip, M. Grace and J.E. Morley, Effect of centrally administered corticotropin releasing factor (CRF) on multiple feeding paradigms. *Neuropharmacology* 22 (1983), 337.

Le Magnen, J., Le processus de discrimination par le rat blanc des stimuli sucrés alimintaires et non altimentaires. *J. Physiol., Paris* 46 (1954), 414-18.

Le Magnen, J., Effets sur la prise alimentaire du rat blanc des administrations post-prandiales d'insuline et le mecanisme des appetits. *J. Physiol., Paris* 48 (1956), 789-802.

Le Marchand, L., L.N. Kolonel, J.H. Hankin and C.N. Yoshizawa, Relationship of alcohol consumption to diet: a population-based study in Hawaii. *Am. J. Clin. Nutr.* 49 (1989), 567-72.

Leathwood, P.D. and D.V.M. Ashley, Behavioral strategies in the regulation of food choice. *Experientia* 44 (1984), Suppl., 171-96.

Leibe, R.L. and J. Hirsch, Diminished energy requirements in reduced-obese patients. *Metabolism* 33 (1984), 164-70.

Leibowitz, S.F and G. Shor-Posner, Brain serotonin and eating behavior. *Appetite* 7 (1986), 1-14.

Leibowitz, S.F., G.F. Weiss and G. Shor-Posner, Hypothalamic serotonin: pharmacological, biochemical, and behavioral analyses of its feeding-suppressive action. *Clin. Neuropharmacol.* 11 (1988), suppl. 1, 51-71.

Lenz, H.J., A. Raedler, H. Greten and M.R. Brown, CRF initiates biological actions within the brain that are observed in response to stress. *Am. J. Physiol.* 252 (1987), R34.

Levine, A.S. *et al.*, Effect of breakfast cereals on short-term food intake. *Am. J. Clin. Nutr.* 50 (1989), 1303-7.

Lissner, L., D.A. Levitsky, B.J. Strupp, H.J. Kalkwarf and D.A. Roe, Dietary fat and the regulation of energy intake in human subjects. *Am. J. Clin. Nutr.* 46 (1987), 886-92.

Luiten, P.G.M., G.J. ter Horst and A.B. Steffens, The hypothalamus, intrinsic connections and outflow pathways to the endocrine system in relation to the control of feeding and metabolism. *Prog. Neurobiol.* 28 (1987), 1-54.

Lundgren, O., Vagal control of the motor functions of the lower esophageal sphincter and the stomach. *Vagal Nerve Function: Behavioral and Methodological Considerations.* Eds. J.G. Kral, T.L. Powley and C. McC.Brooks. Amsterdam, Elsevier, 1983, 185-98.

Malina, R.M., W.W. Spirduso, C. Tate and A.M. Baylor, Age at menarche and selected menstrual characteristics in athletes at different competitive levels and in different sports. *Med. Sci. Sport Exerc.* 10 (1978), 218.

Marcus, R., Cann, C., Madvig, P., *et al.*, Menstrual function and bone mass in elite women distance runners. *Ann. Intern. Med.* 102 (1985), 158.

Marken-Lichtenbelt, W. van, Digestion in an ectothermic herbivore, the green iguana *(Iguana iguana)*: effect of food composition and body temperature. *Physiol. Zool.* 65 (1992), 649-73.

Marshall, J.C. and R.P. Kelch, Low dose pulsatile gonadotropin releasing hormone in anorexia nervosa: a model of human pubertal development. *J. Clin. Endocrinol.* 49(5), 1979, 712.

Marshall, J.C. and R.P. Kelch, Gonadotropin-releasing hormone: Role of pulsatile secretion in the regulation of reproduction. *N. Eng. J. Med.* 31 (1986), 1459.

Mattes, R.D., Effects of aspartame and sucrose on hunger and energy intake in humans. *Physiol. Behav.* 47 (1990), 1037-44.

Mayer, J., Regulation of energy intake and body weight, the glucostatic theory and the lipostatic hypothesis. *Ann. N.Y. Acad. Sci.* 63 (1955), 15-43.

McArthur, J.S., B.A. Bullen, I.Z. Beitins, M. Pagano, T.M. Badger and A. Klibanski, Hypothalamic amenorrhea in runners of normal body distribution. *Endocrine Res. Comm.* 7 (1980), 13.

McHugh, P.R., The control of gastric emptying. *Vagal Nerve Function. Behavioral and Methodological Considerations.* Eds. Kral J.G., T.L. Powly and C. McC. Brooks. Amsterdam, Elsevier, 1983, 221-32.

Meijer, G.A.L., K.R. Westerterp, W.H.M. Saris and F. ten Hoor, Sleeping metabolic rate in relation to body composition and the menstrual cycle. *Am. J. Clin. Nutr.* 55 (1992), 637-40.

Miller, W.C., A.K. Lindeman, J. Wallace and M. Niederpruem, Diet composition, energy intake, and exercise in relation to body fat in men and women. *Am. J. Clin. Nutr.* 52 (1990) 426-30.

Mitchell, J.E., D.E. Laine, J.E. Morley and A.S. Levine, Naloxone but not CCK-8 may attenuate binge-eating behavior in patients with the bulimia syndrome. *Biol. Psychiatry* 21 (1986), 1399-1406.

Mitchell, J.E., R.L. Pyle, E.D. Eckart, D. Hatchukami, C. Pomeroy and R. Zimmerman, A comparison study of antidepressants and structured intensive group psychotherapy in the treatment of bulimia nervosa. *Arch. Gen. Psychiatry* 47 (1990), 149-57.

Mook, D.G., On the organization of satiety. *Appetite* 11 (1988), 27-39.

Morley, J.E., Neuropeptide regulation of appetite and weight. *Endocrine Rev.* 8 (1987), 256-79.

Morton-Boyd, J. and P.A. Jewell, The Soay sheep and their environment: a synthesis. *The Ecology of the Soay Sheep of St Kilda.* London, Athlone Press, 1974.

Nelson, M.E., E.C. Fisher, C.N. Meredith, R.N. Turksoy and W.J. Evans, Diet and bone status in amenorrheic runners. *Am. J. Clin. Nutr.* 43 (1986), 910.

Newman, J.C. and D.A. Booth, Gastro-intestinal and metabolic consequences of a rat's meal on maintenance diet ad libitum. *Physiol. Behav.* 27 (1981), 929-939.

Nieuwenhuijs, R., *Chemoarchitecture of the brain.* New York, Springer Verlag, 1985.

Niijima, A., Nervous regulation of metabolism. *Prog. Neurobiol.* 33 (1989), 135-47.

Nimrod, A. and K.J. Ryan, Aromatization of androgens by human abdominal and breast fat tissue. *J. Clin. Endocrinol. Metab.* 40 (1975), 367.

Norgan, N.G. and J.V.G.A. Durnin, The effect of 6 weeks of overfeeding on the body weight, body composition, and energy metabolism of young men. *Am. J. Clin. Nutr.* 33 (1980), 978-88.

Oatly K., Dissociation of the circadian drinking pattern from eating. *Nature* 229 (1971), 494-96.

Ogden, J. and J. Wardle, Cognitive restraint and sensitivity to cues for hunger and satiety. *Physiol. Behav.* 47 (1990), 477-81.

Olster, D.H. and M. Ferin, Corticotropin-releasing hormone inhibits gonadotropin secretion in the ovariectomized Rhesus monkey. *J. Clin. Endocrinol. Metab.* 65 (1987), 262.

Oomura, Y., Glucose as a regulator of neuronal activity. *Advances in Metabolic Disorders, vol. 10. CNS Regulation of Carbohydrate Metabolism.* Ed. A.J.Szabo. New York, Academic Press, 1983, 31-67.

Parfitt, D.B., K.R. Church and J.L. Cameron, Fasting-induced suppression of pulsatile LH secretion is rapidly reversed by refeeding in rhesus monkeys. *Soc. Neurosci. Abstr.*, 1990, Abstract 168.8.

Porikos, K.P., M.F. Hesser and T.B. Van Itallie, Caloric regulation in normal-weight men maintained on a palatable diet of conventional foods. *Physiol. Behav.* 29 (1982), 293-300.

Porte, D, β-Cells in Type II diabetes mellitus. *Diabetes* 40(1991), 166-80.

Prentice, A. M., A. E. Black, W. A. Coward, H. L. Davies, G. R. Goldberg, R. Murgatroyd, J. Ashford, M. Sawyer and R. G. Whitehead, High levels of energy expenditure in obese women. *Br. Med. J.* 292 (1986), 983-87.

Prentice, A.M., R.G. Whitehead, S.B. Roberts and A.A. Paul, Long-term energy balance in child-bearing Gambian women. *Am. J. Clin. Nutr.* 34 (1981), 2790-99.

Prewitt, T.E., D. Schmeisser, P.E. Bowen, Aye, T.A. Dolecek, Langenberg, T. Cole and L. Brace, Changes in body weight, body composition, and energy intake in women fed high- and low-fat diets. *Am. J. Clin. Nutr.* 54 (1991), 304-10.

Prins, A.J.A., A, de Jong-Nagelsmit, J. Keyser and J.H. Strubbe, Daily rhythms of feeding in the genetically obese and lean Zucker rats *Physiol. Behav.* 38 (1986), 423-26.

Prop, J. and C. Deerenberg, Spring staging in Brent Geese *Branta bernicla*: feeding constraints and the impact of diet on the accumulation of body reserves. *Oecologia* 87 (1991), 19-28.

Quigley, M.E. and S.S.C. Yen, The role of endogenous opiates on LH secretion during the menstrual cycle. *J. Clin. Endocrinol. Metab.* 51 (1980), 179.

Reiter, E.O. and M.M. Grumbach, Neuroendocrine mechanisms and the onset of puberty. *Ann. Rev. Physiol.* 44 (1982), 595.

Revusky, S.H., Effects of thirst level during consumption of flavored water on subsequent preferences. *J. Comp. Physiol. Psychol.* 66 (1968), 777-79.

Richter, E.A., N.B. Ruderman, H. Gavras, E.R. Belur and H. Galbo, Muscle glycogenolysis during exercise: dual control by epinephrine and contractions. *Am. J. Physiol.* 242 *(Endocrinol. Metab.5)*, 1982, E25-E32.

Ritter, R.C., P.G. Slusser and S. Stone, Glucoreceptors controlling feeding and blood glucose: location in the hindbrain. *Science* 213 (1981), 451-53.

Rivier, C. and W. Vale, Modulation of stress-induced ACTH release by corticotropin-releasing factor, catecholamines and vasopressin. *Nature* 305 (1983), 325.

Rivier, C., J. Rivier and W. Vale, Stress-induced inhibition of reproductive functions: role of endogenous corticotropin-releasing factor. *Science* 231 (1986), 607.

Rodin, J., Comparative effects of fructose, aspartame, glucose and water preloads on caloric and macronutrient intake. *Am. J. Clin. Nutr.* 51 (1990), 428-35.

Roger, P.J., J.A. Carlyle, A.J. Hill and J.E. Blundell, Uncoupling sweet taste and calories: comparison of the effect of glucose and three intense sweeteners on hunger and food intake. *Physiol. Behav.* 43 (1988), 547-52.

Rohner-Jeanrenaud, F., E. Bobbioni, W. Ionescu, J.F. Sauters and B. Jeanrenaud, Central nervous system regulation of insulin secretion. *Advances in Metabolic Disorders. vol. 10, CNS regulation of Carbohydrate Metabolism.* Ed. A.J. Szabo. New York, Academic, 1983, 193-220.26.

Rolls, B.J., E.T. Rolls, E.A. Rowe, K. Sweeney, Sensory specific satiety in man. *Physiol. Behav.* 27 (1981), 137-42.

Rolls, B.J., E.A. Rowe and E.T. Rolls, How sensory properties of foods affect human feeding behavior. *Physiol. Behav.* 29 (1982), 409-17.

Rolls, B.J., M. Hetherington and V.J. Burley, The specificity of satiety: the influence of foods of different macronutrient content on the development of satiety. *Physiol. Behav.* 43 (1988a), 143-53.

Rolls, B.J., M. Hetherington and V.J. Burley, Sensory stimulation and energy density in the development of satiety. *Physiol. Behav.* 44 (1988b), 727-33.

Ropert, J.R., M.E. Quigley and S.S.C. Yen, Endogenous opiates modulate pulsatile LH release in humans. *J. Clin. Endocrinol. Metab.* 52 (1981), 584.

Rosenberg, K. and J.V.G.A. Durnin, The effect of alcohol on resting metabolic rate. *Br. J. Nutr.* 40 (1978), 293-98.

Rosenwasser, A.M., R.J. Pelchat and N.T. Adler, Memory for feeding time: possible dependence on coupled circadian oscillators. *Physiol.Behav.* 32 (1984), 25-30.

Rossmanith, W.G. and S.S.C. Yen, Sleep-associated decrease in luteinizing hormone pulse frequency during the early follicular phase of the menstrual cycle: Evidence for an opioidergic mechanisms. *J. Clin. Endocrinol. Metab.* 65 (1987), 715.

Rozin, P., Specific aversions as a component of specific hungers. *J. Comp. Physiol. Psychol.* 64 (1967), 237-42.

Rozin, P., Are carbohydrate and protein intakes separately regulated? *J. Comp. Physiol. Psychol.* 65 (1968), 23-29.

Rozin, P. and J.W. Kalat, Specific hungers and poison avoidance as adaptive specializations of learning. *Psychol. Rev.* 78 (1971), 459-86.

Ruiter, L. de, J.M. Koolhaas and J.H. Strubbe, Brain and behavior: a gordian knot? *Neth. J. Zool.* 35 (1985), 209-37.

Rusak, B., Vertebrate behavioral Rhythms. *Handbook of Behavioral Neurobiology. Vol. 4, Biological Rhythms.* Eds. F.A. King, J. Aschoff. New York, Plenum Press, 1981, 183-205.

Russek, M., Demonstration of the influence of a hepatic glucosensitive mechanism on food intake. *Physiol. Behav.* 1970, 1207-09.

Russel, J. and P.J.V. Beaumont, The endocrinology of anorexia nervosa. *Handbook of Eating Disorders Part 1. Anorexia and Bulimia Nervosa,* Eds. P.J.V. Beumont, G.D. Burrows and R.C. Casper. New York, Elsevier, 1987, 201-24.

Saad, M.F., W.C. Knowler , D.J. Pettitt, R.G. Nelson, D.M. Mott and P.H. Bennett, Insulin and hypertension. Relationship to obesity and glucose intolerance in Pima Indians. *Diabetes* 39 (1990), 1430-35.

Saris, W.H.M. and D. van Dale, Effects of exercise during VLCD diet on metabolic rate, body composition and aerobic power: pooled data of four studies. *Int. J. Obes.* 13 (1989), 169-70.

Sauder, S.E., G.D. Case, N.J. Hopwood, R.P. Kelch and J.C. Marshall, The effects of opiate antagonism on gonadotropin secretion in children and in women with hypothalamic amenorrhea. *Pediatr. Res.* 18 (1984), 322.

Scheurink, A.J.W. and S. Ritter, Sympathoadrenal responses to glucoprivation and lipoprivation in rats. *Physiol. Behav.* 53 (5), 1993, 995-1000.

Scheurink, A.J.W. and A.B. Steffens, Central and peripheral control of sympthoadrenal activity and energy metabolism in rats. *Physiol. Behav.* 48 (1990), 909-20.

References

Scheurink, A.J.W., A.B. Steffens and L. Benthem, Central and peripheral adrenoreceptors affect glucose, free fatty acids and insulin in exercising rats. *Am. J. Physiol.* 225 (*Regul. Int. Comp. Physiol.* 24), 1988, R547-R556.

Scheurink, A.J.W., A.B. Steffens, H. Bouritius, G. Dreteler, R. Bruntink, R. Remie and J. Zaagsma, Adrenal and sympathetic catecholamines in exercising rats. *Am. J. Physiol.* 256 (*Regul. Int. Comp. Physiol.* 25), 1989a, R155-R160.

Scheurink, A.J.W., A.B. Steffens, H. Bouritius, G. Dreteler, R. Bruntink, R. Remie and J. Zaagsma, Sympathoadrenal influence on glucose, FFA and insulin levels in exercising rats. *Am. J. Physiol.* 256 (*Regul. Int. Comp. Physiol.* 25), 1989b, R161-R168.

Scheurink, A.J.W., A.B. Steffens, G. Dreteler, L. Benthem and R. Bruntink, Experience affects exercise-induced changes in catecholamines, glucose and FFA. *Am. J. Physiol.* 256 (*Regul. Int. Comp. Physiol.* 25)*, 1989c, R169-R173.

Scheurink, A.J.W., A.B. Steffens and R.P.A. Gaykema, Hypothalamic adrenoreceptors mediate sympathoadrenal activity in exercising rats. *Am. J. Physiol.* 259 (*Regul. Int. Comp. Physiol.* 28), 1990a, R470-R477.

Scheurink, A.J.W., A.B. Steffens and R.P.A. Gaykema, Paraventricular hypothalamic adrenoreceptors regulate energy metabolism in exercising rats. *Am. J. Physiol.* (*Regul. Int. Comp. Physiol.* 28), 1990b, R478-R484.

Scheurink, A.J.W., T.D. Mundinger, B.E. Dunning, R.C. Veith and G.J. Taborsky Jr., α_2-adrenergic regulation of galanin and norepinephrine release from canine pancreas. *Am. J. Physiol.* 262 (1992), R819-R825.

Scheurink, A.J.W. H. Leuvenink and A.B. Steffens, Metabolic and hormonal responses to hypothalamic administration of norfenfluramine in rats. *Physiol. Behav.* 53 (5), 1993, 889-98.

Schmidt-Nielsen, K., Energy metabolism. *Animal Physiology: Adaptation and Environment*, Second Edition. Cambridgde, UK, Cambridge University Press, 1979, 161.

Schreihofer, D.A., D.B. Parfitt and J.L. Cameron, Evidence that suppression of hypothalamic-pituitary-gonadal axis activity during fasting results from a nutritional signal and not the psychological stress of food deprivation. *Soc. Neurosci. Abstr.*, 1990, Abstract 168.7.

Schulz, S., K. Brück and C. Leitzmann, Thermogenesis in small and large eaters. *J. Therm. Biol.* 8 (1983), 73-5.

Schutz, Y., J.P. Flatt and E. Jéquier, Failure of dietary fat intake to promote fat oxidation: a factor favoring the development of obesity. *Am. J. Clin. Nutr.* 50 (1989), 307-14.

Schwartz, B., D.C. Cumming, E. Riordan, M. Selye, S.S.C. Yen and R.W. Rebar, Exercise associated amenorrhea: A distinct entity?. *Am. J. Obstet. Gynecol.* 141 (1981), 662.

Schwartz, W.J., S.M. Reppert, S. Eagan and M.C. Moore-Ede, In vivo metabolic activity of the suprachiasmatic nucleus: A comparative study. *Brain Res.* 274 (1983), 184-87.

Shillabeer, G. and J.S. Davison, Proglumide a cholecystokinin antagonist increases gastric emptying in rats. *Am. J. Physiol.* 252 (1987), R 353-R360.

Shimazu, T., Neuronal control of intermediate metabolism. *Neuroendocrinology*, Eds. Lightman, S. and B. Everitt. Oxford, UK, Blackwell Scientific, 1986, 304-30.

Silverstone, T. and E. Goodall, Serotonergic mechanisms in human feeding: The pharmacological evidence. *Appetite* 7 (1986), 85-97.

Sims, E.A.H., E. Danforth, E.S. Horton, G.A. Bray, J.A. Glennon and L.B. Salans, Endocrine and metabolic effects of experimental obesity in man. *Recent Progress in Hormone Research* 29 (1973), 457-96.

Sinclair, A.R.E., The resource limitation of tropic levels in tropical grassland ecosystems. *J. Anim. Ecol.* 44 (1975), 497-520.

Smith, G.P. and A.N. Epstein, Increased feeding in response to decreased glucose utilisation in the rat and monkey. *Am. J. Physiol.* 217 (1969), 1083-87.

Spiteri, N.J., Circadian patterning of feeding, drinking and activity during diurnal food access in rats. *Physiol. Behav.* 28 (1982), 139-47.

Spiteri, N.J., A.J.A. Prins, J. Keyser and J.H. Strubbe, Circadian pacemaker control of feeding in the rat at dawn. *Physiol. Behav.* 29 (1982), 1141-45.

Steffens, A.B., The influence of insulin injections and infusions on eating and blood glucose level in the rat. *Physiol. Behav.* 4 (1969a), 823-28.

Steffens, A.B., A method for frequent sampling of blood and continuous infusion of fluids in the rat without disturbing the animal. *Physiol. Behav.* 4 (1969b), 833-36.

Steffens, A.B., Influence of reversible obesity on eating behavior, blood glucose and insulin in the rat. *Am. J. Physiol.* 228 (1975), 1738-44.

Steffens, A.B., and J.H. Strubbe. CNS regulation of glucose secretion. *Advances in Metabolic Disorders, vol. 10, CNS Regulation of Carbohydrate Metabolism.* Ed. A.J. Szabo. New York, Academic Press, 1983, 221-58.

Steffens, A.B., G.J. Mogenson and J.A.F. Stevenson, Blood glucose insulin and free fatty acids after stimulation and lesions of the hypothlamus. *Am. J. Physiol.* 222 (1972), 1446-52.

Steffens, A.B., J.H. Strubbe, B. Balkan and A.J.W. Scheurink, Neuroendocrine mechanisms involved in regulation of body weight, food intake and metabolism. *Neurosci. Biobehav. Rev.* 14 (3), 1990, 305-13.

Stellar, E., The physiology of motivation. *Psychol. Rev.* 61 (1954), 5-22.

Stephan, F.K., Phase shifts of circadian rhythms in activity entrained to food accces. *Physiol. Behav.* 32 (1984), 663-71.

Strauss, J.F., L.A. Schule, M.F. Rosenblum, T. Tanaka, Cholesterol metabolism by ovarian tissue. *Adv. Lipid Res.* 18 (1981), 99.

Strominger, J.L. and J.R. Brobeck, A mechanism of regulation of food intake. *Yale J. Biol. Med.* 25 (1953), 383-90.

Strubbe, J.H., Food intake regulation in the rat. *Exogenous and Endogenous Influences on Metabolic and Neural Control on Feeding.* Eds. A.D.F. Addink and N.Spronk. Oxford, Pergamon Press, 1982, 31-39.

Strubbe, J.H., Central nervous system and insulin secretion. *Neth. J. Med.* 34 (1989), 154-67.

Strubbe, J.H., De hersenen en insulinesecretie. *Integraal* 5 (1990), 20-26.

Strubbe, J.H., and P.R. Bouman, Plasma insulin patterns in the unanaesthetized rat during intracardial infusion and spontaneous ingestion of graded loads of glucose. *Metabolism* 27 (1978), 341-51.

Strubbe, J.H. and J. Gorissen. Meal patterning in the lactating rat. *Physiol. Behav.* 25 (1980), 775-77.

Strubbe, J.H. and C.G. Mein, Increased feeding in response to bilateral injections of insulin antibodies in the VMH. *Physiol. Behav.* 19 (1977), 309-13.

Strubbe, J.H. and A.J.A. Prins, Reduced insulin secretion after short term food deprivation in rats plays a key role in the adaptive interaction and free fatty acid utilization. *Physiol. Behav.* 37 (1987), 441-45.

Strubbe, J.H. and A.B. Steffens, Rapid insulin release after ingestion of a meal in the unanesthetized rat. *Am. J. Physiol.* 229 (1975), 1019-22.

Strubbe, J.H. and A.B. Steffens, Blood glucose levels in portal and peripheral circulation and their relation to food intake in the rat. *Physiol. Behav.* 19 (1977), 303-07.

Strubbe, J.H. and A.B. Steffens, Hormonal changes induced by food intake contribute to body weight regulation and alterations in metabolism. *Ann. d'Endocrinol.* 49 (1988), 105-12.

Strubbe, J.H. and van Wachem, Insulin secretion by the transplanted pancreas during food intake in fasted and fed rats. *Diabetologia* 20 (1981), 228-36.

Strubbe, J.H., A.B. Steffens and L. de Ruiter, Plasma insulin and the time pattern of feeding. *Physiol. Behav.* 18 (1977), 81-86.

Strubbe, J.H., J. Keyser, T. Dijkstra and A.J.A. Prins, Interaction between circadian and caloric control of feeding behavior in the rat. *Physiol. Behav.* 36 (1986), 489-93.

Strubbe, J.H., A.J.A. Prins, J. Bruggink and A.B. Steffens, Daily variation of food induced changes in blood glucose and insulin in the rat and the control by the suprachiasmatic nucleus and the vagus nerve. *J. Autonom. Nerv. Syst.* 20 (1987), 113-19.

Strubbe, J.H., J.G. Wolsink, A.M. Schutte, B. Balkan and A.J.A. Prins, Hepatic-portal and cardiac infusion of CCK-8 and glucagon induce different effects on feeding. *Physiol. Behav.* 46 (1989), 643-46.

Stunkard, A.J., Genes environment and human obesity. *Progress in Obesity Research*, Eds. Y. Oomura, S. Tarui, S. Inoue and T. Shimazu. London, John Libbey, 1990, 669-74.

Sugerman, H.J., Evolution of gastric proccedures for morbid obesity. *Progress in Obesity Research*, Eds Y. Oomura, S. Tarui, S. Inoue and T. Shimazu. London, John Libbey, 1990, 597-603.

Taggart, N., Diet, activity and body-weight. A study of variations in a women. *Br. J. Nutr.* 16 (1962), 223-35.

References

Tempel, D.L., G. Shor-Posner, D. Dwyer and S.F. Leibowitz, Nocturnal patterns of macronutrient intake in freely feeding and food-deprived rats. *Am. J. Physiol.* 256 (1989), R541-R548.

Tordoff, M.G. & M.I. Friedman, Hepatic portal glucose infusions decrease food intake and increase food preference. *Am. J. Physiol.* 251 (1986), R192-R196.

Trayhurn, P. and W.P.T. James, Thermogenesis and obesity. *Mammalian Thermogenesis*, Eds. L. Girardier and M.J. Stock. London, Chapman & Hall, 1983, 234-58.

Treit, D., M.L. Spetcher and J.A. Deutsch, Variety in the flavor of food enhances eating in the rat: a controlled demonstration. *Physiol. Behav.* 30 (1983), 207-11.

Tremblay, A., G. Plourde, J-P. Despres and C. Bouchard, Impact of dietary fat content and fat oxidation on energy intake in humans. *Am. J. Clin. Nutr.* 49 (1989), 799-805.

Turek, F.W., Circadian neural rhythms in mammals. *Ann. Rev. Physiol.* 47 (1985), 49-64.

Van der Walt, L.A., E.N. Wilmsen and T. Jenkins, Unusual sex hormone patterns among desert dwelling hunter gatherers. *J. Clin. Endocrinol. Metab.* 46 (1978), 658.

Van Erp-Baart, A.M.J., W.H.M. Saris, R.A. Binkhorst, J.A. Vos and J.W.H. Elvers, Nationwide survey on nutritional habits in elite athletes, part I: Energy, carbohydrate, protein, and fat intake. *Int. J. Sports Med.* 10 (1989), S3-S10.

Van Es, A.J.H., J.E. Vogt, C. Niessen, J. Veth, L. Rodenburg, V. Teeuwse, J. Dhuyvetter, Deurenberg, J.G.A.J. Hautvast and E. van der Beek, Human energy metabolism below, near and above energy equilibrium. *Br .J. Nutr.* 52 (1984), 429-42.

Van Itallie, T.B. and H.R. Kissileff, Human obesity. A problem in body energy economics. *Handbook of Behavioral Neurobiology, Volume 10, Neurobiology of Food and Fluid Intake*, Ed. E.M. Stricker. New York, Plenum Press, 1990, 207-40.

Van Staveren, W.A., Deurenberg, J. Burema, L.C.P.G.M. de Groot and J.G.A.J. Hautvast, Seasonal variation in food intake, pattern of physical activity and chance in body weight in a group of young adult Dutch women consuming self-selected diets, 1985.

Veldhuis, J.D., A.D. Rogol, E. Samoslik and N.H. Ertel, Role of endogenous opiates in the expression of negative feedback action of androgens and estrogens on pulsatile properties of luteinizing hormone secretion in man. *J. Clin. Invest.* 74 (1984), 47.

Veldhuis, J.D., W.E. Evans, L.M. Demersm, M.O. Thorner, D. Wakat and A.D. Rogol, Altered neuroendocrine regulation of gonadotropin secretion in women distance runners. *J. Clin. Endocrinol. Metab.* 61 (1985), 557.

Verboeket-van de Venne, W.P.H.G. and K.R. Westerterp, Influence of the feeding frequency on nutrient utilization in man. *Eur. J. Clin. Nutr.* 45 (1991), 161-69.

Verhofstad, A.A.J., The adrenal medulla. An immunohistochemical and ontogenetic study on the noradrenalin and adrenalin storing cells of the rat. PhD. Thesis, University Nijmegen, Netherlands, 1984.

Vigersky, R.A., A.E. Andersen, R.H. Thompson and D.L. Loriaux, Hypothalamic dysfunction in secondary amenorrhea associated with simple weight loss. *N. Eng. J. Med.* 297 (1977), 1141.

Villaneuva, A.L., C. Scholsser, D.I. Hoffman and R.D. Rebar, Increased cortisol production in women runners. *J. Clin. Endocrinol. Metab.* 63 (1986), 113.

Vonk R.J., A.B.D. Van Doorn and J.H. Strubbe, Bile secretion and bile composition in the freely moving unanesthetized rat with a permanent biliary drainage: influence of food intake on bile flow. *Clin. Sc. Mol. Med.* 55 (1978), 253-59.

Wardle, J., Hunger and satiety: a multidimensional assessment of responses to caloric loads. *Physiol. Behav.* 40 (1987), 577-82.

Warren, M.P., The effects of exercise on pubertal progression and reproductive function in girls. *J. Clin. Endocrinol. Metab.* 51 (1980), 1150.

Warren, M.P., Effects of undernutrition on reproductive function in humans. *Endo. Rev.* 4(4), 1983, 363.

Warren, M.P., Metabolic factors and the onset of puberty. *The Control of the Onset of Puberty II*, Eds. Grumbach, M.M., P.C. Sizonenko, M.A. Aubert. Baltimore, Williams & Wilkins, 1987.

Warren, M.P., R. Jewelewicz, I. Dyrenfurth, R. Ans, S. Khalafand R.L. Vande Wiele, The significance of weight loss in the evaluation of pituitary response to LH-RH in women with secondary amenorrhea. *J. Clin. Endocrinol. Metab.* 40 (1975), 601.

Webb, and J.F. Annis, Adaptation to overeating in lean and overweight men and women. *Hum Nutr. Clin. Nutr.* 37C (1983), 117-31.

Wendorf, M. and I.D. Goldfine, Archeology of NIDDM, excavation of the "Thrifty" genotype. *Diabetes* 40 (1991), 161-65.

Westerterp, K.R. and D.M. Bryant, Energetics of free existence in swallows and martins (hirundinidae) during breeding: a comparative study using doubly-labeled water. *Oecol. Berl.* 62(1984), 376-81.

Westerterp, K.R., W.H.M. Saris, M. van Es and F. ten Hoor, Use of the doubly-labeled water technique in humans during heavy sustained exercise. *J. Appl. Physiol.* 61 (1986), 2162-67.

Westerterp, K.R., F. Brouns, W.H.M. Saris, and F. ter Hoor, Comparison of doubly labeled water with respirometry at low- and high-activity levels. J. Appl. Physiol. 65 (1988), 53-56.

Westerterp, K.R., G.A.L. Meijer, W.H.M. Saris, P.B. Soeters, Y. Winants and F. ten Hoor, Physical activity and sleeping metabolic rate. *Med. Sci. Sports Exerc.* 23 (1991a), 166-70.

Westerterp, K.R., W.H.M. Saris, P.B. Soeters and F. ten Hoor, Determinants of weight loss after vertical banded gastroplasty. *Int. J. Obes.* 15 (1991b), 529-34.

Westerterp, K.R., G.A.L. Meijer, G.M.E. Janssen, W.H.M. Saris and F. ten Hoor, Long term effect of physical activity on energy balance and body composition. *Br. J. Nutr.* 68(1), 1992, 21-30.

Westerterp, K.R., B. Kayser, F. Brouns, J.P. Herry, and W.H.M. Saris, Energy expenditure climbing Mt. Everest. J. Appl. Physiol. 73 (1992), 1815-19.

Westerterp-Plantenga, M.S., L. Wouters, F. Ten Hoor, Deceleration in cumulative food intake curves, changes in body temperature and diet-induced thermogenesis. *Physiol. Behav.* 48 (1990), 831-6.

Weststrate, J.A., P.J.M. Weys, E.J. Poortvliet, Deurenberg and J.G.A.J. Hautvast, Diurnal variation in postabsorptive resting metabolic rate and diet-induced thermogenesis. *Am. J. Clin. Nutr.* 50 (1989), 908-14.

Wiepkema, P.R., Positive feedbacks at work during feeding. *Behavior* 39 (1971), 266-73.

Wilson, G.T., Cognitive-behavioral treatments of bulimia nervosa: The role of exposure. *The Psychobiology of Bulimia Nervosa.* Eds. K.M. Pirke, W. Vandereycken and D. Ploog. Heidelberg, Springer-Verlag, 1988, 137-45.

Woods, S.C., Central nervous system control of nutrient homeastasis. *Handbook of Physiology, Section I. The Nervous System Vol. 4. Intrinsic Regulatory Systems of the Brain.* Ed. F.Bloom. Am. Physiol. Soc. Bethesda, 365-411, 1986a.

Woods, S.C., G.J. Taborsky Jr., and D. Porte Jr., CNS control of nutrient homeostasis. *Handbook of Physiology, Section I. The Nervous System. Vol. 3. Sensory Processes.* Bethesda, MD: Am. Physiol. Soc., 1984.

Woods, S.C., D. Porte, J.H. Strubbe and A.B. Steffens, The relationships among body fat, feeding and insulin. *Neural and Humoral Control of Food Intake*, Eds. R. Ritter, S. Ritter and C.D. Barnes. Academic Press, Orlando, 315-27, 1986.

Wooley, O.W., S.C. Wooley and R.B. Dunham, Calories and sweet taste: effects of sucrose preference in the obese and non-obese. *Physiol. Behav.* 9 (1972), 765-8.

Wooley, S.C., Physiologic vs cognitive factors in short-term food regulation in the obese and non-obese. *Psychosom. Med.* 34 (1972), 62-8.

Wurtman, R.J. and J.J. Wurtman, Carbohydrate craving, obesity and brain serotonin. *Appetite* 7 (1986), 99-103.

Yager, J., The treatment of eating disorders. *J. Clin. Psychiatry* 49 (1988), 18-25.

Young, J.B. R.M. Rosa and L. Landsberg, Dissociation of sympathetic nervous system and adrenal medullary responses. *Am. J. Physiol.* 247 (*Endocrinol. Metab.* 10), 1984, E35-E40.

Zacharias, L., W.M. Rand and R.J. Wurtman, A prospective study of sexual development in American girls: the statistics of menarche. *Obstet. Gynecol. Surv.* 31 (1976), 325.

Index